TABLE OF CONTENTS

Table of Contents	1
Introduction	2
Chapter 1 - Transforming Technologies	6
Chapter 2 - The XB-45s	16
Chapter 3 - The B-45A Description	41
Chapter 4 - The B-45C Description and Service	84
Chapter 5 - The RB-45C Description	101
Chapter 6 - The B-45A in Service	142
Chapter 7 - The RB-45C in Service	202
Chapter 8 - The B-45's Place in USAF History	233
Chapter 9 - Tables and Data	238
Chapter 10 – Acronyms	248
Chapter 11 - Sources and References	249
Chapter 12 - The B-45 in Color, Part 1, Scale Models	252
Chapter 12 - The B-45 in Color, Part 2, Unit Insignia	256
Chapter 12, The B-45 in Color, Part 3, the Aircraft	260

XB-45 Technical Manual Illustration Courtesy of Alan Griffith

INTRODUCTION

Photo 1: This image shows a 1946 North American Manufacturers Model of the NA-130 at Inglewood, CA. (NMUSAF)

After years of devastating conflict, the last year of World War II (WWII) led all to believe that the war would be won on Allied terms, and a long-lasting world peace would follow. But for some the time from the last year of the War through the Soviet Blockade of Berlin brought the realization that the technology of atomic weapons, and ground-breaking advances in aviation sciences called that peace into question. Although the six years of world–wide conflict had eliminated the two authoritarian forces that had destroyed the pre-war world order, the flawed but fundamentally positive peace was under threat. Communism was rapidly replacing those forces as the primary danger to the current calm. Having already obtained many of the German scientists and engineers who had developed liquid-fueled rockets and ballistic missiles, the Soviets were prioritizing the design of these vehicles to deliver atomic weapons. Communist forces were correctly judged to be the new, immediate, and immense threat. There existed a new sense of urgency to revolutionize the United States Army Air Force (USAAF), not unlike that felt in 1941 and 1942.

The aviation technologies concerned were those based upon the turbojet engine and turbojet-powered combat aircraft. Their implications, based on just over a year of combat experience and nearly two years of intelligence collection, were perceived to be revolutionary. Aircraft designed to successfully implement the benefits of the turbojet's seemingly unlimited increase in aerial capabilities **could and would** fundamentally change aerial combat in all respects. The full impacts of these technologies upon aircraft and those who flew them were still great unknowns. Further, how these impacts would reveal themselves, were also mysteries. Most importantly, what needed to be done in terms of research, engineering, design, testing and production to benefit from them, had yet to be fully discovered.

Planners and contractors at this time commonly called for extensive periods of research and development to unravel at least a few of the great number of unknowns about this new form of aeronautical power called the turbojet. The developmental goal for the US was to master these new technologies and their implications to aircraft which had been fielded by the German Luftwaffe and the Royal Air Force through the second half of 1944. Then we would improve and exploit them to obtain an unbeatable air arm.

By 1943 the USAAF Headquarters staff began the US aeronautical revolution with fighter aircraft developments. The Bell XP-59 had flown in late 1942, the Lockheed XP-80 in January 1944, and two more designs, the Republic F-84, and the North American F-86, were under development. Turbojet-powered transports and dedicated reconnaissance airframes would wait. Due to the growing urgency, new jet-powered bombers would have to be airframes based upon conventional aerodynamic and manufacturing "knowns" learned during the first three to four years of the war. They would, however, still be powered by the radical turbo-jet, but with little true knowledge of the challenges these power plants held in store. Initial design concepts began in 1943 based on Technical Intelligence gained by the RAF since 1941, shared with the USAAF and resulting in the 1943 request for proposals (RFP) such as the Boeing Model 413, a design remarkably similar in layout to the Arado 234V series but some 40% larger.

Four aircraft - the North American XB-45, the Convair XB-46, the Boeing XB-47, and the Martin XB-48 - were designed, built, and flown because of a 1944-initiated USAAF RFP to develop a turbojet-powered bomber. Three of these, the B-45, B-46, and B-48, were essentially conventional straight-wing aerodynamic designs. The Boeing developed XB-47 design was more of a true technological challenger that would also pay a very high price, almost too high, in terms of funding, accidents, and multiple efforts to achieve each of its aerodynamic goals. Two of these four designs initiated by that 1944 USAAF Headquarters letter, the Convair XB-46 and the Martin XB-48, were eliminated almost immediately because they had been built without access to Manhattan-Project held details and could never be made capable to carry the existing US atomic inventory. The Boeing XB-47 was the sole swept-wing design, and the contestant that Headquarters (HQ) USAAF was betting on to truly prove what an advanced turbojet strategic bomber would look like and how it would perform. But it would take longer to prove out as it would be a true developmental aircraft from the ground up. Its performance and aerodynamic advancements if realized, would insure its survival and service.

The North American XB-45, like the Martin and Convair proposals, had also been denied Manhattan-Project access and almost had its service terminated for the same reason as the XB-46 and XB-48. The Air Staff had committed everything to a jet-powered service and had three fighters already preparing to come into wide service. There was however no large turbojet-powered aircraft that could teach the service the mysteries of operations, maintenance, training and logistics for sizable multi-engine turbojet aircraft.

Photo 2: The first XB-45 559479 was photographed over the Muroc Dry Lake area in mid-1947. (NMUSAF)

Photo 3: B-45A-1 7015 conducted extreme cold weather tests in Alaska and Canada by North American Aviation before it was photographed over southern California. Shortly after this flight the aircraft returned to the 3200th Proof Test Group (PTG) at Eglin Air Force Base (AFB). (Courtesy NAA via Tommy Thomason)

Therefore, the service chose a "conventional" design, powered by four revolutionary power plants to drive that airframe. The choice to produce an airframe without fully understanding the aerodynamic and maintenance challenges it would face brought a high price in terms of lost airframes and crews. This decision did however allow the United States Air Force (USAF), an independent service as of September 1947, to become a world-leading turbojet-driven combat arm much, much faster when time was of the essence. The North American B-45, along with the RB-45C series endured to become a truly ground-breaking aircraft setting a significant number of historical firsts for the US and the USAF. This book is the first detailed look at the development and use of this overlooked, but historically significantly aircraft. Enjoy!

Photo 4: North American Aviation (NAA) B-45A-5 assigned to the 47th Bombardment Group (Light) (BG) at Barksdale AFB, LA was photographed from the bomb bay of a Group sister ship over northern Louisiana in late 1949. (Courtesy Knox Bishop)

Notes on Serial and Tail Number Presentation....

During the production life of the B-45, the USAF had yet to standardize the presentation of the serial numbers on the airframe, in the aircraft's data block, or the radio call number (or tail number) on the vertical fin. The few images of B-45A data blocks show the common last two digits of the fiscal year of the airframe's contract followed by a dash. B-45 serial numbers all ended in single or dual numerical digit on its end. On aircraft with single digit ending serials, the dash is followed by that single digit, with no preceding zero (e.g. 47-1). On aircraft with two-digit serials, that dash is followed by those last two digits of the serial (e.g. 47-56).

Tail numbers, on the other hand, which were highly visible had four digits. The first digit was the last numeral of the fiscal year in which the aircraft was purchased, either a 7 for Fiscal Year or FY 1947, or an 8 for FY 1948. Two zeros then followed that year digit, and the sole digit of the serial if it ended in a single number. For example, serial B-45A-1 47-002's "tail" number was 7002. Serial numbers which ended with two numerical digits began with the last digit of the fiscal year in which the aircraft was purchased, then a single zero followed by the last two digits of the serial number. Thus serial 47-056's tail number was 7056.

BUZZ numbers added another layer of confusion. BUZZ numbers were painted onto USAF aircraft to positively identify aircraft that had flown at extremely low altitudes over civilians or civilian facilities. That form of aviation fun and daring was known as BUZZing, usually in an area near an air base or range. The numbers consisted of two letters indicating the specific role of aircraft and the second the type. The three digits were the last three of the aircraft's serial number. This five-digit identification was initially painted on the underside of the port wing, and later solely on both sides of the aft fuselage. The B-45A-1's Buzz Number showed the aircraft type identifier "BE", a dash and three digits: two zeros followed by the single serial number digit, e.g., BE-002, or the BE aircraft identifier, a dash, one zero and the last two digits of the serial number, i.e., BE-010 for aircraft 7010. Four-digit numbers, which mirror the layout of the "tail number" on each aircraft (e.g., **7015**), will be used to identify specific aircraft in this book for clarity.

A note on Nomenclature....

Weapons that derive their destructive energy from the fission or fusion of an atomic nucleus are both technically "nuclear" weapons. However, early fission weapons were popularly known as "atomic" or "A" bombs. This nomenclature carried over into official technical and historical documents of the day and therefore the Marks (Mk) 5, 7 and 8 weapons carried by the B-45 are referred to as atomic weapons in this book.

The B-45 never carried fusion weapons, popularly known as "hydrogen" or "H" bombs from their use of the hydrogen isotopes Deuterium and Tritium. After the nature of how "atomic" and "hydrogen" weapons worked became common knowledge, it became customary to call fission weapons "nuclear" weapons, and fusion weapons "thermonuclear" weapons.

Photo 5: B-45C 8001 was photographed during a pre-delivery acceptance flight. (The Gerald Balzer Collection via Tony Landis from HQ AFMC/HO)

Photo: 6: XB-45 number 1, 559479, was photographed over Muroc AFB in late 1947. (NARA II via Tommy Thomason)

CHAPTER 1 – TRANSFORMING TECHNOLOGIES

Trepidation is defined as a feeling of fear or anxiety over something that may happen. The period from 1942 through 1945 was filled with trepidation that rushed the creation of the B-45 and expedited the advancement of two technologies, the turbojet engine, and atomic weapons. Each would fundamentally impact the B-45 development and operational use.

Trepidation....

By the end of 1942 Axis offensive operations had secured most of Europe from the English Channel and North Sea to portions of the Soviet Union from Taganrog on the Sea of Azov, then northward to the outskirts of Moscow and on to Leningrad in the far North; as well as the Balkans; portions of the Middle East and North Africa. In the Far East, the Imperial Japanese military forces had secured Manchuria, half of China, Southeast Asia to the border of India, south over the Dutch East Indies and Timor, and into the western Pacific as far as a line from the Solomon Islands to the Gilbert Islands, then due north to Kiska in the Aleutian chain.

These stunning gains in the space of less than two years brought serious foreboding to the Allies and American military leaders about what could happen in the not-too-distant future. The trepidations of 1939 through 1942 drove a tremendous development in technologies of the day that the idea of an Allied victory in two to three years might be possible. Pushing these technological limits did indeed produce exceptional aircraft like the P-51 Mustang, the B-29 Superfortress, the C-54 Skymaster, and the F6F Hellcat; all of which were momentous advancements in aviation.

By late 1944 however, a newly recognized geo-political development brought about a far greater and more serious level of trepidation at USAAF Headquarters. The Soviets quietly initiated and controlled subtle coups in a significant number of former German-occupied countries, such as Romania, Bulgaria, Communist elements in Greece, Albania, Yugoslavia, and Hungary. Poland, the collection of countries which were merged to form Czechoslovakia, Lithuania, Latvia, and Estonia were under Soviet control by the end of 1945. The Soviet presence in Eastern Europe was initially thought to be temporary and that it would exist only until Germany could be totally defeated and dealt with by all the Allies. But it soon became clear that the Soviet's true intention was the firm and permanent occupation of the half of Germany and all of Europe which they had conquered.

In the Far East, the pre-World War II civil war in China, suspended by the war against Japan, resumed immediately. The foes were the same, the Kuomintang led by Chiang Kai-shek, and the Chinese Communist Party led by Mao Zedong. In the eastern hemisphere the "western Allies" were now faced with the first of many conflicts from India to the Far East in the rebellions against colonialism. Most were viewed as Communist-dominated and controlled by dictatorial and authoritarian leaders. Having just destroyed two such powers over a six-year World War at the costs of millions of lives, the prospect of a new global conflict was very, very real.

While the US at the time had the sole weaponized atomic devices, some knew it was only a matter of time before the possession of that power would spread. This could mean that some of the anti-colonial conflicts could become serious. There were many in both the U.S. government and the U.S. military who understood that potential atomic conflicts would not only be over the control and philosophy of governments, but be unbelievably devastating as well.

These were the apprehensions that influenced the decisions made by USAAF Headquarters throughout the development and initial service periods of the NAA American B-45. While the continued technological developments of existing weapons were good, no longer would they be enough to assure security. The immediate concern was the risk of being technologically behind by slow development of the revolutionary turbojet engine. Second and much more threatening was the realization that the conflicts against authoritarianism and dictatorship would continue despite VE Day and VJ Day and might have an atomic aspect. The turbojet was very definitely in play for the B-45, atomic weapons were not due to their then-current physical size and weight. Let us look first at turbojet technologies as without this type of engine, there would have been no B-45.

Turbojets....

Heralded by an increasing number of intelligence reports about highly advanced and unusually powered combat aircraft encountered in Europe, beginning in late 1941 and continuing steadily through 1944, a technological concern was emerging. The word turbojet, formerly spoken in a whisper, was now stated loudly about Germany, and glimpses were becoming clear images. With British assistance, the US had done some research in turbojet power beginning in late 1941 including flying the XP-59 in October of 1942. But the USAAF Headquarters staff was still largely convinced that existing and proven technologies should not be risked or replaced by unproven concepts.

Intelligence was rapidly providing proof that reports of exceptional speed, and little or no means of defense against these propeller-less aircraft were valid. Soon it was confirmed that the threat was based upon turbojet technology. Though these aircraft were few in number, the concern that the arrival of greater numbers would prolong the war and rapidly we would be as far behind our enemies as in 1942. U.S. leadership at the highest level was also concerned that if the U.S. military failed to develop turbojet technology, allies and adversaries would perceive the nation unwilling or unable to lead the world that would follow World War II (WWII). The new technology was still in development in the United States and USAAF General Henry "Hap" Arnold was determined to accelerate that process and make it operational.

The turbojet engines that powered the B-45 family (the J35 initially and later the J47) were air-breathing gas-turbine engines. The gas-turbine had an air inlet or intake, a compressor (with multiple stages or fan-like disks mounted onto a common shaft), combustion chambers, and a turbine (another disk which endured extremely high temperatures and pressures), which drove the common shaft that rotated the compressor stages.

Air was drawn into the rotating compressor stages via the intake and then compressed to a higher pressure before entering the combustion chambers. Fuel, initially aviation gasoline, then pure kerosene known as Jet Propulsion 1 or JP-1) was mixed with the compressed air and ignited into a continuous burning flame within the combustion chambers. These chambers surrounded the shaft extending between the turbine and the compressor section. The combustion gases left the combustion chambers under very high pressure and temperature and flowed across the turbine blades, which in turn drove the compressor section via the compressor shaft.

The turbine exhaust, which still contained considerable energy, then forcibly expanded in the propelling nozzle, also known as the tail pipe. There it began to expand through the tail pipe and applied pressure upon the nozzle cone and the funnel-shaped plug mounted on the aft side of the turbine. These pressures produced the forward propelling force known as thrust (the opposing forward force or reaction to the hot air mass exiting to the rear) which is the power of a turbojet engine.

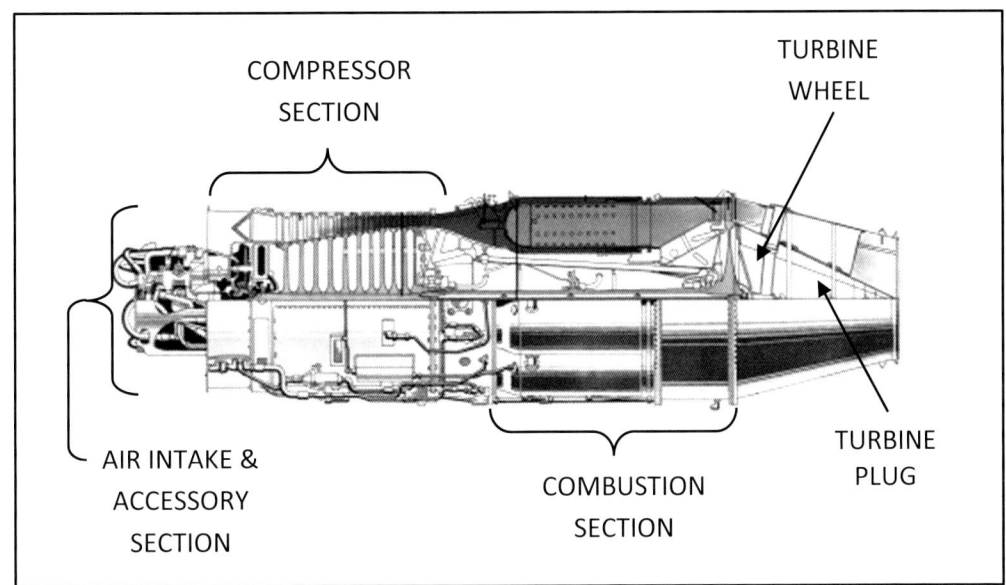

Drawing 1: This drawing shows the major components of the General Electric J35 turbojet engine. (Author)

Operational turbojets through the end of WW II featured one of two different compressor section designs. The designs differed in how the compressor directed the air flow that entered via the intake through the compressor stage or stages, and into the combustion chamber or chambers. The first design was known as the axial flow compressor. Air from the intake entered and travelled through the compressor stages parallel to the turbine shaft that powered the compressor, and into the combustion chambers along the axis of the engine, then through the turbine and out via the exhaust nozzle or tail pipe. This design was based on a paper published in 1926 by A.A. Griffin in the United Kingdom.

The axial flow compressor's advantages were higher efficiency and larger air-mass flow rates, particularly in relation to its diameter. This was because the design caused almost no impact on the directional flow of the intake air through the compressor stages and the combustion chambers, leading to a smooth and efficient airflow throughout the engine. However, to produce more thrust, additional compressor stages and/or turbine stages would be required. This would lead to additional length, weight, complexity, and cost.

The second design was known as a centrifugal compressor. It featured a single or double compressor stage and the intake air often entered that stage at an angle of approximately 45^0, meaning an immediate change of direction as it passed through the compressor. Once in the compressor the air was redirected at least 90^0 or more toward the outside of the engine casing by centrifugal force from a large impellor. The compressed air would be redirected for a third time into the combustion chambers for ignition. After combustion, the air flowed to the turbine and out of the engine via the exhaust nozzle or tail pipe. In some centrifugal flow designs the intake air was redirected upward by 90^0, then forward by another 90^0 turn for combustion in the burner cans. Leaving the chambers, it was turned once more by 180^0 to flow aft through the turbine (**see Photos 5 and 12**).

The advantages of the centrifugal flow compressor were compactness, ease of design, and relative simplicity, as often it only required one compressor disc, and therefore much lower production costs. Further it was much shorter in length than an axial flow compressor turbojet, and for that reason the design is still often used to power rotary wing aircraft today. One disadvantage was the significant and frequent change in air flow direction which was not efficient at all. This led to lower thrust generation capacity, higher fuel consumption and poorer high-altitude capability. The most important disadvantage was that centrifugal flow engine was larger in diameter and the only method to increase thrust was to increase pressure by increasing the diameter and therefore the weight of the impellor. In addition to the large weight increase, material properties were also limiting the size of the impellor disk. The large diameter of the engine caused drag induced on the corresponding airframe design to be significant.

Developmental History….

Efforts to develop an operational and maintainable turbojet started independently in a number of European nations and the United States prior to WWII. The United Kingdom (UK) and Germany would be the most successful by the end of the War. Gyorgy Jendrassik in Hungary, and the General Electric Company (GE) in the US were early turbojet or turboshaft pioneers, but except apart from GE, all work was halted at the beginning of the War. German sources reported to the US Navy after the war that it was only the Imperial Japanese Navy that showed any strong interest in turbojets. And this did not manifest itself until 1942 after the Japanese Naval Attaché saw the performance of the Messerschmitt Me-262 and was quick to recognize what the aircraft meant.

Japan actively pursued three engines. Their first turbojet was by the First Air Technical Arsenal, and it employed a centrifugal compressor. It produced 770 lbs. of thrust and was to be used in the twin-engine "anti-invasion" bomber Kikka, loosely based on the Me-262 design. The Kikka was later designed to use the Ishikawajima Ne-10 centrifugal-flow turbojet, and the Ne-12, which added a four-stage axial compressor to the front of the Ne-10. Tests of this power plant soon revealed that it would not produce the thrust required, and the project was temporarily stalled. It was then decided to produce a new axial flow turbojet based on the German BMW 003, the Ishikawajima Ne-20, finally built in 1945. The engine weighed 1,000 lbs. empty and produced only 1,000 lbs. of thrust. At least one successful flight was logged in the Kikka and the US Navy came to believe that the low weight of the Kikka would offset the low power of the Ne-20 to such a degree that the aircraft would be reasonably effective against Allied propeller driven fighters. The Ne-20 was intended to power the Oka 43 "Baka" Suicide Attacker, giving the aircraft considerably more safety on the out-bound flight and more range. As all these were heavily based on German design and engineering, no further development was pursued by the Allies using what they recovered at various locations in Japan after the end of the war.

While three independently developed turbojet engine projects initiated the turbojet revolution, it was the work of Frank Whittle and Hans von Ohain that revolutionized aircraft propulsion by recognizing that the combination of a compressor and a turbine was uniquely suited as a power plant for flight. Sir Frank Whittle's **(Photo 1)** axial flow design began at RAF College Cranwell in 1928 and was patented in 1935. But it was not until 12 April 1937 that his first turbojet power plant, the Power Jets WU, became the world's first operating turbojet. It was a centrifugal design and included an internal fuel pump for liquid fuel. However, Whittle was unable to interest the UK Government in his creation and its development continued at a slow pace. The beginning of World War II in September 1939 began to change the UK government's attitude towards Whittle's invention. Whittle's first operational turbojet designated the W.1 (also a centrifugal flow design) was built under contract by the British Thomson-Houston (B.T.H) company in 1940. At the same time Gloster Aircraft designed the E.29/39, the first British aircraft built to be powered by turbojet power. A second Whittle design, built by B.T.H., the W.2 **(Photo 2)** was quickly fitted to the E29/39 **(Photo 3)**. It made the UK's first turbojet powered flight on 15 May 1941. After two improvement generations they emerged as the Rolls Royce Derwent centrifugal flow turbojet which powered the Gloster Meteor in the UK's air defense against the Fieseler Fi 103 Vergeltungswaffe 1 or Vengeance Weapon 1.

Photo 1: Frank Whittle posed for this photo during the War when he was an RAF Group Commander. (Wikipedia CC BE 03) **Photo 2**: The Whittle W.2B shown here on display at the Midland Air Museum in the UK powered the Gloster E29/39. (Wikipedia CC BE 03) **Photo 3**: The Gloster E.28/39 W4046/G was the first turbojet powered aircraft to fly in the UK. (Wikipedia CC BE 03)

Hans von Ohain **(Photo 4)** started to develop the turbojet engine effort in Germany during the same time period of Frank Whittle's work in the UK. While their turbojets were developed at roughly the same time, they were independent achievements. Neither Whittle nor von Ohain appear to have had any knowledge of the work of the other. Von Ohain's most significant achievement was the first flight of a turbojet-powered aircraft, powered by the Heinkel HeS 3 (also a centrifugal flow engine) **(Photo 5)**, in the Heinkel He 178 **(Photo 6)**.

Photo 4: Hans von Ohaim, photographed in the mid-1970s, developed the first operational turbojet in Germany just prior to WW II. (Wikipedia CC BE 03) **Photo 5**: This image of the Heinkel HeS 3, taken at the *Deutsches* Museum in Munich, Germany, shows the engine which powered Germany's and the world's first jet-powered aircraft in 1937. (Wikipedia CC BE 03) **Photo 6**: A wartime image of the Heinkel He 178 shows the first turbojet powered aircraft in pre-War Germany. (Wikipedia CC BE 03)

There were two other war-time turbojet efforts in Germany. While not the first, they were the most successful operational turbojets flown by the Luftwaffe during the War. One was a Junkers *Motorenwerke* project which led to the Jumo 109-004 **(Photo 7)**, the axial flow turbojet that powered the Messerschmitt Me-262. The exhaust area of the Jumo 109-004 featured a movable plug **(Photo 8)** called a *zwiebel* (German for onion due to its side-view appearance) that could travel fore and aft some 40 cm or 16 inches to vary the exhaust nozzle area. Narrowing the exhaust flow through the nozzle aft of the turbine created a more thrust-producing jet flow at the aft end of the nozzle.

The other was a project that the *Beyerische Motoren Werke* (BMW) obtained when they absorbed the *Bramo Motorenwerke* of Spandau, Germany. It was extensively re-designed through 1942 and the engine was designated the BMW 109-003A-1 or simply the BMW 003A-1. It was an axial flow turbojet with 7 compressor stages and a single turbine stage. A major innovation was its annular combustion chamber, which did away with the separate and multiple chambers or "burner cans". It was simply a continuous lining \
]and casing in a ring (the annulus) that enclosed the area between the compressor section and the turbine. The advantages were much more uniform combustion, shorter size (therefore lighter), and less surface area. By 1944 the axial flow BMW 003C **(Photo 9)** was producing over 1,700 pounds of thrust and would be used to power the four-engine Arado Ar 234C-3 and Heinkel He 162 *Spatz* or Sparrow fighter **(Photo 10)** in 1945.

Photo 7: This Junkers *Motoren Jumo* 004 turbojet was photographed at the Wings of Eagles Museum, Elmira, NY. (Courtesy Paul Cooper) Photo 8: This image shows a cut-a-way view of the *zwiebel* or onion, which moved fore and aft inside the exhaust nozzle to vary the diameter of the exhaust nozzle. (Wikipedia CC BE 03)

Photo 9: The BMW 003C powered the He-162 and the Ar 234C-3. This cut-a-way is on display at The *Militärhistorisches Museum der Bundeswehr - Flugplatz Berlin-Gatow* Airfield. (Wikipedia CC BE 03) Photo 10: The He 162 shown was obtained during Operation Lusty in 1945 and was photographed on the ramp at Wright Field, near Dayton, OH. (Wikipedia CC BE 03)

A third set of turbojet efforts began in the United States in 1940 by various companies including Lockheed, Northrop, and GE. More importantly, the National Advisory Committee for Aeronautics (NACA) together with GE began exploring designs for high pressure compressor systems before the War and had developed an 8-stage turbine-driven axial flow turbojet design by the end of 1940.

By the spring of 1941, at the urging of the British Minister of Aircraft Production Lord Beaverbrook, General Hap Arnold **(Photo 11)** directed the USAAF to begin a dedicated effort to build copies of the Whittle W.2B turbojet in the US as an insurance policy if the UK were invaded by Germany. Westinghouse, Allis-Chalmers, and GE were the favored contractors due to their long and successful history in steam-driven turbines, as well as expertise and production of Turbo Superchargers for internal combustion engines powering the B-17, P-38, P-47 etc.... In July 1941 Arnold selected GE to be the sole contractor on the program (known as the GE Type I Supercharger for security) to copy the Whittle engine and be the primary USAAF source for turbojet research and engines in the near term.

In late 1941, General Arnold contracted with Bell aircraft to design and produce a "pursuit type aircraft" to use the GE Type I-A engine **(Photo 12)** as its power plant. This effort became the Bell XP-59 **(Photo 13)**. Drawings of the Whittle W.2B were sent to the GE turbo-supercharger department at Lynn, Massachusetts and the project began under the leadership of Reginald Standerwick. The first GE Type I-A engine ran in March 1942 after only 8 weeks of work. Following several improvements efforts it became the GE I-16, two of which powered the first flight of the Bell XP-59 on October 1, 1942. Work on this power plant design continued through four additional refinements and it was designated as the J31-GE-4 by the USAAF in late 1944. The design was also the basis of the centrifugal flow I-40, which as the GE J33, powered the Lockheed P-80 and T-33 families.

11: USAAF General Hap Arnold, shown here just prior to his retirement, was fundamentally responsible for US turbojet engine development. (Wikipedia CC BE 03) Photo 12: This GE Type I-A turbojet engine is shown on display at the National Air and Space Museum's Udvar-Hazy Center. (Wikipedia CC BE 03) Photo 13: The Bell XP-59A was the USAAF's first turbojet powered aircraft. (Alan Griffith)

Arnold chose the Whittle engine over the NACA design although he was aware of the better performance potential from the axial flow compressor. But time was of the essence. The Whittle design was immediately available and already proven to be airworthy. While he did not choose the axial flow design to power the XP-59, he did not discard it. The steam turbine division of GE at Schenectady, NY, was directed to develop an axial flow turboshaft engine using an axial flow compressor design based on work with NACA. The GE TG-100 **(Photo 14)** program, led by Glen Warren, was designated the GE T31 turboprop and powered the US Navy (USN) Ryan XFR-1 on its December 21, 1945 first flight. Continuing issues with the gear box connecting the turboshaft to the propeller led to its demise.

Photo 14: The United States Air Force Museum (USAFM) once displayed this TG-100 (T31) turbo-shaft at the Wright-Patterson AFB, OH Museum. (Author)

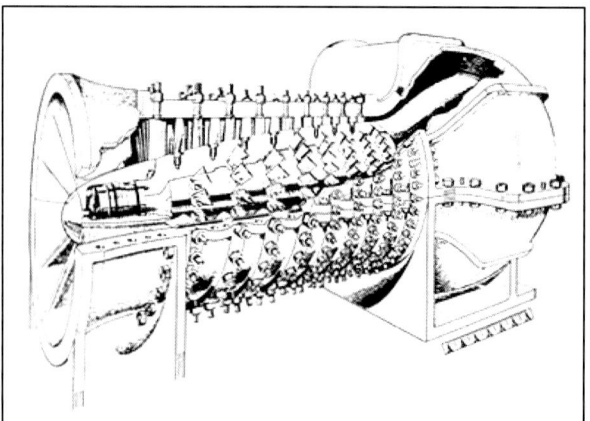

Drawing 2: This is a NACA drawing of their 8-stage compressor axial flow turbine which became the basis for a joint effort between NACA and GE which resulted in the GE TG180/J35. (NASA)

GE encountered serious issues meeting USAAF and USN production demands for the J35 and the USAAF transferred production to the Chevrolet Division of General Motors (GM), and shortly thereafter to GM's Allison Division in Indianapolis, IN. While a serious financial blow to GE, it forced the company to concentrate on solving its production shortfalls, thus laying the groundwork for the Company to produce the J47, one of the most successful turbojets on record. The final turbojet to play a role in the B-45's history was a development of the J35, known at GE as the TG-190 program and lead by Neil Burgess. It was similar in design and components to the J35, but fitted with one additional axial compressor stage, for a total of twelve, and was 6 inches smaller in diameter. Designated the J47 **(Photo 16)** by the USAAF, the turbojet developed 5,000 to over 7,000 pounds of thrust by using an afterburner. It could burn JP-1(kerosene), JP-2 (like JP-1 but with a higher freezing temperature), JP-4 (a 50-50 mix of kerosene and gasoline) and military aviation fuel (octanes 80/87, 100/130, and 115/145). It first flew on 20 May 1948 in the first production F-86A-1 (47-605).

Photo 15: The GE J35, shown on display at the National Museum of Naval Aviation, powered the first twenty-two B-45A airframes built by North American. (Courtesy Stephen Kahl)

The J47 went on to power all B-45A-5s, the B-45C, and the RB-45C. It then powered all series in the Boeing B-47 Stratojet family; the NAA F-86A, D, E, and F; the North American FJ-2 Fury; and the prototypes of the Martin XB-51; Republic XF-91 and the Chase XC-123A. It was added as supplemental power mounted in underwing podded-nacelles to the Convair B-36D and later series; the Boeing KB-50J and K; and the KC-97L. The J47 design was the first axial-flow engine approved in the U.S. for commercial airline use.

Photo 16: The J47 powered most of the USAF combat aircraft from 1949 through the mid-1950s, and over 36,000 were produced. (GE Engines)

Atomic Weapons….

The United States atomic weapons program was began on 19 January 1942 by President Franklin Delano Roosevelt (FDR) following a letter from Albert Einstein. This letter alerted the President to the importance of research on nuclear chain reactions and the possibility that such research might lead to developing powerful bombs. Einstein noted that Germany had stopped the sale of uranium and German physicists were engaged in uranium research. Further, FDR also received a National Academy of Sciences report determining that an atomic powered bomb was feasible. President Roosevelt instructed the Army to take responsibility for the construction of an atomic weapons complex, which was delegated to the U.S. Army Corps of Engineers on 17 June 1942. On 13 August 1942, the Army Corps of Engineers established the Manhattan District to develop and build the atomic bomb with facilities at Oak Ridge, Tennessee and Hanford, Washington.

On 19 September 1942, Brigadier General Leslie R. Groves, head of the Manhattan District, selected Oak Ridge, Tennessee as the site for facilities to produce nuclear materials. Isotope separation of uranium235 took place in the gaseous diffusion plant built in the K-25 area of the site, in the electromagnetic plant in the Y-12 area, and in the liquid thermal diffusion plant. Pilot pile reactor and plutonium isotope separation facilities were built and put into operation at the X-10 area. On 25 November 1943, Groves selected Los Alamos, New Mexico, as the site for a separate scientific laboratory to design an atomic bomb. On 16 January 1943, Groves selected Hanford, Washington, as the future full-scale site for plutonium production and isotope separation facilities. Three reactors, B, D, and F were built immediately.

The release of a fantastic amount of energy that was an atomic bomb required the near instantaneous creation of a critical mass of highly refined uranium or plutonium. There were clearly two ways to do this, one quite simple and the other very complicated. The simple way was to divide the critical mass into two parts: one was a set of rings made of uranium likened to a roll of Lifesavers; the other was a slug of uranium the same diameter as the hole in the ring. The roll of Lifesavers was fired by an explosive charge down a gun-barrel-like steel tube onto the slug, resulting in a critical mass. The much more complicated method was to use shaped explosive charges surrounding a plutonium sphere to compress it into a critical mass around an initiator.

The simple way, a so-called gun-type mechanism, was used to design the "Little Boy" later designated the Mk I. It was so straightforward that it was not tested as a bomb, per se, before it was dropped on Hiroshima, Japan on 6 August 1945. The complex mechanism, termed an implosion device, was tested because there was some uncertainty as to whether the essential exact timing of the shaped charges would function as designed. On 16 July 1945 Los Alamos scientists successfully tested one in the "Trinity shot" at Alamogordo, New Mexico. On 8 August 1945, the device called "*Fat Man*", was dropped on Nagasaki, Japan. Five days later, Japan surrendered. The "*Fat Man*" device was designated the Mk III about one year later. Basically, Oak Ridge remained the only uranium processing facility, and the Hanford, Washington facility was the sole location for plutonium processing. Los Alamos, now somewhat supported by a facility at Kirtland USAAF field at Albuquerque, NM, continued to be the sole weapons development facility.

The gun-type device was relatively inefficient in converting the uranium into energy, a significant shortcoming in view of the amount of processing it took to refine ore containing uranium to the purity needed. In part for this reason, only a limited number of Mk I's were built. Most were to be utilized by the Navy as a placeholder for their atomic weapon delivery capability from aircraft carriers using the North American AJ and the Lockheed P2V Neptune. Since both were bombers, they were big enough to carry the heavy Mk 1, and since a gun device was smaller than the implosion device, the Mk 1 was small enough to fit in their bomb bay.

Beginning in 1946, a formal program to "productionize" the "*Fat Man*" resulted in standardized drawings for the device and a standardized procedure to make new weapons. The device was then formally designated the Mk III atomic weapon. It would be two more years before enough uranium and plutonium were available to make new weapons and during that period here were fewer than two dozen Mk IIIs for the Strategic Air Command's SILVERPLATE B-29s to carry. The weapon was huge, five feet in diameter, 128 inches long, and weighed almost 11,000 lbs. Its dimensions alone made it unacceptable as a weapon for the B-45. Quantity production began in April 1947. By April 1949 there were 120 Mk IIIs to arm Strategic Air Command (SAC) AC SILVERPLATE B-29s and SADDLETREE B-50s.

On 1 January 1947 in accordance with the Atomic Energy Act of 1946, all atomic energy activities including those related to atomic weapons were transferred to the newly created the civilian Atomic Energy Commission (AEC). Beginning in 1947 two new production reactors were authorized for the Hanford site. Over the next five years as the Cold War intensified, the AEC greatly expanded the nation's atomic weapons complex. New facilities included three additions to the Oak Ridge gaseous diffusion complex; new gaseous diffusion plants at Paducah, Kentucky, and Portsmouth, Ohio; three additional plutonium production reactors at Hanford, Washington; five heavy-water reactors for producing tritium and plutonium at a site on the Savannah River in South Carolina; a reactor testing station near Idaho Falls, Idaho; a feed materials production center at Fernald, Ohio; component and assembly plants at Rocky Flats, Colorado, and Amarillo, Texas; a second weapons laboratory at Livermore, California; and a continental testing site near Las Vegas, Nevada.

Early Atomic Testing….

Between the summer of 1945 and late spring of 1948, there had been two atomic weapon tests. The first was the 16 July 1945 "Trinity" test in central New Mexico. The second test involved the detonation of one air-dropped and one submerged "*FAT MAN*" weapon as Operation CROSSROADS, on 1 July and 25 July 1946. It was held at the northeast lagoon, Bikini Atoll in the southwest Pacific to investigate the effect of atomic weapons on warships. *Trinity* and CROSSROADS had demonstrated their immense power. However, all the services and the atomic design and engineering community were dissatisfied with them otherwise. The "*LITTLE BOY*" was extremely inefficient, and the "*FAT MAN*" not far removed from a laboratory subject.

Operation SANDSTONE....

The first atomic weapon <u>development</u> test series by the AEC was Operation SANDSTONE. Operational planning began at Los Alamos in March of 1946 with some concepts that had been proven in 1945. The main goals of SANDSTONE that impacted the B-45's future were to test "levitated" cores or "pits", and to test composite cores. Levitation meant that instead of being immediately inside the tamper (a metal sphere that separated the explosive shell and the atomic pit), there would be an air gap between the tamper and the core, which would be suspended inside on strong, but small and very symmetrical wires. This would allow the tamper to gain more momentum before striking the core. The principle was to generate as much compression on the core or pit as possible to generate more atomic fission reactions. Two of the three planned detonations would use levitated cores.

The motivation behind the composite core was to make better use of the available fissionable material. The use of uranium-235 in an implosion weapon rather than the inefficient gun type "*LITTLE BOY*" was obvious development. However, while plutonium was more expensive and harder to produce than uranium-235, it fissioned faster, because it made better use of the neutrons its fission produces. On the other hand, the slower reaction of uranium-235 permitted the assembly of supercritical masses, making it theoretically possible to produce weapons with higher yields. By July 1945, both materials were being developed in a composite core containing 3.25 kilograms (7.2 lbs.) of plutonium and 6.5 kilograms (14 lbs.) of uranium-235. The composite cores became available in 1946 with double the yield of Trinity with much less fissile material required.

SANDSTONE was a complete success with detonations using basic Mk IIIs with the new levitated and composite cores installed. Detonations were conducted on 15 April, 1 May, and 15 May 1948. They proved a new design for the atomic cores or "pits" for atomic weapons. Even before the third weapon was exploded, the Director of Los Alamos National Lab immediately suspended the production of WW II vintage pits and directed that new weapons using the new cores and other improvements be available in the national stockpile by mid-1952. This decision meant that two problems were solved. First, serious concerns about the relatively small amount of fissile material available for weapons production was no longer an issue. Now hundreds or thousands of weapons were possible to produce. Second, new weapon designs using these new pits were now possible that significantly reduced the size and weight of atomic weapons. And for the first time, by 1949, weapons designers were developing designs that could be carried within two to three years by the B-45A to targets throughout most of Soviet occupied Europe.

The AEC had been already working on an improved weapon for the USAF and USN before SANDSTONE, known as the Mk IV. It now had a levitated and composite core, the ability to insert and remove the atomic core or pit once the aircraft was airborne and no longer over populated areas. It also required far less atomic material, about 7 lbs. of plutonium and 14.4 lbs. of uranium, half of what a MK III required. Further it had much better arming and firing components and circuitry, and better ground handling. It entered the inventory in March of 1949 and was retired by the end of May 1953. However, this weapon was still a 60-inch diameter bomb and another B-29 weapon, and in the future a B-47 and B-36 weapon.

The next weapon would be key to the B45's future. The Mk 5, an implosion fission weapon had a higher aspect ratio, not to mention being smaller. The AEC's goal was a lighter, smaller bomb with the same yield as the Mk IV or better. By August 1948 it had begun direct conferences and studies with the USAF on exactly what the service wanted. A September 1948 meeting settled on a 50-inch diameter, 6,000 lb. weapon agreed to by the US Navy (USN) and the Armed Forces Special Weapons Project. The Mk 5 design and dimensions were fixed by September 1949 for delivery in 1951. The final design of the Mk 5 in 1951 was 43 and ¾ inches in diameter, 132 inches in length, and a weight of 3,175 lbs., with a yield of 81 kilotons (KT), an explosive power of 81,000 tons of TNT using a levitated pit and composite core. These dimensions allowed the Mk 5 to fit into the bomb bay of a B-45 and it was among the initial weapons the Tornado carried in its atomic tour of duty in Europe beginning in June of 1952. Early models of the Mk 5 required the bomb/nav to enter the B-45A-5 bomb bay to insert or remove the pit in the weapon. Later models were equipped with an automatic In-Flight-Insertion/Extraction (IFI/IFE) system to insert or remove the pit that was controlled from the bomb/nav's station. The Mk 5 was carried by the B-45, B-29, B-50, B-36, B-47, and B-52 USAF bombers, and the AJ, and A3D carrier-based bombers of the USN. The A2J was not far in its development when it was cancelled and never carried the Mk 5, nor did the Martin P6M turbojet-powered seaplane patrol bomber which never carried it for the same reasons.

Photo 17: An early Mk 5 with manual IFE/IFI wis on dislay at the National Museum of the United States Air Museum. The two light recttagles were antenna covers for radar fuses. (Author) Photo 18: The two doors in front of the Mk 5 opened the core of the weapon for access to manually insert or remove the pit. (George Cully collection)

An even smaller diameter weapon resulting from SANDSTONE, to be carried by fighter bombers and light bombers, was the Los Alamos Scientific Lab and Douglas Aircraft Company Mk 7 weapon. The Mk 7 was compatible with a variety of pits and cores, which in various combinations could produce a yield between a single KT to 80 KT. The weapon weighed 1,700 lbs. and between 1,700 and 1,800 weapons were produced beginning in 1951. Its diameter was 31 inches and length 182 inches. It carried a core able to withstand the aerodynamic forces of being carried externally by tactical fighter bombers and the improved explosive sphere detonation system like the Mk 5's. The Mk 7, a very aerodynamically clean weapon, was considered by the AF Special Weapons Command (AFSWC) to be the most ballistically perfect tactical atomic weapon ever developed. Part of this ballistic performance was that the Mk 7 Mod 1 and later was spin stabilized using three spin tabs on the tail fins which caused a slow roll rate between one-third and one revolution per second. The roll greatly reduced dispersion (or uncontrolled errors in its flight path) and improved the circular error probable for the weapon. Its case, designed and built by Douglas Aircraft, was designed be carried by high speed (but below Mach 1) fighters such as the F-84E, F, and G, and the Navy's F2H-2 Banshee, at high or low altitudes. The weapon featured a smaller diameter casing and more aerodynamically efficient body than nearly all other implosion fission weapons.

All models of the weapon had an automatic IFI/IFE system controlled by the pilot, but the entire aft portion of the weapon had to be removed before the pit could be loaded onto the IFI/IFE gear. Of course, these tail sections had to be secured on or near the ramp or each aircraft, to prevent wind damage to the tail sections or other aircraft and equipment nearby. The tail section was then re-attached prior to take-off.

Photo 19: This Mk 7 weapon, on display at the National Museum of the USAF (NMUSAF) shows the center tail fin was retractable to provide takeoff rotation clearance for fighters carrying it under a wing. (NMUSAF)

The Unique Special Weapon….

Only five Mk 1 weapons, the gun-type weapon used on Hiroshima, had been built up to February 1950, and all had been retired by 1951. Following CROSSROADS, interest in the mission of destroying submerged or buried hardened targets was reemerging. The AEC, Los Alamos National Labs and the USN Bureau of Ordinance were realizing that the type of targets the RAF had dropped a significant number of Barnes Wallis-designed Tall Boys and Grand Slams on, were still there and likely to be put back into use by the Soviets and their allies. At first, the USAF opposed the effort over the amount of fissile material that must be committed to building gun-type weapons, but the Tactical Air Command's target research confirmed the findings of the other parties and began to push for similar weapons. The fact that the new weapon, designated the Mk 8, could be carried internally by bombers and externally by fighter bombers was a great advantage. Naval targets such as

Photo 20: The Mk 8 could be carried externally by fighter-bombers or internally by bombardment aircraft. It was designed to destroy or severely damage buried or other deep targets. (US Department of Defense)

Sub-pens and certain harbor structures as well as underground command posts, dams, bridge abutments, or other buried or underground factories simply could not be breached or dug out by implosion weapon air burst. As atomic planners began to look at more extensive target lists, deeply buried or highly reinforced structures had to be left off and a new gun-type weapon became more urgent.

The Mk 8 was essentially a much-refined Mk I LITTLE BOY design with greatly improved weight, ballistics, and size. It could be dropped at very high attitudes by long-range bombers, or at low-level by tactical attack and fighter aircraft. It was the first US atomic weapon to have post-impact-detonation fusing and a structure that could withstand the impact of high "G" decelerations that could set off the barrel-mounted charge too early. The weapon weighed less than 3,300 pounds, entered the U.S. stockpile in April 1952, and rapidly became a penetration weapon that would be valuable in a European conflict. The weapon could penetrate 30 to 40 feet of sand, 45 to 60 feet in loam, 85 to 100 feet in plastic clay.

The final 60-inch Atomic Weapon from Los Alamos

The Los Alamos Lab introduced another implosion weapon at this time, the Mk 6. This again was a 60-inch diameter weapon but was designed and built with a much lighter exterior case and internal components. The weapon also had a new 60 detonation-point circuitry that significantly increased the amount of explosive pressure that was delivered to the fissile "pit". More pressure on the "pit" meant more atomic-isotope fissile and atomic yield or power. Production began in July of 1951. It initially weighed 8,500 lbs., later reduced to 7,600 lbs. The Mk 6 became the main SAC fission weapon for many years arming B-29s, B-50s and the B-47. It was the last of the 60-inch diameter fission weapons designed and produced by the AEC.

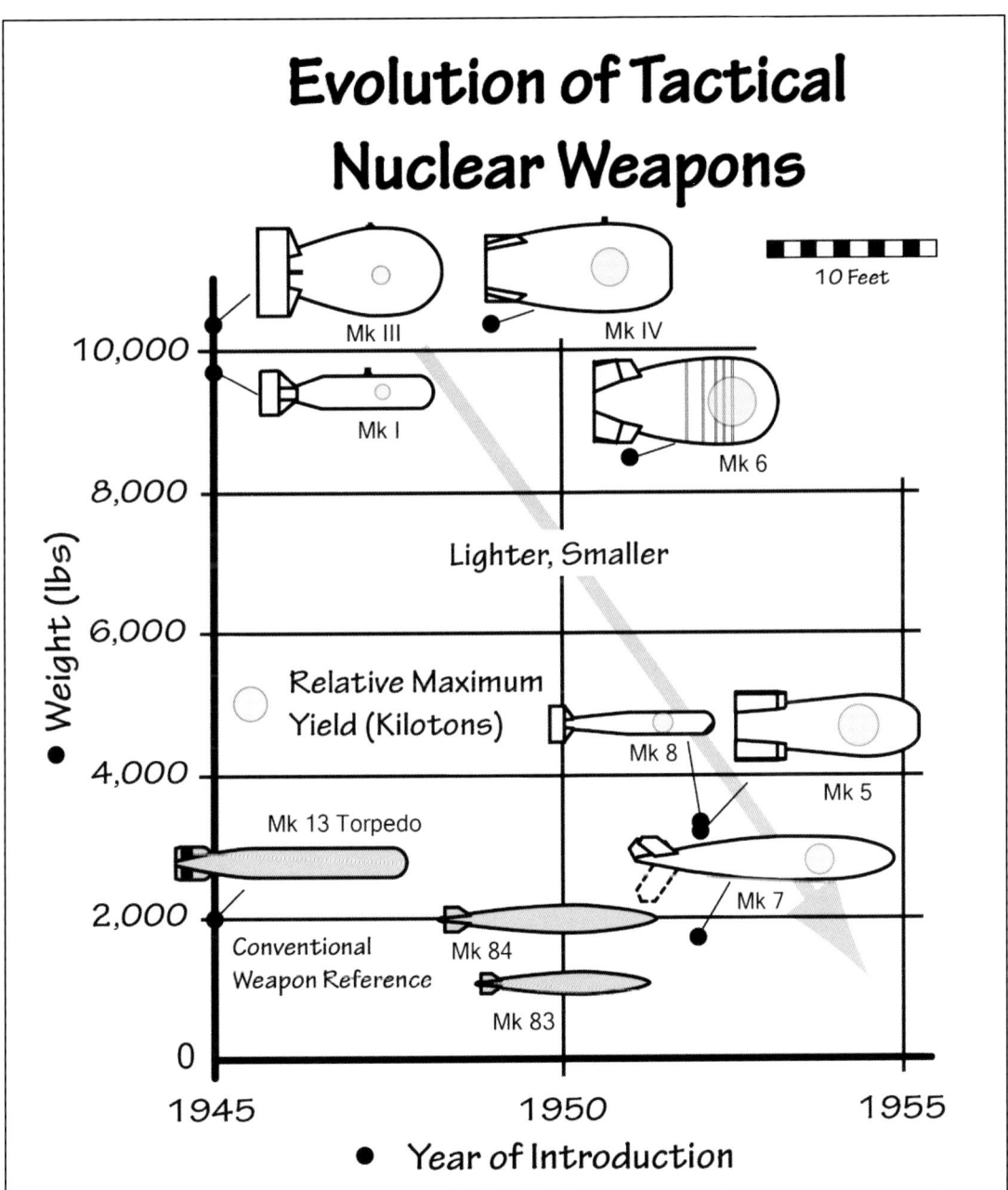

Drawing 3: This drawing captures the time-frame of stockpile entry, the relative yield between weapons, and the respective weights of the fission weapons in the US stockpile between August of 1945 and 1952. (Tommy Thomason)

CHAPTER 2 – THE XB-45s

The Start of the US Military Jet Age….

On 18 April 1942, GE's first turbojet, the I-A began testing producing 1,250 lbs. of thrust. By June 1943, it had developed into the USAAF's first production turbojet, the GE I-40, or the USAAF J33, centrifugal-flow engine, producing 4,200 lbs. The J33 was destined to power Lockheed's first production jet fighter, the XP-80 Shooting Star. Turbojet design and performance improvement continued and by 21 April 1944, the axial-flow turbine GE Model TG-180 went to test. It produced 4,000 lbs. of thrust significantly more efficiently than any other US- or UK-designed engine. With the I-40/J33 turbojet already flying in the Lockheed F-80, work remained slow on the TG-180, later the USAF J35, which first flew in February 1946.

GE, having experienced extremely severe issues with production and dependability of the TG-180/J35, sought a new project to reclaim the company's creditability with the USAF and the service's confidence in the company's ability to produce a more dependable and better performing engine. The new project, the TG-190, later known as the J47, was a "bet the company" effort with the commitment that if the TG-190 and/or its production failed, GE would immediately leave the turbojet engine business. With J33 and J35 production gone, GE had to make a key decision to make Lynn, MA the R&D center, and the Lockland Plant the production center for an engine that might find itself the power plant for almost every USAF and many USN jets for the next decade.

The TG-190, which was the final turbojet that played a role in the B-45's history, was a development of the TG-180. The program was led by Neil Burgess and the engine was similar in design and components to the TG-180, but fitted with one additional axial compressor stage, for a total of twelve, and was 6 inches smaller in diameter. The turbojet developed 5,220 to over 7,000 lbs. of thrust by using an afterburner. It first flew on 20 May 1948 in the first production F-86A-1 (47-605).

Effective Military Aircraft powered by turbojets….

An aircraft capable of putting the turbojet to effective use was the second major challenge to the USAAF from Pearl Harbor to the surrender on Tokyo Bay. It is also likely that General Arnold was briefed initially in early 1942 of Luftwaffe development of a multi-engine turbojet powered bomber or reconnaissance aircraft. To anyone directly associated with the turbojet, an aircraft powered by these near science fiction like devices might mean success or failure in conflict. The Luftwaffe development aircraft first scrutinized by the Allies was being designed by the Arado Company. The jet was first reported through an informant in the Arado design office in Potsdam, Poland as a "stratospheric bomber" known as the Ar 234. The first known plans were dated in December 1941. By May of 1943, Human Intelligence (HUMINT) reported that Arado was building a limited number of jet-powered prototypes at its plant at Brandenburg, Germany about 38 miles west, southwest of Berlin. The first photo reconnaissance image was made on 9 March 1944 at Brandenburg of a twin-engine prototype and at this time the Allies were reasonably certain that a bomber and reconnaissance version were being developed and both close to being operational. HUMINT again reported that an air and ground personnel training program had been established for KG 76 crews at Alt Lonnewitz Airfield, some 22 miles south of Templehof Airfield in Berlin.

The Blitz, or Rechlin 46, as it was known to Allied Intelligence, was next seen at Rechlin/Larz Airfield, 70 miles northwest of Berlin on 25 August 1944. A complete set of plans for the aircraft were obtained by US Army Ground Forces in September 1944. The Allies declared the aircraft was fully operational on 12 October 1944 following numerous combat reports in September of 1944. The Rechlin 46 was also confirmed as the Arado Ar 234 at the same time. Production of the Ar 234 was confirmed at Kupper-bei-Sagan Airfield, in Poland 50 miles southeast of Cottbus, Germany. By this time, the Ar 234s were considered a highly capable aircraft beyond interception in some missions but vulnerable to ground-based AAA in others. The only saving grace was that there were so few of them; the same advantage the Allies experienced with the Messerschmitt Me 262. Combat radio detection and radar and spotter's reports indicated that the aircraft had completed visually-unobserved high altitude reconnaissance missions at 30,000 feet and higher over the UK in early May 1945.

Photo 1: The Ar 234C-3 featured a pair of BMW 003C axial-flow turbojets under each wing in a very low drag nacelle. (NMUSAF) Photo 2: This image clearly shows the similarity of the XB-45 and the four-engine Arado Blitz, in this case an Arado 234 V19. (Author's Collection)

The "twinned" nacelle design was extremely streamlined and a great aid in dealing with asymmetrical thrust emergencies (the loss of one or more engines on one side of the aircraft). It also made the distribution of various loads inside the aircraft much easier while maintaining the Center of Gravity (CG) (the balance point of an aircraft in flight) within a proper "range" relative to the center of lift while in flight. The CG was a vital point calculated for every flight considering the empty CG and the aircraft's load in terms of weapons, cameras, fuel, crew, etc. and its distribution along the longitudinal axis of the aircraft. The pilot insured that the center of gravity was within limits of the load and in the event, it was disposable or used up in flight, without it. If the CG were too far forward, the aircraft could not be rotated for takeoff or flare for landing. If it was too far aft, the aircraft could be unstable longitudinally and therefore difficult if not impossible to fly. The Ar 234 nacelles, because they concentrated weight at the require CG, allowed large loads to be carried in the cockpit area, just ahead, above, and aft of the wing, ordinance on the center of the fuselage and engines, and a heavy set of cameras in the aft fuselage.

Significant intelligence was available on the Ar 234 from mid-1943, enough to convince HQ USAAF, that a serious effort was required to create a similar combat aircraft. By late 1943, HQ USAAF released some of the intelligence reports, drawings, and data on the twin engine Ar 234B to industry (the B model was twin-engine, while the C model was a four–engine aircraft). Sources at the GE Engine Division confirmed to the Author that Allied intelligence on the Jumo, BMW and Arado designs were shared with GE during design phase of the TG-180 and the NAA Model NA-130. An informal RFP based on an airframe powered by GE TG-180 engines was the first official document to express a formal interest in obtaining a combat-capable turbojet-powered bomber aircraft. The TG-180 was specified because it was the only power plant which would be available in the near future that could deliver the thrust to carry the combat load and range the USAAF felt was required. NAA initiated a design study for a four-engine turbojet powered bomber on 19 November 1943.

The first official NAA illustrations (**Drawings 1 and 2**), along with drawings and specifications were submitted to the HQ USAAF in response to their 1943 request. The illustrations outlined a NAA four-engine bomber powered by TG-180 engines. There is a definite influence from the Ar 232C Blitz, whose Intel reports were released to Industry throughout 1944. However, the nacelle intakes had yet to feature individual intakes for each engine that were separate all the way to the front of the nacelle. In fact, they matched those on early three views of the NAA Model NA-130 in response to the HQ USAAF letter. Another strong similarity in the two designs was the "twinned" nacelles under the wing each housing two TG-180 turbojet engines like the Ar 234C design.

Drawings 1 and 2: These are the first official Company illustrations of the North American Model NA-130, the company designation for North American's submission to meet the USAAF's need for a turbojet medium bomber. (Alan Griffith)

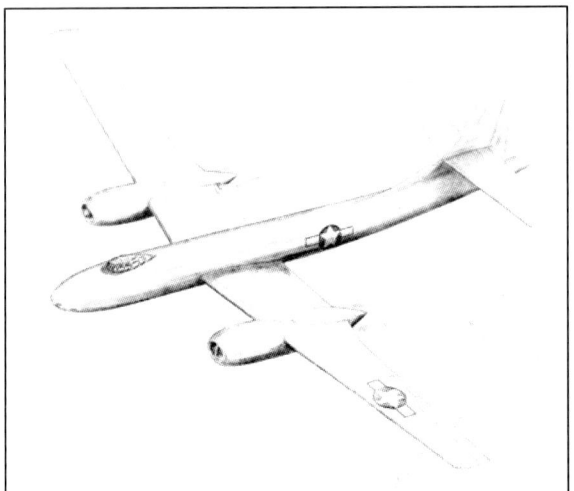 

Drawing 3: This is the conceptual design of the Boeing Model 413 that was submitted to the USAAF in January of 1944. This design was dropped by HQ USAAF shortly after receiving word in favor of a much more advanced proposal powered by six J35s. This new layout would be heavily influenced by NACA studies involving highly swept wing designs and would ultimately result in the XB-47. (Alan Griffith) Drawing 4: This illustration portrays a very early rendering of the North American Model NA-130 design, likely completed in early 1944. Note the engine intakes each feature two vertical dividers producing three intakes into the nacelles, two for engine air intakes and one for accessory section cooling air. They match those found on the Boeing B-413. (Steve Ginter Collection)

It is not known who the other recipients were of the USAAF letter, or how many replied, but The Boeing Aircraft Company did reply, and a graphic of their design exists (**Drawing 3**). The aircraft known as the Boeing Model 413 was like an enlarged version of the Arado 234C not unlike the NAA Model NA-130, having been completed after intelligence was received on the four-engine version of the aircraft in mid-1944. The Ar 234C was 42' 21.5" in length, had a span of 47' 3.25", and a height of l3' 7". The NA-130 was approximately 74' long, 89" in span and 25' high. The Boeing Model 413 was the largest of the three with a length of 109 ', a span of 139.5 feet, and a height of 32 feet.

Four engines were mandatory to come anywhere close to USAAF requirements, so Boeing quite properly anticipated that design feature. The design was in essence approximately 100 percent larger than a production Ar 234C with wings to lift a larger fuselage for fuel and payload. A crew of at least three was planned by location and size of glazed panels. The engine intakes were "twinned" like the Ar 234C but featured different intakes from those on the production Ar 234Cs.

Also received by HQ USAAF in early 1944 were conceptual drawings from NAA of the NA-130 being powered by the Allison Model 500 turbo-prop engines, in 1947 redesignated Allison T40s. **Drawing 5** shows a heavily armed wing tip ground attack aircraft, while **Drawing 6** shows a much more streamlined high altitude and speed bomber. The Model 500 power plant featured a pair of T38 axial-flow turbojet engines mounted side-by-side to a common frame with separate accessory sections for each engine. Combustion air entered through intakes in the front of each T38 and then routed to the individual compressor sections and turbine sections. Each T38 had a separate turbine shaft which extended from the front of each compressor section and led to a common reduction gear box. The separate T38s also had the ability to drive both counter-rotating propellers on its Model 500 if needed. It would be interesting to know if any Allied intelligence reports on the DB-021 PTL (turbo prop) to be used on the Ar 234 ever surfaced and were passed to NAA. Perhaps that was the origin of this NA-130 variant.

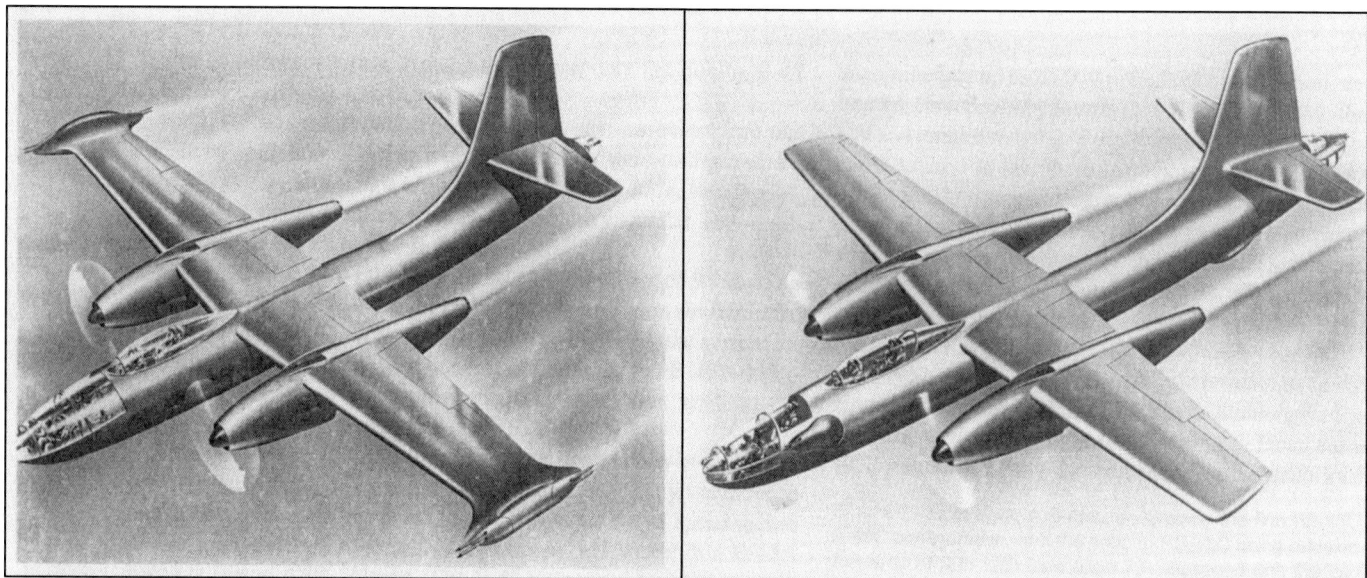

Drawing 5: This drawing shows a conceptual drawing of two Allison 500 counter-rotating turboprop engines powering a North American ground attack aircraft based on the NA-130 medium bomber design. The aircraft featured a mechanical twin gun manned tail turret for self-defense, and wing tip mounted forward-firing 20mm cannons for ground attack missions. Bombs and fuel could be carried in the fuselage bomb bay as planned for the turbojet NA-130 design. (Alan Griffith) **Drawing 6:** This drawing shows a conceptual drawing of two Allison 500 counter-rotating turboprop engines powering a North American design for a high altitude and high-speed medium bomber based on the NA-130 medium bomber design. The layout was basically the same as the ground attack variant, but this aircraft had no wing tip armament, with somewhat less weight and drag. However, it did lose a bit in both categories due to the pair of nose-mounted turreted 20mm self-defense weapons. It was fitted with a ventral radome to house a bomb/nav radar in a good location for drag reduction. It could also carry weapons and/or fuel in the fuselage bomb bay. Each design called for a crew of four, pilot, co-pilot, bomb/nav and tail gunner. (Alan Griffith)

Photos 6 and 7 show a later model Allison T40 on display at the NMUSAF. The engine was never used on a NAA design because it flew well beyond the availability date of the early NA-130s. It did however power the Republic XF-84H with its supersonic propeller in 1955.

Photo 3: This image shows a side view of the T40 without the engine's twin drive shafts and common reduction gear box. The compressor section of each T38 power plant featured a 14-stage section, hence the long length of that section. (Author) **Photo 4:** This forward view of the T40 shows where each of the two drive-shafts powered by the T38 turbine shafts left the twin engine just forward of the compressor section. As with all the twin drive shaft reduction gear box engines driving counter rotating props, the T40 failed because that component was not durable or reliable. (Gayle Hays)

On 25 August 1944, USAAF Air Materiel Command (AMC) issued a letter to NAA granting authority for procurement of preliminary engineering, wind tunnel models, data, and mockup of the Model NA-130. The Model NA-130 mockup was to be shipped to Wright-Patterson and NAA requested an increase of the letter contract for the project. The increase was approved and Letter Contract No. W33-038-ac-5126 dated 8 September 1944 was issued for a total of $226,311.90, or 30% of the price estimated in the North American's proposal.

Following the War, Arado, Boeing and NAA were joined by Avro and Avro Canada in using the "twinned" nacelle. Photo 5: This image shows an Avro Tudor on approach, (Author's Collection) Photo 6: This photo caught the Avro Ashton Mk 1, an airliner prototype used for high altitude testing by the RAF. (Author's Collection) Photo 7: The Avro Canada a C102 Jetliner, another prototype jet transport for Trans Canada Airlines also used a variant of the Arado 'twinned" nacelle" at the approximate location of the Arado's use. However, none went into production. (Author's Collection)

The 1944 Headquarters, USAAF Bomber Letters....

Encouraged by the response to the 1943 inquiries, and continuing alarm from more detailed intelligence reports about the capabilities of the Arado turbojet, on 17 November 1944 HQ USAAF requested the aircraft industry to submit proposals for various turbojet-powered bombers, with gross weights ranging from 80,000 to more than 200,000 pounds. Power would be from one of the GE family of engines starting with the 2,300 horsepower TG-100 axial flow turbo-prop engine through the 3,280 lb. thrust TG-180 to the 5,000-lb. thrust TG-190. The designs were to have a payload capacity of 8,000 pounds with a combat radius of 1,000 miles at a top airspeed of 500 miles per hour or greater. The maximum payload capability was the ability to carry a single 22,000-lb. Grand Slam bomb over a shorter distance.

The request effectively called for two designs, a light, and a medium bomber, based on range. There were few firm design requirements with two exceptions. First, the aircraft had to be powered by one of the GE engines; and second the "light" bomber had to feature a "conventional" design. This meant that design and engineering should be, to the greatest degree possible, based on aircraft and technologies already in service. It provided little emphasis on research into the realm of higher speeds, altitudes, pressures and temperatures in which the aircraft would routinely operate. Further it did not take into consideration that turbine-powered aerodynamics had barely been defined when the bomber proposals were requested. Would a "conventional" aircraft design really meet the need? No one in the United States had that answer.

NAA had submitted their proposal, the Model NA-130 (Drawing 7) in late spring 1944, and on September 8, 1944 in Contract Letter AC-5126, the USAAF contracted with NAA for the development and testing of three experimental aircraft, designated the XB-45. Company President James H. Dutch Kindelberger chose First Vice President J. Lee Atwood as the Technical Director of the program. He in turn chose Harold E. Dale to head the design. The aircraft was a very conventional design, meaning it was entirely based on aerodynamic and construction standards used for aircraft like the Martin B-26 Marauder and the North American B-25. The XB-45's strength and performance standards were increased from these light bombers to meet the anticipated multi-engine turbojet bomber requirements, but they were estimates. The aircraft soon evolved into a four-position pressurized light bomber, also powered by four J35s and featuring an easy to construct fuselage, upper-fuselage mounted straight wing, with retractable landing gear and a dihedral tail plane.

The North American Model NA-130 project continued with AMC and USAAF emphasis. On 10 February 1945 a fixed price contract, W33-038-ac-5126 was made with North American Aviation for additional wind tunnel models, tests, mockups and data for the now designated XB-45 aircraft for $506,861.62. Following a successful mock-up inspection at Inglewood, CA between 8 and 13 March 1945, a final construction contract for experimental XB-45 airframes was authorized later on 29 March 1945. On 4 December 1945 it was requested that the Chief, Procurement Division, HQ AMC execute a contract for the procurement of three XB-45 test aircraft as soon as possible. On 3 May 1946 Supplemental Agreement No. 1 to fixed price contract No. W33-038-ac-5126 was executed between the War Department (USAAF) and North American Aviation, Inc., for three USAAF Model XB-45 Airplanes, spare parts, special tools, tests, data, and certain service in connection therewith for a total cost of $12,723,075.58 in 1945 dollars, and $182,067,211.55 in 2021 dollars.

Soviet Hegemony in Europe....

By mid-1946 it was clear the Soviet intention to occupy all the area it had retaken from the *Wehrmacht* during World War II was fact, and a heightened sense of urgency gripped the War Department. Faced with the very real possibility of a new conflict in a variety of locations in Europe, the USAAF took immediate action to speed the delivery of the new light and medium bomber aircraft. Back in November of 1944, the USAAF originally planned to hold a formal competition between the contractors working on projects to satisfy the USAAF's 1944 "Bomber Aircraft Proposals Letters". Only four manufactures had replied, North American with the XB-45 (**Photo 8**); Convair with the XB-46 (**Photo 9**); Boeing with the B-47 (**Photo 10**); and Martin with the XB-48 (**Photo 11**). All four were told to begin design efforts immediately and all were soon very surprised at the depth of turbojet and turboshaft development in the US at the time. This indicates just how tightly the GE and Bell efforts on the I-16 and XP-59 project had been held. Throughout 1944 and 1945 they were read into USAAF and Allied intelligence reports on turbojet-powered Luftwaffe and RAF aircraft development and combat performance in Europe, and at least Boeing and NAA had preliminary designs in mind.

The USAAF decided to skip the normal contractor competition, design, review, and flight test steps, and directly choose from the four submitted aircraft, the XB-45, XB-46, XB-47, and XB-48. Headquarters would review the four fully constructed submissions to include flight tests and chose two winners beginning in the summer of 1946. The North American Aviation XB-45 and Convair XB-46 were selected as the light bomber candidates immediately. Both were conventional designs to be powered by four TG-180s. From a design and shape standpoint the XB-46 was graceful beyond words and possessed fine flying qualities. But it was much heavier, and the finesse of its design caused great concern about its ability to carry the significant amount of equipment associated with the radar and communications. Its overall performance, again due to its weight, was significantly below that of the XB-45 and it was cancelled after five months of flight test.

Boeing had been directly involved in modifying the MANHATTAN PROJECT B-29s to carry the "*LITTLE BOY*" and "*FAT MAN*" atomic weapons during the war and had a reasonable idea of what the requirements and capabilities to carry current weapons were. Boeing had submitted their Model B-413 (**Drawing 3**) to the USAAF in mid-1944 it was a conventional design. After learning of the swept wing designs in Germany however, Boeing went back to the drawing board. The USAAF opted to appraise the new XB-47 and XB-48 in the future, which might be two or more years down the road. The medium bomber completion was then between The Boeing Aircraft Company XB-47 (**Photo 10**) and the Glenn L. Martin Company XB-48 (**Photo 9**). Atomic weapons capacity was a major reason why the XB-47 was considered as an atomic bomber candidate from the very beginning, although only a very few at HQ USAAF knew that fact.

The existing contract with NAA was amended within one week and a new Letter Contract was approved on approved on 29 August 1946. A fixed price contract between the USAAF and North American Aviation, Contract W33-038-ac-15569 which called for an initial lot of 96 B-45As (NAA NA-147) plus a non-flying static test airframe, at a cost of $73.9 million was and signed on 20 January 1947.

Photo 8: The first North American XB-45 45-59479 was photographed over southern California in late 1947. (NARA II via Tommy Thomason)

Photo 9: The striking Convair XB-46 45-59582 was photographed over the San Bernardino Mountains shortly after its first flight on 2 April 1947. (NMUSAF)

The new XB-47 was an exceptional design from the beginning, while it would demand a tremendous amount of effort and funding to resolve the seemingly enumerable setbacks, it put Boeing at the lead in bombardment design and service for decades. The XB-47 had the performance, range, payload and reconnaissance capabilities beyond all others and it was a given that the aircraft would win the medium bomber competition.

Boeing chose the NACA-developed highly swept and very vertically flexible wing; lighter but stronger than any wing yet designed. The aircraft had a bomb bay that could accommodate the MK III weapon and a fuselage with room for an advanced bomb/nav radar. The XB-47 was selected as the USAAF's new medium bomber on 17 April 1946 and its first flight was on 17 December 1947. The first combat capable B-47B Stratojets entered service at MacDill AFB, FL in October 1952 and over 2,000 B-47 airframes were produced.

Photo 10: The first XB-47 46-065 was photographed on the ramp at Boeing Field, Seattle, WA in late 1947. (NMUSAF)

The XB-48 was more conventional than the XB-47, which was intended to be a groundbreaking aircraft from start. Both would be powered as prototypes by six GE J35s, with the XB-48 the first 6-engined turbojet bomber to fly. It was over 12,000 pounds too heavy and its three-engine nacelles per wing produced significant drag. Therefore, its performance was considerably below that of even the XB-45 or XB-46. The major reason the XB-48 was eliminated so early was the Martin Company's complete lack of atomic weapon design data and no preparation for that mission, again a result of Manhattan Project and AEC policy. The XB-48 did feature one novelty, the tandem main landing gear housed in the fuselage that had been tested successfully on a Martin B-26 about a year earlier. It was ultimately adopted onto the XB-47 with great success. The XB-48 program was finally cancelled in its entirety on 31 March 1949.

Photo 11: The Martin XB-48 45-59585 was photographed over the Chesapeake Bay in the summer of 1947. (NMUSAF)

The Inglewood Plant and Mines Field....

The three XB-45s were designed and built at the NAA plant at Mines Field near Inglewood and El Segundo, CA. This plant had built North American B-25s, including the "Doolittle Raiders" and the early versions of the P-51 Mustang and its Allison-powered export versions that were flown by the RAF. This work had all been accomplished on the same floor space and outdoor ramp areas that were used to manufacture and assemble the first three XB-45s and the first 22 B-45A-1s. Mines Field became the Municipal Airport of Los Angeles in 1930, and Los Angeles Airport in 1941. At that point it was leased by the USAAF from the city to support the aircraft production and delivery work of North American Aviation and Douglas Aircraft, as well as USAAF training. The airfield became the Los Angeles International Airport in 1949, labeled LAX by the Civil Aeronautics Board, and is today one of the largest and busiest international airline operating locations in the world.

The Inglewood plant, adjacent to the south side of Los Angeles Airport, would handle the production of the Navion four seat private aircraft, the F-82 Twin Mustang long range escort, the FJ Fury program, the AJ-1 atomic-capable carrier-based bomber, and experimental design and engineering that led to such aircraft as the X-15 Mach 6 research aircraft and the Mach 3 XB-70 Valkyrie bomber. The first two XB-45 aircraft **45-59479** and **45-59480** were true prototypes, while the third **45-59481** was a pre-production airframe used to development production methods and plan machine and tooling layouts needed for production aircraft. The following images of **45-59479**, taken in 1946, show the construction lines at the Mines Field plant which mirror those used throughout the industry, and some of the major components of the aircraft.

Photo 12: A WW II image shows the indoor production line of the North American plant at Mines Field, showing the same floor space used for the initial B-45 production some five years later. (Library of Congress) **Photo 13:** The third XB-45, 45-59481, sits on the production line at North American's facility at Inglewood, CA. Note the "twinned" engine nacelles whose intakes have yet again been modified from those used on the Ar 234C. (Gerald Balzer Collection via Tony Landis from HQ AFMC/HO)

The upper area of the assembly **(Photo 13)** was designed to house the aircraft's Aerial Observer Bombardier (AOB) radar and would ultimately be covered by a metal-framed transparent canopy. The turret-like feature in the upper level held the rotation gear for the radar antenna. The area below, cradled by the wooden supports, housed the large bomb/nav radar antenna and was ultimately covered by a fiberglass radome. None of the XB-45s carried the AN/APQ-24 bomb/nav radar, and few of the early B-45A bombers would have it installed until after 1951 when the B-45As were being prepared for deployment to the UK and service with North Atlantic Treaty Organization (NATO) in Europe.

Photo 14: The extreme nose of the XB-45 is shown on its fabrication jig. The dome in the central shelf would serve as a mount for the AN/APN-21 Bombing and Navigation Radar if fitted. (Alan Griffith)

The two triangular-shaped vertical mounts **(Photo 15)** would hold the AOB instrument panel. The starboard side was the plotting table for the AOB and to the right in the image was the AOB's seat. This seat would pivot counter counter-clockwise to make it possible for the AOB to get to the narrow passageway that led to the access hatch on the port side of the fuselage. The cockpit **(Photo 16)** was very basic in this view toward the starboard side. Note how far forward beneath the windscreen frame the pilot's instrument panel was located. This led to the aft windscreen frame obstructing the pilot's lateral line of sight. The annotated items in the image are: item 1 the checklists; item 2 the flare chute; item 3 the emergency hydraulic pump for the co-pilot; item 4 internally operated flight control locks; item 5 the pilot's cabin decompression control; item 6 the pilot's throttles; and item 7 the co-pilot's emergency fuel shutoff switch under a plastic guard.

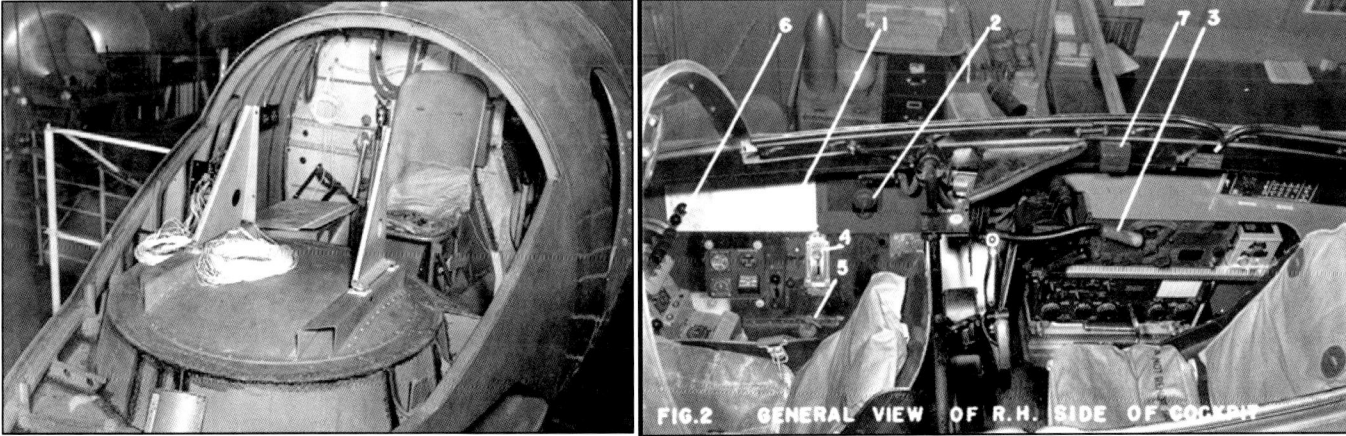

Photo 15: This image looks aft showing the AOB station which was covered with Plexiglas glazing on production aircraft. Note the bundle of aluminum wire laid on the AN/APN-21 radome mount. Some five miles of such wiring was used to connect controls, indicators, and components throughout the aircraft. (Alan Griffith) Photo 16: The cockpit view looks to starboard. Note that the aircraft did not feature ejection seats and that the canopy was fixed in place and could not be jettisoned. (Alan Griffith)

Photo 17: This image shows the starboard side of the forward fuselage. The opening was installed to fit a flight controls maintenance access hatch. The oval opening on the aft port side of the bulkhead housed a hatch to the bomb bay for inflight access. (Alan Griffith) Photo 18: This image shows the port fuselage as the exterior skin was held in place by spring-loaded clamps called Cleco fasteners. They fitted into pre-drilled holes through the skin and into the formers, clamping components together until they were removed individually, and replaced with rivets. (Alan Griffith)

Photo 19: The image shows the starboard side of the gunner's station at the very aft end of the fuselage. The hatch in Photo 18 is the gunner's access hatch and emergency escape route. The structures just forward of the hatch were supports for an airbrake assembly that gave the gunner a small area of smooth air to clear the aircraft in an emergency bail-out. No operational turrets were installed in the XB-45s, although 45-59479 was used to test aerodynamic shapes for the Emerson E-1 through E-3 turrets. (Alan Griffith) Photo 20: This image shows the empennage of the first XB-45 and the large leading-edge fillet between the vertical stabilizer and aft fuselage. The horizontal stabilizers are the original short span units. (Alan Griffith)

Photo 21: In this image the entire wing center section (a one piece assembly) has been mated to the center fuselage which also housed the XB-45's two bomb bays. The aft fairings for each pair of engine nacelles extended a short way aft of the wing and flap sections. (Alan Griffith)

The wing **(Photo 21)** featured three major components with a center section which spanned the bomb bays and extended to the outboard sides of the engine nacelles. The port and starboard outboard wing sections were bolted to the center section just outboard of the engine nacelles. Each wing outboard of the fuselage held eight self-sealing tanks of multiple cells and three spars. Self-sealing tanks had liners on the inside of the metal tank structure that rapidly plugged and sealed a hole from hostile fire or structural damage, preventing fuel loss and fuel or fumes from entering the aircraft structure. The aircraft's structure was a semi-monocoque design, which combined a tensile-stressed or flexible skin and a compressive structure made up of longerons and ribs or frames. This resulted in a much lighter, but stronger, structure to share and absorb the loads and forces the aircraft would endure in flight. Less weight in turn meant better performance and less fuel consumption. **Photo 22** shows the stringers riveted to the skin for strength and forming. These panels were then riveted to the ribs to enclose and shape the airfoil of the wing.

Photo 22: The image shows the internal structure of the port side outboard wing panel with ribs and spars. (Alan Griffith) Photo 23: The monocoque aluminum skin was riveted to stringers which in turn were securely fastened to the wing's internal structure. (Alan Griffith)

The forward engine bay doors were very large and heavy, and were bolted closed (**Photo 24**). Note how fine the junction between the engine compressor casing forward flange and the nacelle is. However, plumbing and mounts made engine changes lengthy endeavors even in good weather. The torpedo-shaped cowls at the center of each engine protected the accessory section of each engine. The accessories located here included the electrically power engine starters. Each circular engine air intake featured a slimmer band at its front, creating a narrower area between the cowl and the accessory section cowl of the engine. The intake duct then expanded in diameter as it extended aft over the compressor section. This was another NAA design that created a "venturi" effect for the intake air. This effect accelerated both the speed and mass of the intake air entering the compressors. This increased pressure and mass would mean a larger volume of air for the compressor to pressurize and direct to the combustion chambers. More volume produced in this manner meant more thrust from the engine without an additional fuel burn.

Photo 24: J35 engines number 1 (to the right) and 2 (to the left) are shown mounted in the port nacelle, looking aft. (Alan Griffith)

Photo 25: This nearly head-on view of the first XB-45, 45-59479 at Mines Field in late 1946 shows its overall layout. The nacelle anti-glare panels have only been applied on the inboard side of each nacelle. The North American Navions shown in the background were also built at this plant and note the North American P-82 immediately behind the XB-45. The image has a second element of historical aviation as well: the C-54 at the upper center of the image was one of two owned by the Santa Fe Railway Company to fly fresh flowers and meat from southern California to Chicago, IL. It was the only such operation by a railroad company in US history. (NARA II via Tommy Thomason)

Photo 26: The first XB-45 45-59479 was photographed on the North American ramp at Mines Field in late 1946, shortly before its surface move to Muroc AAF. (NARA II via Tommy Thomason)

First Flight....

NAA completed the first XB-45 in September 1946. The next five months consisted of system and TG-180 ground tests at Mines Field. It was then disassembled for shipment via truck **(Photo 27)** to the USAAF flight test facility at Muroc AAF, north of the San Bernardino Mountains and due east of Rosamond, CA. It arrived there in early February 1947 and was reassembled at the North American facility, a WW II wooden hangar known as Bldg. 4901 **(Photo 28)** in the desert next to the dry lake runway and prepared for its first flight.

Photo 27: The first XB-45 travelled by truck and trailers to Muroc Field in early February 1947. Multilane highways were still extremely rare in southern California at the time. (The Boeing Aircraft Company via John Fredrickson).

Photo 28: The first XB-45 45-59479 undergoes reassembly at Building 4901, Muroc AAF in February, 1947, just a month before the aircraft's first flight. The hangar was located on the area that was known as "North Base" of Muroc AAF. (The BBA Collection via the Author)

During this time at Muroc, a large flight test airspeed Pitot tube and yaw indicator were installed on a slender boom mounted in the leading edge of the port wing tip. By this time, the anti-glare panels on the nacelle upper surfaces had also been extended to cover their entire upper portions in the pattern common to all B-45s.

Photo 29: North American's George Krebs (left) was the Pilot and Paul Brewer (right) the Flight Test Engineer on 45-59479's first flight on 17 March 1946, seen here in a publicity image taken at the plant at Mines Field. (San Diego Aerospace Museum, SDAM, via Craig Kaston)   Photo 30: XB-45 45-59479 was imaged on the ramp beside Bldg. 4901 just a few days before its first flight at Muroc Army Air Field. First flight pilot George Krebs is the second on the left facing away from the camera, while flight test engineer Paul Brewer, the second from the right on the ramp, discusses a flight data recorder. (The Boeing Aircraft Company via John Fredrickson).

Photo 31: Muroc AAF Fire Crews were present at all test operations such as this engine test run just outside of Bldg. 4901 at North Base prior to the first flight of 45-59479. (The Boeing Aircraft Company via John Fredrickson) Photo 32: Another 17 March 1946 image shows the JP-1 fueling of the number 1 XB-45 in front of Bldg. 4901 on North Base before its first flight. (Steve Ginter collection)

Photo 33: It is possible that this image of North American's Flight Test Engineer Paul Brewer (left) and Flight Test Pilot George Krebs (right) suiting up was the taken on 17 March 1947, the day of XB-45 45-59479's first flight at Muroc. (SDAM via Craig Kaston) Photo 34: XB-45 45-59479 taxis its take-off on 17 March 1947 with Krebs and Brewer at the controls. (The Gerald Balzer Collection via Tony Landis from HQ AFMC/HO)

Unfortunately, confirmed images of **45-59479**'s first flight or even the three taxi tests made that day are exceedingly rare. **Photo 33** is the only one obtained by the author that is confirmed to show **45-59479** taxying to the dry lake edge just prior to its first take off. Even though it shows the pre-take off taxi, it also shows some configuration details of **45-59479** on its first flight. First flight images are ones that show only one large flight-test Pitot tube mounted on the port wing near the wing tip, and no translucent panels in the nose or canopy shade in the cockpit. There are several images **(Photos 35 through 37)** taken within a few weeks or months of that first flight which reveal the numerous small but significant appearance changes to the aircraft during early flight test.

Photo 35: XB-45 45-58479 touches down on Muroc dry lake during an early test flight. There is no sunshade in the cockpit canopy and the bomb/nav station is not covered with transparent material. However, a second flight-test Pitot tube has been installed on the starboard wing tip. (NARA II via Tommy Thomason) Photo 36: The first XB-45's single nose wheel just touches down as the elevator was near full up to hold it off the lakebed surface. Portions of North Base facilities are in the background. (Steve Ginter Collection)

Photo 37: By the summer of 1947, test work on the Mojave Desert had forced a few changes on the aircraft for crew safety and comfort. An opaque oval area was installed on the cockpit canopy immediately above the pilot and co-pilot seats to provide these stations with some protection against the direct sunlight. Little could be done for cabin air flow due to the fixed canopy, so a set of translucent nose canopy sections were installed as a source of sunshade while still allowing some illumination for the crew access hatch and companionway. (SDAM via Craig Kaston)

Other changes shown in **Photo 36** were that production B-45s would not see operational tail turrets installed until the late spring of 1952, so no cooling fixtures were installed. **Photo 38** shows good detail of the single nose wheel and the outboard side of the very large diameter main wheel and tire. The tail cone **(Photo 39)** was a fixed metal, non-functional cone as the planned, manned turret had yet to be fully designed and ground tested. **Photo 40** shows a good plan view of the starboard side of XB-45 **45-59479**

Photo 38: This image shows the single nose gear wheel and the size of the main gear wheel. (NARA II via Tommy Thomason) Photo 39: This image shows an aft view of the port side of the first XB-45 with no equipment installed in the tail turret station. (NARA II via Tommy Thomason)

Photo 40: This starboard side image also shows how aerodynamically clean the XB-45 was and would remain so throughout its operational career. The vane-equipped device on the starboard flight Pitot tube was one of a pair of flight test yaw indicators. Data from these instruments was collected on tape recorders housed in the nose and bomb bay. (SDAM via Craig Kaston)

The first flight lasted approximately one hour, and was limited to a max speed of 250 miles per hour due to a malfunctioning main landing gear wheel well up-lock. George Krebs and Paul Brewer reported that the XB-45 was a very smooth and controllable aircraft with very responsive controls. They considered it an easy aircraft to fly, handled asymmetric power very, very well and had no adverse stall characteristics at all. On 26 March after the gear door locks were repaired the XB-45 achieved 516 miles per hour on its second flight. As with any new design, problems and shortcomings became immediately apparent. First, the emergency escape means had little chance of success. It was awkward to exit the seats, and then thread along the companionway to get to the access hatch. There escape would bring them very close to the port nacelle in the very high speed-slip stream. Paul Brewer in later flights experimented with exiting the aircraft via the forward bomb bay using dummies and sandbags, but this was impossible as well. Anything released in the bomb bay tumbled violently, pushed into the slipstream briefly then up again into the bay and hard into the aft bulkhead. Engineers would design and install a set of air deflectors in front of a redesigned access hatch on production aircraft, but the answer was ejection seats which would take another year to be installed into production aircraft. None of the XB-45s would ever have them fitted.

Photo 41: This summer 1947 image shows the extended landing gear and the overall size of the main gear wheel well doors. Located between the fuselage and the main gear struts they could be used as speed brakes. Note the single nose gear wheel and tire, and the original short span horizontal stabilizer. (NARA II via Tommy Thomason)

The XB-45's flight controls were powered by two separate and independent 3,000 lbs. per square inch (psi) hydraulic systems with artificial "feel" provided to the control column and rudder pedals. The problem was that there was no method to transfer hydraulic power from one system to another in case of a system failure. If one side failed, that side lost its aileron and flaps and there was no back-up boosted system that could manage the flight controls on that side at high speeds. The fix in production aircraft was a system that would link the two independent systems. In an emergency, either portion of the system could then provide pressure to control surfaces on both sides of the aircraft. Another control issue was slowing the aircraft from cruise airspeeds to pattern speeds for approach and landing. The aircraft was not designed with airbrakes or spoilers and being a clean design, it did not lose airspeed even at idle thrust rapidly. The fix was a solely procedural one developed by the flight test crew to extend the main gear wheel well doors that almost doubled the drag on the aircraft. Fortunately, they were strong, well attached, and could withstand high airspeeds. The increased drag proved sufficient to establish a reasonable rate of decent in a manner that even inexperienced crews could handle.

Learning the Jet....

The spool-up delay of the TG-180s also meant that high power settings had to be maintained all the way down final approach along with full flaps. This was needed so the aircraft could generate enough thrust quickly enough to make a safe missed approach and go-around. Even with full flaps, approach speeds where over 130 miles an hour, higher than pilots flying propeller planes were used to. Landing rollouts were proportionally longer as well. These lengths were much more than matched by take-off distances; in fact, they were doubled. The B-45 would require airfields of at least 9,000 to 10,000 feet in length; a rare and expensive commodity in 1947.

The next major issue that **45-59479** uncovered was a lateral stability problem at altitude and cruise speeds. The aircraft produced an unpredictable but consistent yaw to either side. It was not severe enough to dangerously affect overall aircraft stability, but it would be significant problem for the bomb/nav system and created drag. The fix was a seven-foot extension on the fixed portion of the horizontal stabilizer, 3.5 feet on each side **(Photos 42 and 43)**. This change would be retrofitted to the first XB-45 and fitted to the following two XB-45s and all other B-45s on the production line. There were no accidents associated with this problem.

These images illustrate the change to the horizontal stabilizer to correct the lateral stability issue. Photo 42: The original short version horizontal stabs are shown. (NARA II via Tommy Thomason) Photo 43: The 3.5-foot extension on the port side is shown fitted to a B-45A-1. Note that the elevator was not lengthened in the modification. (Alan Griffith)

The XB-45 did nonetheless have exceptional performance for its time. Its ceiling was in excess of 40,000 feet, and it could carry a payload of 8,000 pounds at nearly 500 miles per hour. Its maximum bomb load was 22,000 pounds. While XB-45 **45-59479** could carry no weapons, the ballast required to reach its design gross weight to equal its designed max loads was carried in the two bomb bays. With both bomb bay tanks installed and full the XB-45 had a ferry range of over 2,900 nautical miles. However, some problems could not be solved using the prototype aircraft.

One was cockpit layout. **Photo 45** highlights the canopy framework obstruction problem on the XB-45. The pilot's head is exactly beside the aft frame of the windscreen, obstructing his peripheral vision and making formation flying and maneuvering within the airfield pattern extremely difficult. For example, it was difficult for the pilot to properly judge when to make the turn from base leg to final approach because his view of the runway was obstructed by the canopy frame. As mentioned above, the aircraft had no ejection seats, and the fix would be a significant redesign of the cockpit configuration with black powder-powered ejection seats and a new jettisonable canopy in production B-45As.

More XB-45s....

The second North American XB-45 **45-59480 (Photos 44 through 46)** was completed on 11 December 1947 at Mines Field and was bailed to NAA until 16 August 1948. The delay in USAF acceptance was due to ongoing pressurization system issues in both **45-59479** and **45-**

59480. Between 16 August and 10 November 1948 **45-59480** was sent to Wright-Patterson AFB, OH (WPAFB) for installation of AN/APN-3 Short-Range Navigation Equipment (SHORAN) as a substitute for the AN/APQ-24 bomb/nav radar system. AMC was very concerned that with the rapid approach of service-ready B-45As, and the Air Force needed to know for sure which conventional weapons the B-45 would be capable of using. **45-59480** and the other aircraft would be assigned to a weapons suitability project at Muroc AFB, and the aircraft returned to Muroc on 9 February 1949 to begin this Project.

Photo 44: This image shows the second XB-45, 45-59480, on the runway at Mines Field in late 1947. It was fitted with the larger span horizontal stabilizer during production. (SDAM via Craig Kaston)

However, its maintenance workload was extremely high due primarily to engine issues and the aircraft completed only one project sortie between 18 August and 18 November 1949. Its low in-commission rate made it a poor test airframe and on 17 October 1949 it was returned to WPAFB to have the AN/APN-3 SHORAN equipment removed. On 15 May 1950 the aircraft was assigned as a ground instructional airframe with the 3750th Technical Training Wing, Sheppard Technical Training Center, at Sheppard AFB, TX. The B-45 weapons compatibility tests were not completed for nearly three more years.

Photo 45: The second XB-45 45-59480 was photographed during a test flight shortly after its arrival at Muroc Air Force Base in 1948. It was first flown fitted with the wide span horizontal stabilizer. (NARA II via Tommy Thomason)

Photo 46: 45-59480 was temporarily assigned to Wright Field at Wright-Patterson AFB, H between October 1949 and May 1950 when this image was taken. During this time period 45-59480 was having its SHORAN equipment removed prior to assignment as a ground trainer at Sheppard AFB, TX. (NMUSAF)

The third XB-45, **45-59481** (**Photo 47** and **48**) was accepted by the USAF on 16 June 1948 and immediately bailed back to NAA at Inglewood from 17 June 1948 to 28 October 1948. It was then permanently transferred to the 2750th Air Base Wing at Patterson Field, WPAFB for modifications and test work. She was then assigned to ARDC and the Wright Air Development Center (WADC) at Wright Field, WPAFB on 16 May 1951. On 24 January 1952 she was assigned to the 4901st Support Wing, AFSWC, KAFB, under ARDC until September 1952. At KAFB the aircraft was involved in getting AFSWC ready to receive and test turbojet bombers in dropping developmental atomic weapons shapes at the Salton Sea Range and live atomic weapons at the NTS. Its time at KAFB was spent developing modification procedures for installing the specialized racks and equipment to carry the new small-diameter atomic weapons developed by AEC. The work at KAFB proved that the B-45 airframe could indeed carry these new weapons.

Photo 47: The third XB-45 45-59481 was photographed prior to its first flight at Mines Field in June 1948; the aircraft was fitted with the wide span horizontal stabilizer during construction. (SDAM via Craig Kaston)

Photo 48: The port side of 45-59481 was photographed at Mines Field in June 1948; and she was assigned to Wright-Patterson AFB and Muroc AFB from October 1948. (SDAM via Craig Kaston)

The Details of Flight Testing for a new Design....

With three aircraft available, even with engine issues and a high maintenance workload per flight hour, flight testing on the XB-45 went quickly, beginning with the first flight on 17 March 1947. Aircraft **45-59479** and **45-59480** flew almost all of the Flight Test load for the XB-45 with **45-59481** intended to fly weapons systems and radar tests at WPAFB, Muroc AFB and KAFB. But that aircraft was unable to complete them due to maintenance issues. Both NAA, USAAF and USAF crews flew flight test missions.

The Phase I Flight Test Program gives a detailed look at what important data was desired and what specific flight test performance goals were set for the initial XB-45 test program. There were four phases: Ground Tests; Initial Flights; Preliminary Flight Tests; and Final Compliance Tests. Ground tests would establish the final overall weight of the aircraft and C.G. location, essential for all future flights. The structure of the aircraft would be tested next for control surface deflection, torsional and bonding stiffness, and vibration and flutter with and without hydraulic boost. All door operations for the landing gear, bomb bay, nose gear, and cabin access would be checked to determine loads and door deflections in operation. All cabins would be pressurized to verify load capacities and check for structural defects. The wing and fuselage would have its torsional and bending stiffness checked.

Once structural tests were complete, systems and power plant checks and tests were then completed. All engine installation, controls, fuel, and oil systems were checked and approved. Each engine was individually started and shutdown; operated at various power settings and under various normal and emergency situations to insure proper operations in all anticipated situations. These same checks were then completed with both engines operating in each nacelle separately; followed by completing the checks again with both engines in each nacelle operating; and finally, with all engines operating. Temperatures were repeatedly checked, and carbon monoxide build up checked in all crew positions. Pressurization and conditioning units; hydraulic flight control boost mechanisms in normal and auxiliary operation; and landing gear wheel assemblies where all thoroughly checked. Flaps and bomb bay doors function and controls were checked.

Following the completion of checks of all radio and electrical equipment on board the aircraft, including data collection, taxi tests began. They were conducted on smooth surfaces at low and high speeds, into and with the wind and in crosswinds on both sides of the aircraft, included a check of longitudinal stability. Brakes and braking systems; the steering mechanism; hydraulic systems; vibration and shimmy damper; carbon monoxide levels at crew stations; engine functioning and cooling were checked during taxi tests. Additionally, high speed taxi tests would test steering, handling, nose wheel shimmy and nose wheel lift off and settling were checked.

Initial Flights Summary....

The preliminary flight tests were the main goal of the First Phase testing. They included: a check of longitudinal stability to obtain data that would allow unaccelerated flight in a neutral C.G. position; dynamic longitudinal, lateral, and directional stability; collecting sufficient data to allow accelerated flight in a neutral C.G. position; aileron effectiveness; directional and lateral control and stability in normal and asymmetric power operation; and effectiveness of trim tab devices. Next came power plant installation tests in level, climb and high-speed flight situations. Take-off and landing tests would follow to determine basic take-off and landing lengths; flight control and aircraft effectiveness; cooling and functional checks of the power plants; and acceleration and stress measurements. Preliminary structural demonstrations came last before the military's Final Compliance Tests. These tests, made at the design gross weight of the XB-45 (82,600 pounds), included dives, climbs, rolls, and accelerated level turns at various "G" loads to establish the minimum load bearing capability of the airframe. With these tests complete the final critical tests were executed to ensure the aircraft complied with power plant, performance, and structural integrity demonstration requirements. With some waivers, the XB-45 passed all three Phases of its Flight Test Program on 13 April 1948.

Research and Development at Muroc....

During **45-59479's** last year at Muroc, assigned to the 2759th Air Base Group, it conducted two important research and development (R&D) projects. The first was flight tests of the **Aerojet XLR13-AJ-1** through **5** liquid fueled rocket-powered ATO devices for the B-45. Only two airfields in the country at the time, MacDill AFB, Tampa, FL and Smokey Hill AFB, Salina, KS would allow an aircraft like the B-45 to safety operate from them without ATO. It was a serious operational short coming of critical proportions particularly if the XB-47 and B-36 faced similar restrictions. The objective of the project was providing a jettisonable booster that would allow a light bomber to take-off and clear a 50-

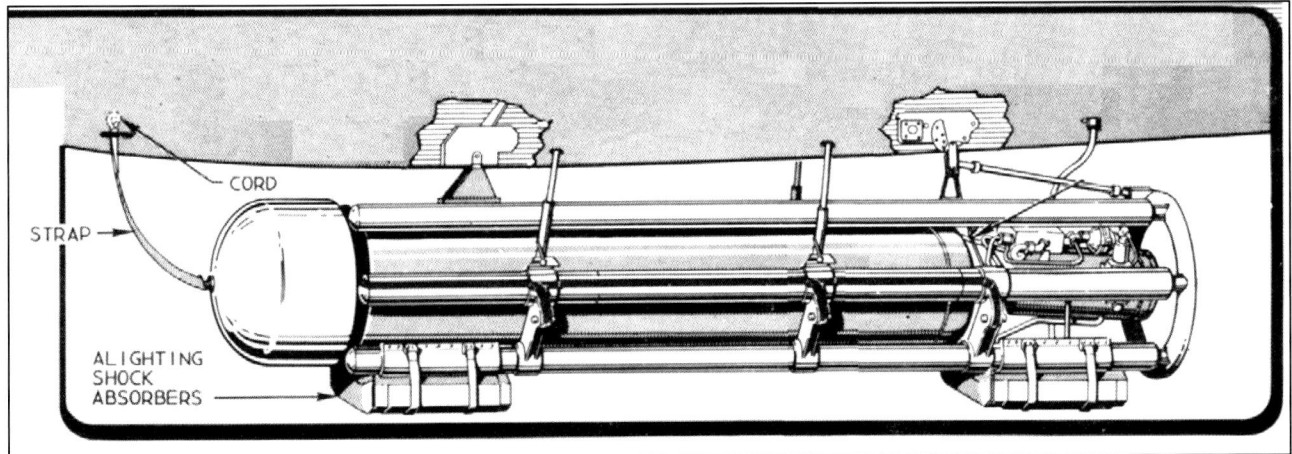

Drawing 7: This illustration shows the basic layout of the Aerojet XLR13-AJ-5 ATO pod. A single pod was suspended under each nacelle and was nnnnnnnnnnnnnnjettisoned about 40 seconds after ignition. (NMUSAF)

foot obstacle and not exceed a take-off distance of about 7,500 feet. Approximately 4,000 lbs. of thrust was generated by each unit over 30 seconds giving the aircraft the equivalent thrust of two additional J35s at full power.

Two pods were carried, one under the centerline of each engine nacelle. The propellant components were based on German R&D with the *Wasserfall Ferngelenkte FlaRakete* (Waterfall Remote-Controlled Anti-Aircraft Rocket designed and built by *Flak-Versuchskommando Nord*, Peenemünde, Germany). The propellant was composed of Red Fuming Nitric Acid (RFNA) oxidizer and dinitrogen tetroxide N_2O_4 fuel, both pressurized in their tanks by nitrogen gas. This propellant combination was hypergolic, producing a violent reaction or explosion on contact. was extremely dangerous to handle, store and use. Severe injuries or death would result from skin contact, inhalation, oral intake, by flame or an explosion of either component.

Photo 49: A German *Wasserfall Ferngelenkte FlaRakete* was photographed launching from the Wehrmacht rocket test facility at Peenemünde in late 1944. (Wikipedia CC BE 03)

Development began in June, 1948 by Aerojet Corporation. The XLR13-AJ-1 consisted of a strong steel tubular structure that held a circular ceramic chamber with two internal tanks inside the circular metal structure of the rack **(Drawing 7)**. These tanks, one for each propellant component, were pressurized by gaseous nitrogen from tanks enclosed by the metal streamlined cap at the front of the rack. Since the fuel components were hypergolic, no igniters were needed. The single ceramic nozzle angled down about 10 degrees and was at the very aft end of the open tubular structure. A pair of mounting lugs and four sway braces were mounted to the external structure of the pod's frame. The unit was jettisonable, but not re-usable as it had no parachute recovery system. The major shortcoming in liquid propellants was that they were extremely toxic through inhalation or contact and absorption through the skin. In the late 1940's protection suits and handling gear were in their infancy and handling of the components meant extreme risk at all times.

The fit of the XLR13-AJ-1 on the XB-45 was confirmed at WPAFB in the fall of 1947. Many questions were discovered and solved there; however no test launches were conducted at WPAFB because of the lack of long runways and the possibility of severe injuries to both base and the local civilian population due to insufficient clear area for those launches. Those would occur at Muroc AFB in 1948 where some 15 launches were conducted, all successful, with no incidents. The final assessment on the Red Fuming Nitric Acid (RFNA) oxidizer and dinitrogen tetroxide N_2O_4 fuel boosters was ruled negative primarily due to the safety issues. This propellant combination was extremely dangerous to handle, store and use.

Despite the continuing hazardous nature of these units, they did solve the take-off issues, saving about 2,500 feet of runway length on a max take-off weight of just over 110,000-lbs. fully loaded with up to 22,000 pound max bomb loads. No structural changes were made to the units and apparently only the XLR-13-1 units were tested with live launches. Muroc was the only facility that had the clear land and unencumbered surrounding area that could withstand any form of accident without putting significant numbers of military personnel and civilian residents in danger. Next the handling to include transportation from storage to a remote area of the flight line and loading was extremely dangerous. And finally, little or no effective personal protective gear had been developed or was on the horizon to protect the propellant handlers from even minor incidents.

Throughout the 1950s both the oxidizer, RFNA, and an aniline-based fuel were continuously developed. The oxidizer was mixed with varying amounts of nitrogen tetroxide (N_2O_4) for stabilization purposes until it was discovered that N_2O_4 was the best available storable oxidizer. Aniline-based fuels were slowly replaced by hydrazine-based fuels until unsymmetrical di-methyl hydrazine was developed and determined to be the most powerful and storable liquid rocket fuel. These successful components were fielded in the first storable-propellant Intercontinental Ballistic Missile (ICBM), the LGM-25C Titan II, designed and produced by the Glenn L. Martin Company. The Titan II was also the highly successful launch platform for the National Aeronautics and Space Administration's Gemini Program during the US's efforts for putting a man on the moon.

Photo 50: An Aerojet Corporation XLR13-AJ-1 is suspended under the port side nacelle of XB-45 45-59479 on the north-south ramp at WPAFB's Wright Field, Dayton, OH. (HQ AFMC/HO via Tony Landis)

Photo 51: XB-45 45-59479 was photographed at Muroc AFB, CA in late 1948 carrying two XLR-13-AJ-1 ATO pods. (Courtesy the Boeing Aircraft Company)

Photo 52: The Strategic Air Command had one more very closely held and unique need for ATO. It operated a single squadron unit based at Ladd Air Force Base near Fairbanks, Alaska initially known as the 46th Reconnaissance Squadron (RS), and later redesignated as the 72nd RS. The Aerojet Booster was successfully tested and mounted on the unit's B-29s for use in winter to ensure they could conduct their radiation monitoring of the USSR year-round. (HQ AFMC/HO via Tony Landis) Photo 53: A B-29 likely assigned to the 46th or 72nd RS takes off boosted by two Aerojet ALR13-AJ-1 ATOs at Muroc AFB. (HQ AFMC via Tony Landis)

Photo 54: An early test flight of XB-45 45-59479 carrying two XLR-13-1 from Muroc AFB was 100% successful, as were all the XB-45 /XLR-13 tests. It was the handling of the propellants for fueling and its hazards to those in the near areas of use of the ATOs that ultimately wrote an end to liquid fuel ATO's later in the B-45 story in 1951. (NMUSAF)

Project Number 2 at Muroc....

Prior to its transfer to WPAFB, XB-45 **45-59479** carried out a second R&D project at Muroc. It carried out flight tests on design shapes of a new Emerson Electric Company Model A-1 through A-3 visually aimed twin .50 caliber tail turret. (**Drawing 8 and Photo 54**). The radar's receiving antenna was to be mounted between the two .50 caliber guns in the center of the turret, the sight to be placed centered on a shelf inside the gunner's station just aft off the gunner, and the electronics to be placed at various locations throughout the aft unpressurized fuselage. The non-functional turret shape mounted on **45-59479** was made of wood, and could be manually adjusted on the ground to test yaw and pitch effects on the aircraft at various turret positions during different airspeeds,

Photo 55: This image shows the original design of the Emerson A-3 self-defense rear tail turret intended to be controlled for ranging and visually controlled for direction. The flight test dummy shown mounted on 45-59479 was not powered and aerodynamic tests required a manual positioning before take-off for data collection. The blister in the upper center of the turret between the two .50 caliber machine guns was to be the location of the range only radar. (The Steve Ginter Collection) **Photo 56:** After some two and one-half years of effort by Emerson, the radar was declared a failure, and without it, the turret would have been a failure as well. Before production of the B-45A-1s began, North American designed a streamlined tail cone to reduce the drag aft of the gunner's enclosure. All B-45As, B-45Cs and RB-45Cs were fitted with this enclosure at production. It was only removed on B-45A-5s fitted with an Emerson A-6 turret for service in Europe and RB-45Cs fitted with a Bell M-7 for combat duty in the Pacific Theater and later service in Europe. (The Boeing Aircraft Company via John Fredrickson)

However, the radar system was never successful on the turret and the project was ended in mid-1949. The original tail cone (**Drawing 9**) developed for the first flight was then produced and installed on all B-45s produced. That cone would remain installed unless that particular airframe was modified to carry later Emerson or Bell turrets. The work was not wasted however, as the design was very similar to the Emerson turret installed on the B-47B a few years later.

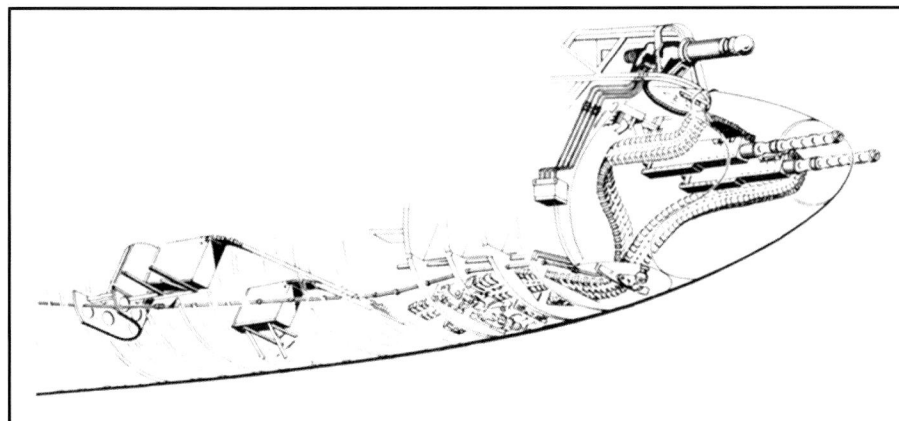 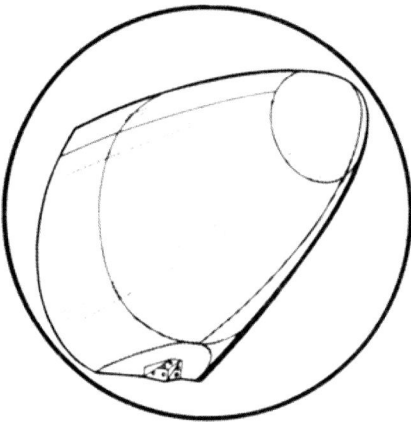

Drawing 8: A technical order drawing of the Emerson A-1 turret tested on the first XB-45 shows the optical sight above the turret and the twin .50 caliber guns. The ammunition boxes were in the center of the aft fuselage for center of gravity purposes. (NMUSAF) Drawing 9: A tech order drawing shows the NAA designed and manufactured tail cone for production aircraft. (NMUSAF)

Finale at Wright-Patterson....

45-59479 remained at Muroc AFB until April 1949 when she was transferred to the 2750th Air Base Group, WPAFB. Two months later she was taken up on a test flight to determine the cause of communications static during precipitation. The aircraft commander was Colonel Al Boyd, the legendary Commander of Air Force Fight Test. Immediately after take-off, the landing gear failed to fully retract. They climbed to 13,000 feet and spent some time trying to retract or lock the gear down. During that period they lost air conditioning, the electrical system, and the number 4 engine. The choice was a crash landing or bailing out, and with no ejection seats, there was little choice. After a bounced touch and go failed to bring the gear down, Boyd set up for the crash landing but as the flaps lowered, the starboard inboard flap hinges failed and it extended over the starboard wing. The aircraft rolled inverted at less than 400 feet and 150 miles per hour, but Boyd rolled the aircraft upright and with three-engines at full power, got it around the pattern. He touched down again on sod and skidded for over a mile, coming to a stop just feet from a large ditch that would have destroyed the aircraft and likely killed the crew. Following the investigation, the airframe was scrapped, but the nose fuselage section forward of the bomb bay was transferred to the 4140th Air Base Unit, WPAFB as a traveling display **(Photo 59)**. The nose stayed in the unit's service until 1960.

Photo 57: The nose of XB-45 number 1, 559479, was assigned to the 4140th Air Forces Base Unit, Wright-Patterson AFB, OH as a traveling display. This unit eventually became the Air Force Orientation Group (AFOG) based at Gentile Air Force Station, Dayton, OH until 1992 (USAF via Dave Menard)

Summary....

The XB-45s were successful test aircraft uncovering serious problems in flight control systems, cockpit layout, and poor performance of the J35 in terms of acceleration and reliability. The jets had no bomb/nav radar so the performance as a bomber remained largely unknown. They also highlighted issues with supply and R&D due to funding and time. Existing supply and transportation systems had little or no experience in handling and transporting sensitive electronic components or fragile canopy and windscreen sections. All of this had to be learned and taught, and it would be some two years before such military and corporate programs were functional.

Headquarters wanted operational jet bombers on the ramp as soon as possible, and conventional means and methods of maintenance, training, communications and analysis were quite often failures on the XB-45. The early J35s were short-lived engines featuring time between overhaul of approximately 20 hours. This would improve of course, but over a period of years, not months. Few companies had the experience or knowledge in manufacturing, inspecting or overhauling high technology components such as compressor and turbine blades or shafts. Dealing with and developing the metallurgy and lubrication means to meet and deal with the extremely variable temperatures and pressures of the turbojet components were also on very new ground.

Finally, these issues had not been adequately considered in the various USAAF and USAF orders and funding. It was assumed that "conventional" reciprocating engine aircraft procedures, requirements, and materials would be sufficient, and extraordinary additional funding not required. They were not. But the XB-45s could only present a preview of the issues an operational aircraft was going to face, and did so with distinction.

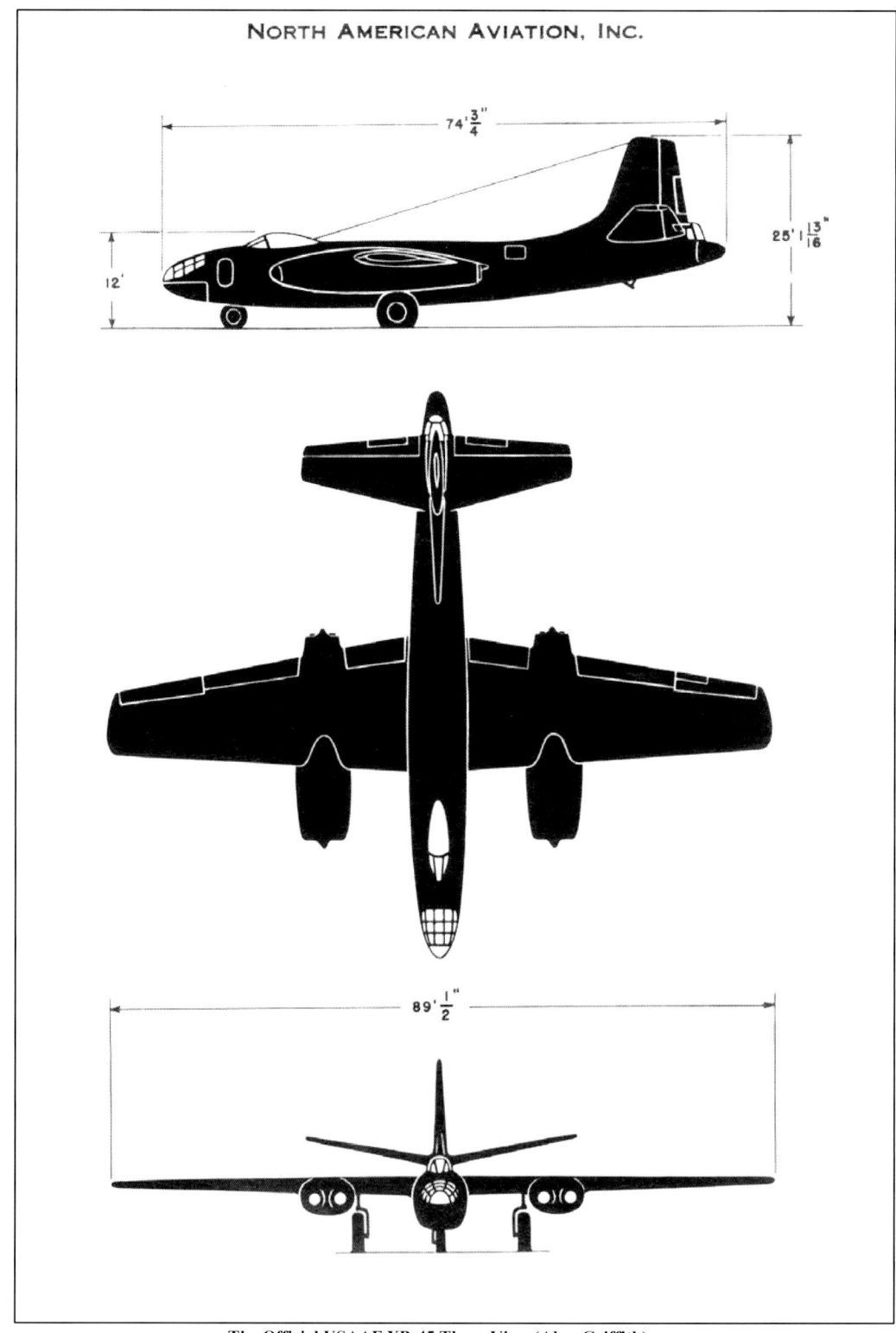

The Official USAAF XB-45 Three View (Alan Griffith)

CHAPTER 3 – THE B-45A DESCRIPTION

Turbulence and Engine Doubt....

The USAAF contracted for production of B-45As well before the first XB-45 was completed at the North American plant at Mines Field. It was a practice not uncommon during the War, and there was no hesitation with a turbine-powered. Contact number W33-038 AC-15569 was signed on 9 August 1946 procuring 96 NAA Model 147s, designated the B-45A. A 97th airframe, a non-airworthy structural test item, was also purchased.

Photo 1: The first B-45A, 7001, was imaged at Muroc Army Air Field in April 1948 and clearly shows the ejection seat headrests, the new canopy, and windscreen. The redesigned Aerial Observer Bombardier (AOB) station, now known as the Bombardier/Navigator or Bomb/Nav station transparency is also well defined. Also shown is the new dual wheel nose gear. The image highlights just how clean and streamlined the first production model of the B-45A was. (USAF via author's collection)

The first B-45A, accepted on 7 April 1948 at Inglewood, was a USAF light bomber (based on range) powered by four General Electric J35 turbojets (two J35-9 mounted in left side of each nacelle and two J35-11s in each right-hand position in each nacelle) which were rated at 4,000 lbs. thrust. This production airframe spanned 89 feet, was 75 feet 4 inches long, and weighed 45,694 pounds empty. As with the XB-45s, the B-45As started production at the North American plant at Mines Field. The standard crew was four: the pilot, co-pilot, the AOB now redesignated as the bombardier/navigator or bomb/nav, and the gunner. Beginning with serial number 7023, NAA began installing four GE J47 turbojets (two J47-7 or two J47-13 in each left-hand position in each nacelle and two J47-9 or J47-15 in each right-hand position) in the remaining 76 B-45As. Aircraft powered by GE J35s would be designated as B-45A-1s; the J47-powered aircraft were designated as the B-45A-5s. The J47 fitted the existing B-45 nacelles with little redesign and each was rated at 5,220 lbs. of thrust, adding 4,440-lbs. of much-needed additional thrust.

For GE, the J35 was a near corporate disaster. While the engine was in general a success in terms of performance, its reliability, durability, and serviceability were initially all poor. Further, GE had serious problems in meeting production demands. USAF then decided to move its production from GE to the Chevrolet Motor Division of General Motors in Tonawanda, NY. In 1947, after Chevrolet returned to automobile production, J35 production was moved to the Allison Engine Company of General Motors in Indianapolis IN. Allison went on to build 14,000 J35s in addition to over 6,600 GE-designed J33s for the F-80 and T-33. GE Engines under the leadership of Harold D. Kelsey, used this setback to leave centrifugal turbojet designs and concentrate on the more promising axial flow compressor engines. He directed the design of a new axial flow engine of 5,000 pounds of thrust using the same basic dimensions of the J35, but with much greater fuel efficiency.

(The TG-180 and TG-190 had been known by those monikers within GE since about 1944. In late 1947, the new Department of Defense moved to standardize aircraft engine designations regardless of manufacturer. The TG-180 was redesignated the J35, and the TG-190 became the J47.) The GE J47 became one of the most successful jet turbines in history. It produced over 1,000 pounds more thrust than the GE J35, was more reliable, and much easier to build. Since the USAF still had considerable doubts that GE could resolve their engine production issues, GE addressed that issue by not only expanding into two facilities, but leasing the largest manufacturing facility under one roof in the world from the War Assets Administration, the former Wright Aircraft Engine plant at Lockland OH near Cincinnati. As GE's Lockland Engine Plant, it was proposed to be the sole source of the J47 that was critical to the F-86, FJ-2, B-36, B-45, and B-47 production programs as well as the B-46, B-51, and B-49 development efforts. With the added demand stemming from the Korean War, GE delivered over 35,000 J47 turbojets from this plant by 1956; it ultimately become known as the GE Evendale Jet Engine Plant, still in operation to this day.

Production....

The first 11 B-45A-5s (7023 through 7033) were also built at the NAA plant at Mines Field in Inglewood, CA. As the images of the XB-45s in Chapter 2 show, the ramp space at Mines was limited, especially for larger aircraft such as the B-45 turbojet-powered aircraft. The Tornado was larger than the World War II B-25s built there (Chapter 2, Photo 12) in open air production which an immediate necessity in 1947. Fortunately, the Inglewood weather, being near the Pacific, was reasonable except for the summer temperature and the line was relocated outdoors. Despite these immediate positives, NAA knew that a better solution was required. The physical plant was still too small to handle two large and one smaller aircraft lines and the electrical power, hydraulic and pneumatic demands were also too high. Further, the local road and rail capacity was uncertain. A facilities search began throughout Southern California to locate a physical plant that included an appropriate airfield and ramps, sufficient talented workforce; a large enough enclosed facility if possible, sufficient supply road and rail networks, machine tool operations area, and a large enough raw material supply area. Fortunately, with the help of the War Assets Administration, a very suitable facility was found at the Long Beach Airport, the plant formally used by the Douglas Aircraft Company to build the C-47 during World War II. The search, relocation effort, re-start of production, and first delivery from Long Beach took less than one year.

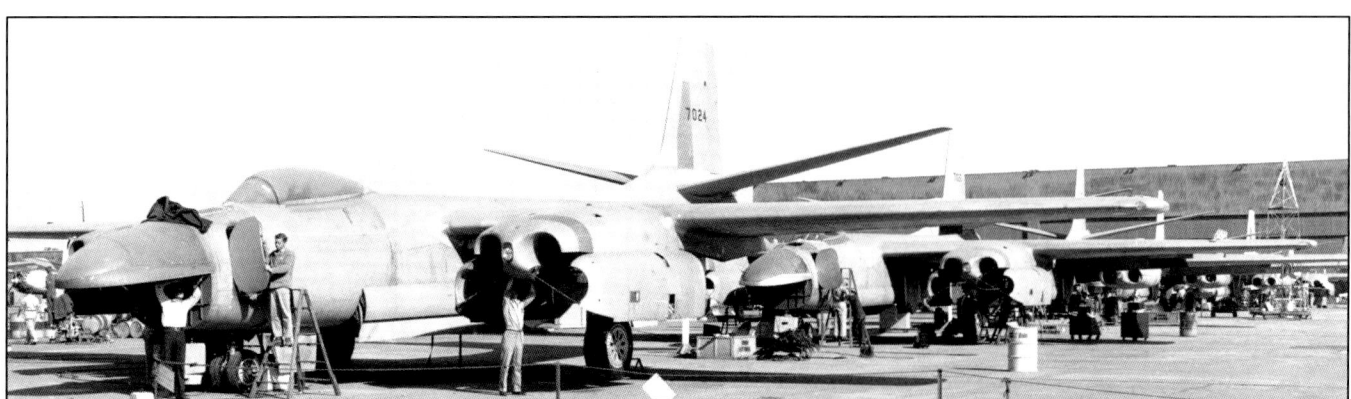

Photo 2: This image was taken in April of 1948 shortly after the B-45A-5 assembly line was moved outdoors at the Mines Field location of the Los Angeles North American Plant. (Alan Griffith)

Photo 3: This image shows the assembly crew positioning two of the four GE J47 turbojets beneath the port nacelle where they will be hoisted by crane and cables routed through openings in the upper nacelles. The engines were then bolted into position within the engine nacelle. (Alan Griffith)

Photo 4: All four engines have now been installed and pumps, valves, hoses, and accessories are being installed inside each nacelle. (Alan Griffith)

Early Changes from the XB-45s....

In the year since the first flight of XB-45 45-59479, several major changes had been made to the design to fix serious shortfalls in the cockpit area. First, the cockpit was redesigned and lengthened to approximately 21.3 feet (an extension of about 4.5 feet) to accommodate the ejection seats, flight controls for the copilot, and additional navigational and communication equipment. The pilot's station was moved aft to eliminate the interference of the windshield frame with his field of view. The canopy could now be jettisoned in an ejection, but still could not be opened on the ramp; so this provided no solution to cockpit heat accumulation. Another major safety improvement was the installation of airbrake panels just forward of the forward crew access hatch and the tail gunner's entry hatch. These, and a new wider forward crew hatch, were designed to improve crew clearance of the fuselage and nacelles in the event of an emergency bail out (**Photos 5 and 6**).

Photos 5: The seven-panel airbrake installed in front of the wider forward crew entry door is shown in the deployed position. **Photo 6:** The tail gunner's airbrake was three panels immediately below the starboard elevator and stabilizer. (Courtesy the Boeing Aircraft Company)

Another highly visible change on the B-45As from the XB-45s was the lower edge of the transparency cover over the bomb/nav station and above the fiberglass radome over the future AN/APQ-24 antenna. The forward end nose dipped slightly down, while the rest of the edge was now almost horizontal along its whole length. To help cure an issue with ground steering and nose gear vibration and shimmy, a new dual wheel nose gear replaced the previously used single wheel nose gear. This helped cure an issue with ground steering and nose gear vibration and shimmy. While an improvement, steering still took a great deal of attention and was best done at low speeds.

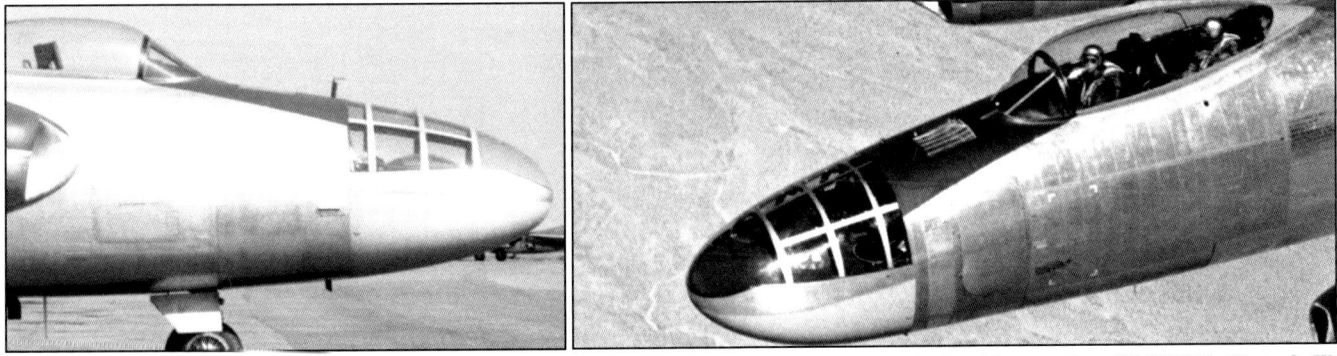

Photo 7: This image shows the new fiberglass radome and the metal framed fiberglass canopy over the bomb/nav station. (NMUSAF) **Photo 8:** This image shows the port side nose, bomb/nav station astrodome cover, and cockpit. (NMUSAF)

Photo 9: A B-45A-1 photographed at Muroc AFB shows that the entire nose cockpit was widened, and the shape flattened somewhat to improve air flow over the area. (NMUSAF)

Photo 10: The first B-45A-1 is shown on the desert just a few yards from the North Base ramp at Muroc Air Force Base in April of 1948. (The Gerald Balzer Collection via Tony Landis from HQ AFMC/HO.)

The flight test programs for the XB-45s were conducted rapidly and with little observable trouble from the standpoint of performance and handling qualities. However, to those USAAF and NAA program directors there were serious issues of design, supply, maintenance, and funding to be dealt with. In the case of the B-45A, the first hard lesson came quickly and tragically. Less than six months after its completion at Mines Field and acceptance by the USAF, still bailed to NAA for Flight Testing, B-45A-1 **7001** was lost. Approximately 112 miles Northwest of Muroc Air Force Base - on fire and out of control - it crashed near Alpaugh, CA at the Southern end of California's Central Valley. Unable to use the newly installed ejection seats nor escape through the forward cabin access hatch, both crew members were killed upon impact. The crew consisted of North American's exceptional Flight Test Engineer Paul Brewer, who had crewed the first flight of the first XB-45 on 17 March 1947, and Nicholas Packard, the mission's Test Pilot.

Photo 11: Paul Brewer, North American's exceptional Flight Test Engineer on the ramp at Muroc North Base. (Craig Kaston via Steve Ginter) **Photo 12:** B-45A-1 7001 was lost on 20 September 1948. (Gerald Balzer via Tony Landis from HQ AFMC/HO)

The cause of the accident had nothing to do with the instability problems that led to the 7.5-foot extortion of the horizontal stabilizer of **7001** and all other B-45s. Instead, there were two factors, neither of which was ever completely solved in the life of the B-45 family. The first was persistent fuel leaks in the wing tanks immediately above and beside the nacelle engine bays. Sealant had yet to be developed that fuel would not dissolve it, or be adversely affected by the frequent changes of ambient temperature during a flight, or the high temperatures of the engine bays. The design or other construction techniques known in the late 1940s to seal the fuel tanks were inadequate in this installation.

The second factor was the leaking fuel collected and soaked the heat reduction blankets installed around the surfaces of the engine exhaust ducts which extended from just aft of the wing leading edge all the way aft to the exhaust nozzles. (**Drawing 1**) With the ducts heated to around 600° to 650° F, the fuel gave off fumes which exploded and blew at least three of the heavy and large engine nacelle doors away from the lower aft portions of the nacelles. (**Drawing 2**) These nacelle cowls collided with the horizontal stabilizers, tearing one off, breaking the other in half and causing the aircraft to instantly become uncontrollable. The ejection seats were dis-functional and neither crew member could make their way to the forward escape door and safely exit the aircraft. The image below shows the aircraft's BUZZ number derived from the aircraft's serial number. It reads BH-001 indicating 7001. There can be no question that 7001 was the aircraft lost on 20 September 1948 becoming the first B-45 to be lost, and was not the first XB-45 45-59479 as continues reported.

Photo 13: The wreckage of 7001 near Alpaugh, CA on 20 September 1948 clearly shows the BUZZ number of the aircraft. (Steve Ginter Collection)

The fire issues were never fully resolved on the B-45. The partial fix was a new design of exhaust duct blankets using new materials in early 1951. Frequent re-sealing and more frequent and more in-depth inspections of the tanks became part of the maintenance routine. This increase in ground man-hour per flight hour cost, aided in dealing with the problem but did not totally eliminate the problem.

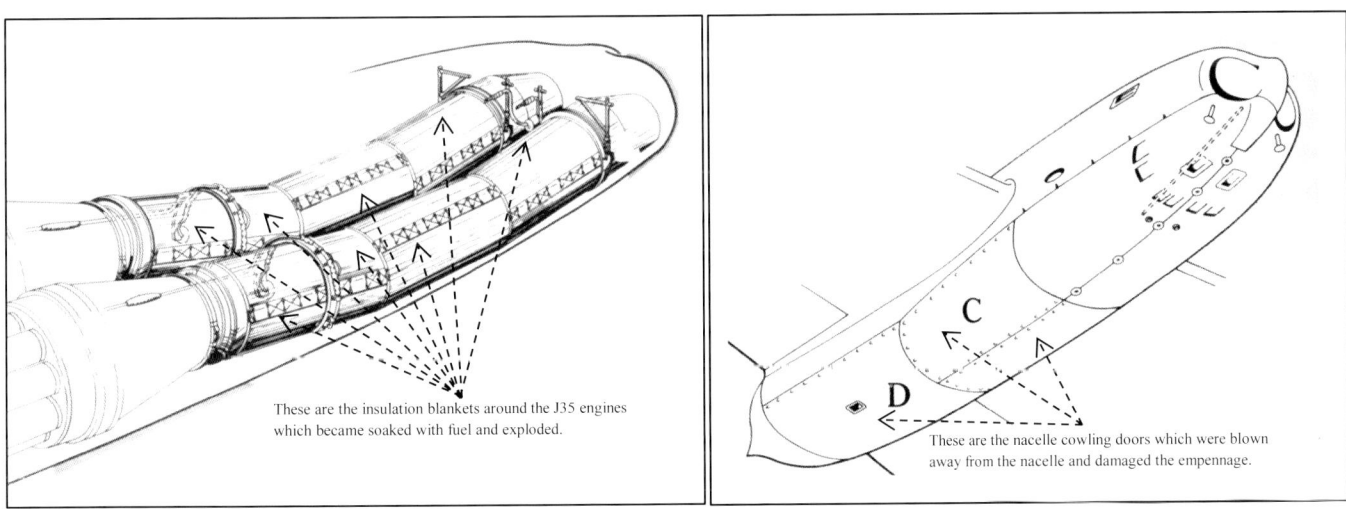

Drawing 1: The blankets, seen here wrapped around the tail pipes engines one and two in the port nacelle generally held the temperatures in these areas to around 600* and 650* F. They became soaked with fuel and exploded. (NMUSAF) Drawing 2: The explosion was massive and completely separated the nacelle door labeled D and likely both doors labeled C. (NMUSAF via the Author)

Nevertheless production went ahead and the search for other serious problems continued. The first 22 B-45As were powered by versions of the GE J35 turbojet, itself seriously behind in both development and production. Within the office of the Chief of Staff, General Hoyt Vanderberg at HQ USAF, the question of the continued production and service of the B-45 had been raised. Development of the aircraft's bomb/nav radar, the AN/APQ-24, was behind schedule. As a result, it was likely that the majority of the fleet would not be equipped with it or the new A-1 Fire Control System, which interfaced with the AN/APQ-24 to control the release of weapons in a selected secquence. The E-4 Auto-Pilot, which provided the AN/APQ-24 with altitude and airspeed information allowing it to accurately aim bombs on a target, had yet to be delivered. The failure to complete pre-production testing of the AN/APT-5 Electronic Countermeasures (ECM) system needed to jam the radars of interceptors was the final major electronics shortfall of the B-45A. HQ USAF remained concerned that GE would be unable to resolve production and design issues for the J47, critical to the success of the B-45, as well as the F-86.

Changes at the end of the decade....

On 22 Aug 1947 the Commander, USAAF General Carl Spaatz had ordered that an additional 43 B-45C models be ordered giving the USAAF another Group of longer range and heavier load carrying bombers. This order was increased to 51 aircraft on 1 June 1948. However, as a result of President Truman's budgetary constraints and General Vandenberg's lack of support, by 12 January 1949 the 51 B-45C aircraft order was cancelled, and the order book was modified to read 22 B-45A-1s, and 74 B-45A-5s (for a total of 96 B-45As), 10 B-45Cs for Research and Development and 33 B-45Cs converted on the production line into RB-45C-1s. The B-45A-1 and B-45-5s would go to the Tactical Air Command (TAC), then still part of the Continental Air Command (CAC). The B-45Cs would be assigned to the Air Research and Development Command (ARDC), and the RB-45C-1s would be assigned to the Strategic Air Command to replace their RB-29s on the most dangerous missions over denied territory.

Photo 14: The eleventh B-45A-1 cruises over the Sierra Nevada Mountains north of Muroc AFB in June of 1948. The new and as yet unframed cockpit canopy significantly improved visibility from the cockpit. The ejection seat headrests are visible inside. (NARA II via Tommy Thomason)

Photo 15: B-45A-1 7010 is christened "Tornado" by Julie London for the North American public affairs division. Julie London was a very popular and beautiful actress, and jazz and pop singer. USAF test pilot Capt. Chuck Yeager is in the pilot's seat. (North American Aviation via Steve Ginter)

Structure of the B-45-1, B-45A-5 and the B-45C....

The B-45A was a conventionally designed aircraft as outlined in Chapter 2. As with the XB-45s, it consisted of a semi-monocoque fuselage, wings, and nacelle structures. The shape was set by vertical or offset formers, to which longitudinal stringers and equipment were attached to the strongest formers. Each of these structural components had to be identified by a unique part number, to insure that they were installed in the proper location and order, and for the essential identification of items that had to be replaced due to damage, corrosion, and service life expiration. These parts were described on large, detailed sets of engineering drawings that define every item, structural or system oriented, of the aircraft.

The identification of the location of structure, components, etc. was carried over from shipbuilding. Longitudinal locations on the drawings were identified by using station numbers. Station numbers were the horizontal distance, in inches, from a "0" reference datum point at the front end of the airframe. (Most were even far beyond the front end of the aircraft so changes in the nose could be made without changing a considerable numbers of station numbers behind the nose). Interestingly, the B-45 began with negative numbers to well behind the nose. Outboard locations, known as butt (for buttock) lines, were measured from a reference line along the exact center of the fuselage. Again numbers in inch dimensions were used to label the width of aircraft structures. Height numbers, known as water lines, was measured from a horizontal reference line usually located below the fuselage so all distances were positive. Again, the measurements were in inches. Sets of these location reference drawings, usually bound in volumes called Structural Repair Technical Orders, were issued in quantity to all units who operated the aircraft. Posting changes was a critical, but all too often unheralded task of the maintenance force. The B-45As, the B-45C and the RB-45C were built to the same structural plans with some changes in the later models in the strengths of the materials used. The following is a set of individual drawings of key structural drawings for the B-45A, B-45AC, and 90% of the RB-45C. Drawings for later models that differ will be found in the "Description" Chapter for that model.

 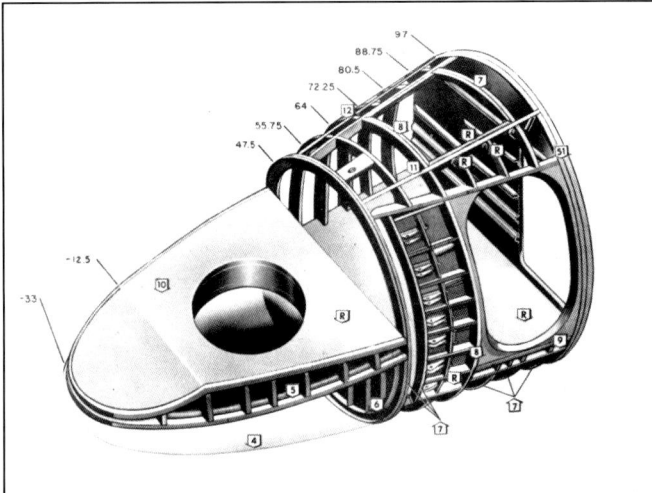

Drawing 3: Major structural components of the B-45. (NMUSAF) Drawing 4: B-45A and B-45C nose exterior. (NMUSAF)

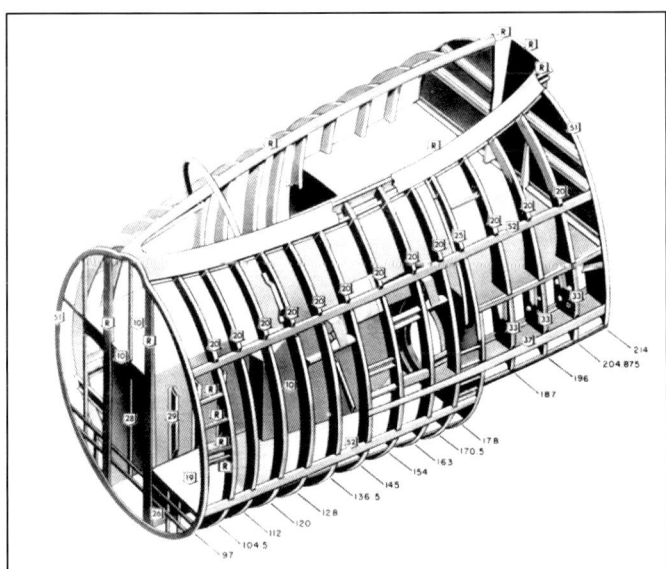 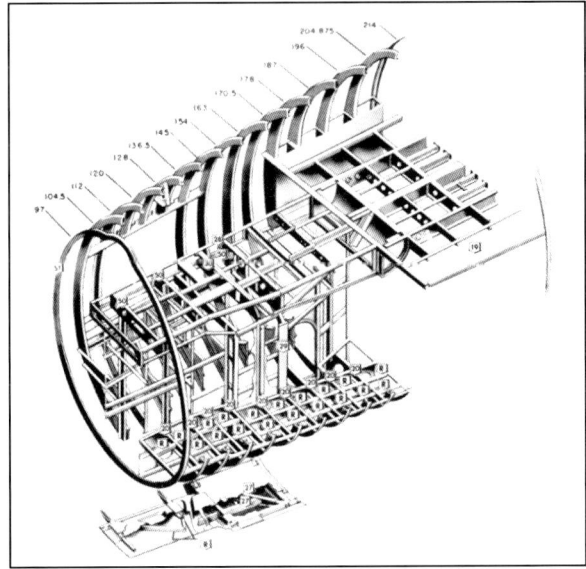

Drawing 5: Is a drawing of the B-45A and B-45C cockpit stations interior and exterior structure. The forward opening was the forward bulkhead for the bomb/nav compartment. On the inboard port side was a small flat floorboard that was the beginning to the pilots' passageway that began to slope downward as it traveled aft to a pressure hatch in the forward bomb bay bulkhead. This passageway allowed the pilot and co-pilot to climb up to the cockpit floor, reach their ejection seats, and strap in. A canvas safety seat was located at aft end of the passageway to secure maintenance or other mission-oriented personnel during flight. (NMUSAF) Drawing 6: Is the B-45A and B-45C cockpit stations interior. (NMUSAF)

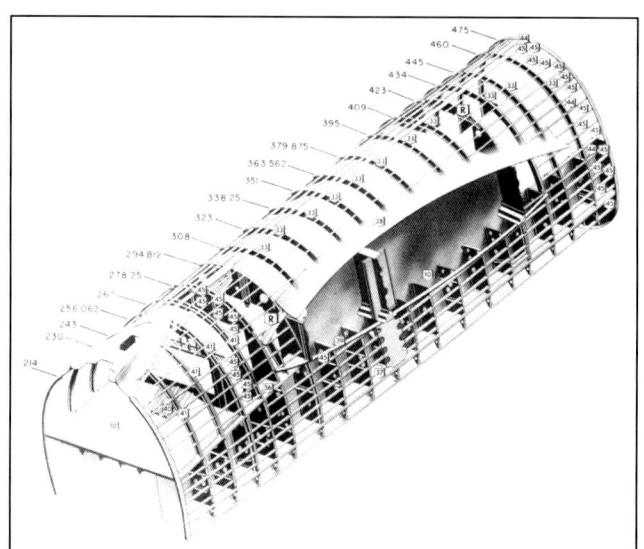 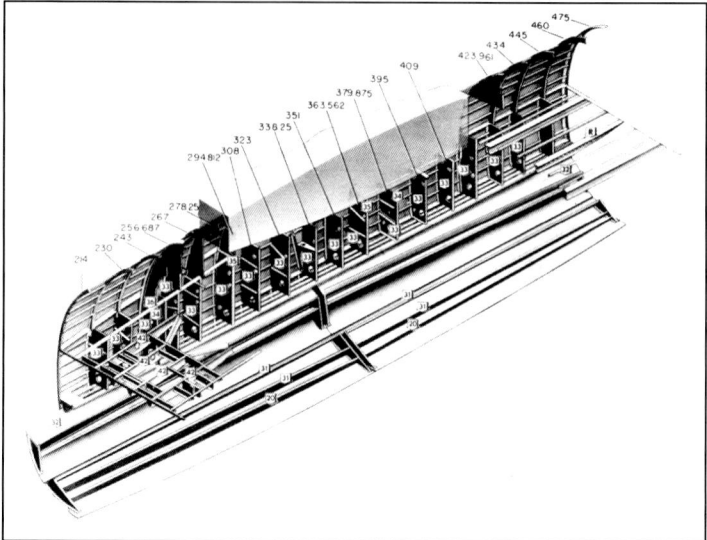

Drawing 7: Is the B-45A and B-45C bomb bay exterior (NMUSAF) **Drawing 8:** Is the B-45A and B-45C bomb bay interior and four bomb bay doors on each side. (NMUSAF)

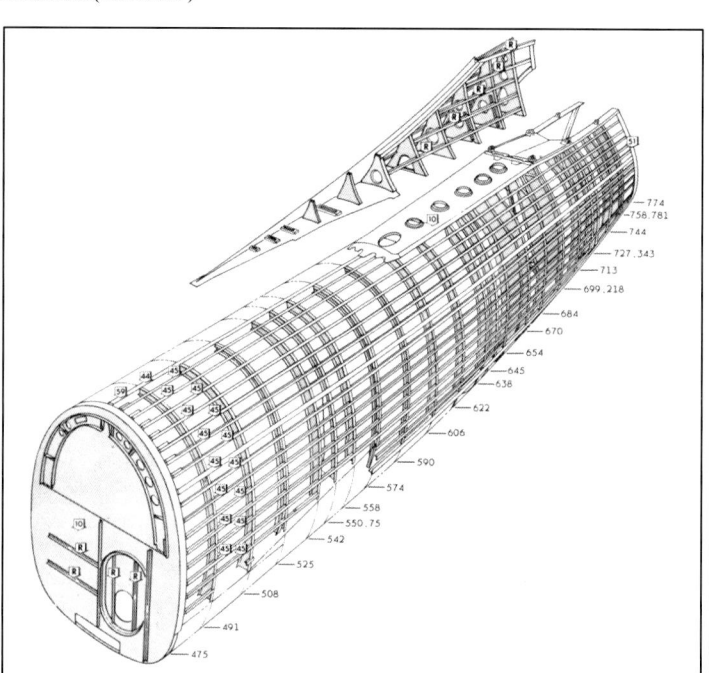 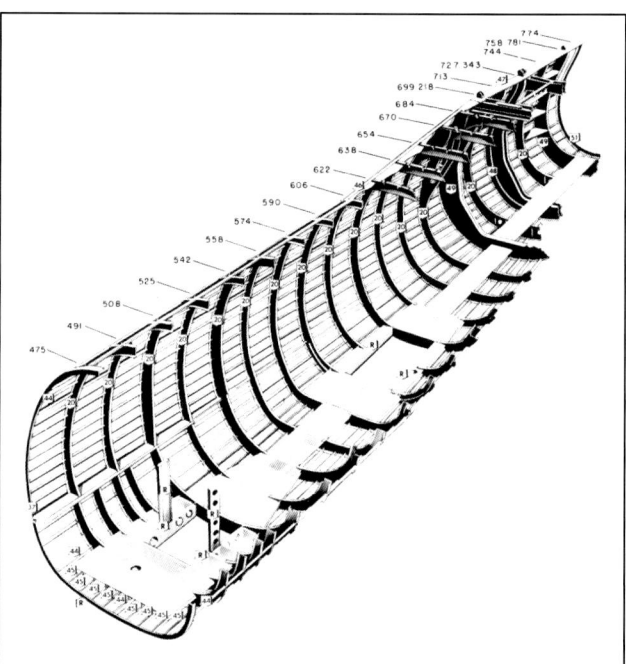

Drawing 9: Is the B-45A and B-45C aft fuselage exterior. (NMUSAF) **Drawing 10:** Is the B-45A and B-45C aft fuselage interior. (NMUSAF)

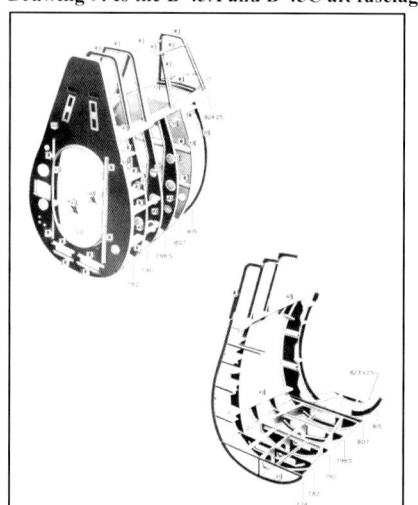 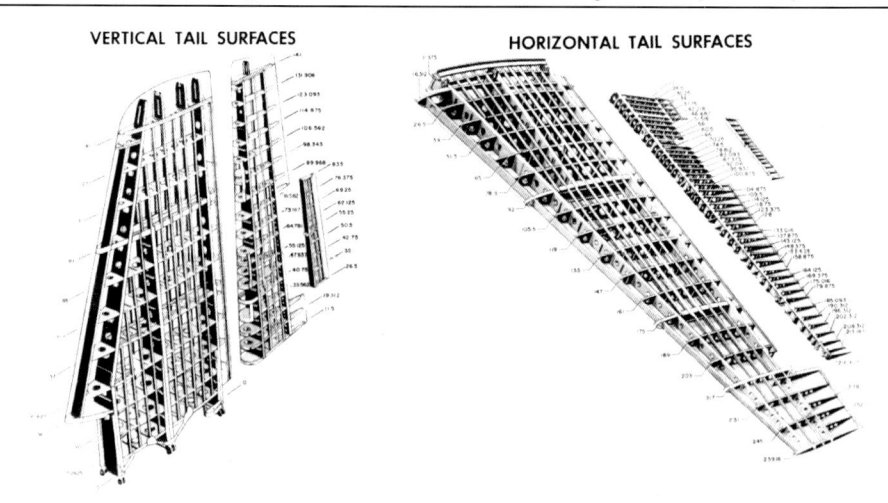

Drawing 11: This is B-45A and B-45C production gunner's station. **Drawing 12:** Are the B-45A and B-45C vertical, horizontal stabilizers, elevators and rudder. (Both NMUSAF)

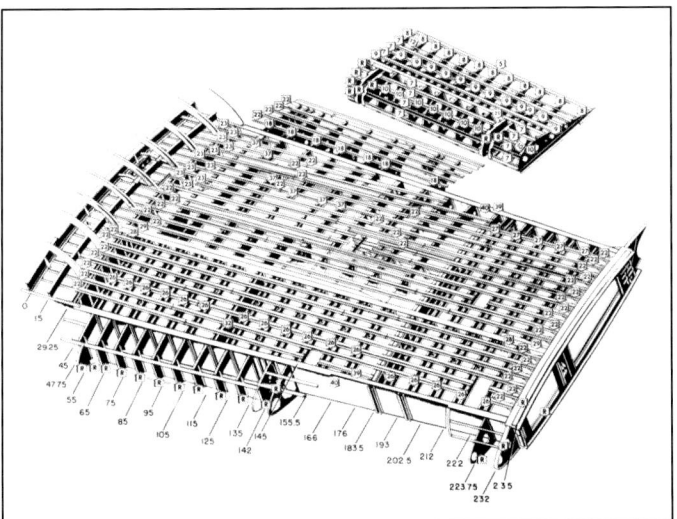 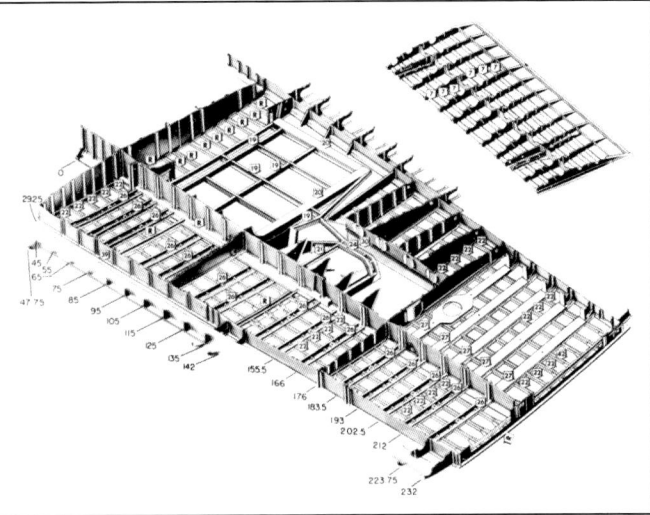

Drawing 13: Is the B-45A and B-45C center wing upper surface including the portion of the upper fuselage structure over the wing. This drawing shows only the port half of the structure as both the port and starboard sides were built as a single unit and installed as a single structure. (NMUSAF) Drawing 14: Is the B-45A and B-45C center wing lower section. Note: This drawing shows only the port half of the structure as both the port and starboard sides were built as a single unit and installed as a single structure. (NMUSAF)

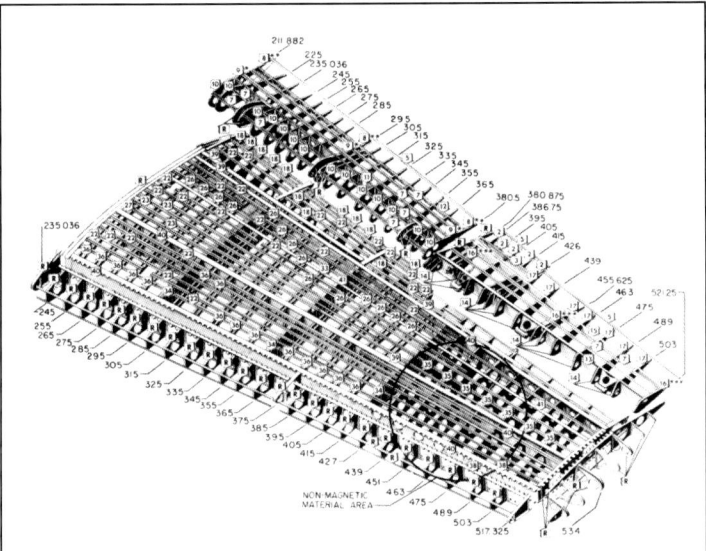 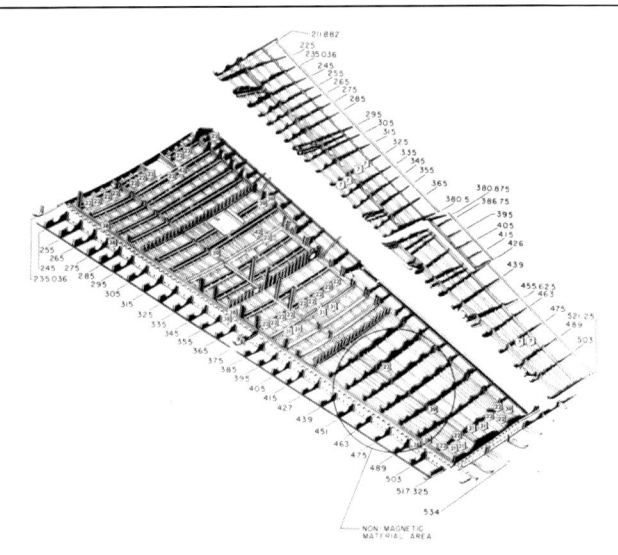

Drawing 15: Is the B-45A and B-45C upper outboard wing section. Note the non-magnetic material area caution where the flux gate compass was installed. (The flux gate compass was an electromagnetic device which directly sensed the horizontal component of the earth's magnetic field. The heading could be displayed remotely and used by an electronic autopilot. The flux gate compass had to be kept as level as possible by mounting it on gimbals or by using a fluid suspension means). (NMUSAF) Drawing 16: Is the B-45A and B-45C the lower outboard wing section with the flux gate compass installation. (NMUSAF)

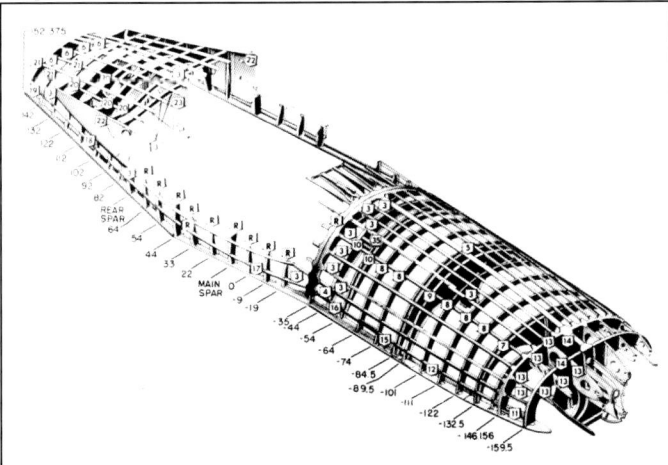 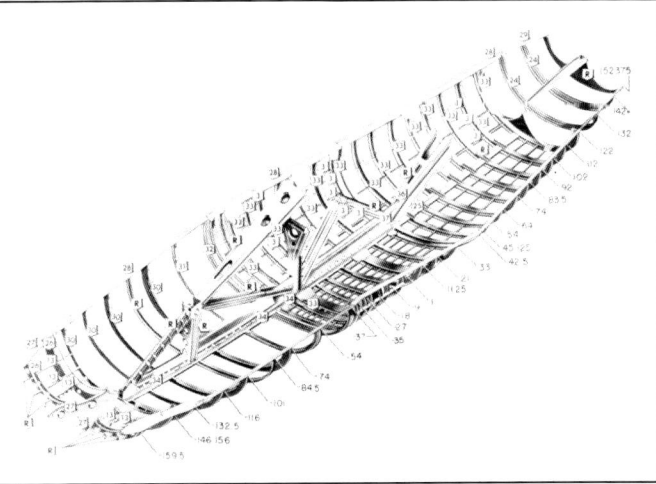

Drawing 17: This drawing shows B-45A and B-45C upper nacelles. The nacelles were identical on both sides of the aircraft. (NMUSAF) Drawing 18: B-45A and B-45C shows the lower nacelle structure. (NMUSAF)

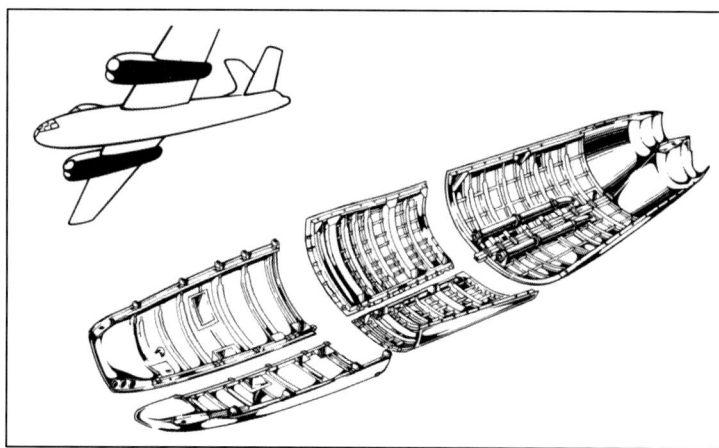 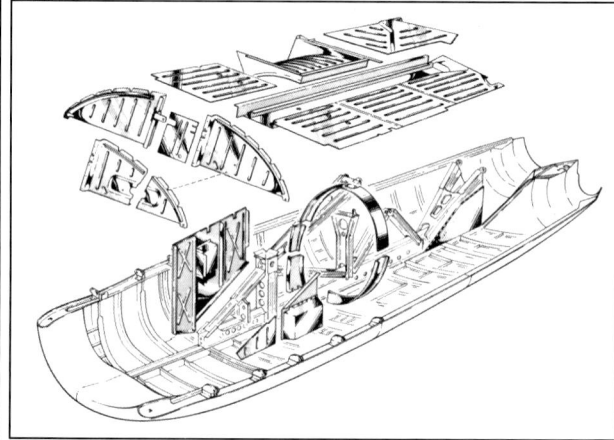

Drawing 19: Shows the B-45A and B-45C the internal lower cowling doors. (NMUSAF) Drawing 20: Shows the B-45A and B-45C nacelle internal mounts and firewalls. (NMUSAF)

Surface Skins for the B-45A-1, B-45A-5 and B-45C....

The following drawings show the skin sections for the B-45A-1, B-45A-5 and B-45C aircraft. The thickness (in decimal fractions of an inch) and the aluminum or other materials are shown next to each section:

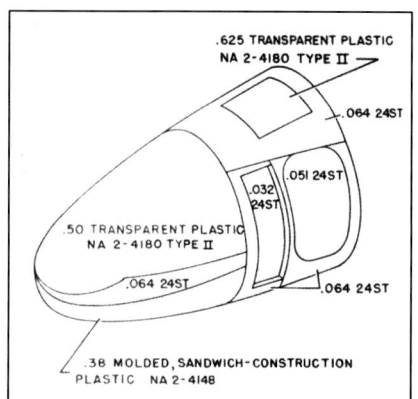 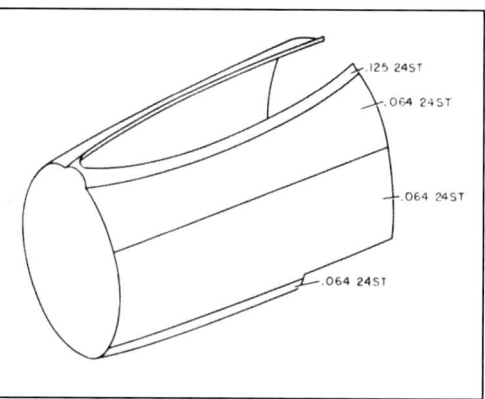 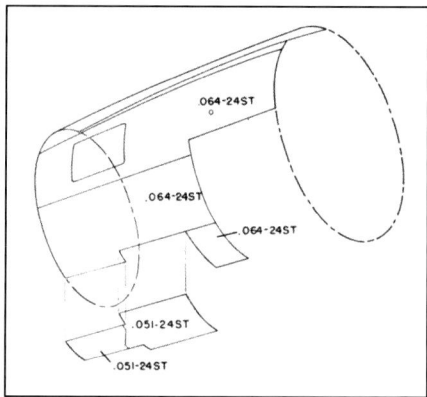

Drawing 21: Shows the B-45A and B-45C nose section skins. (NMUSAF) Drawing 22: Shows the B-45A and B-45C cockpit stations skins. (NMUSAF) Drawing 23: Shows the B-45 and B-45C starboard side and lower cockpit stations skins. (NMUSAF)

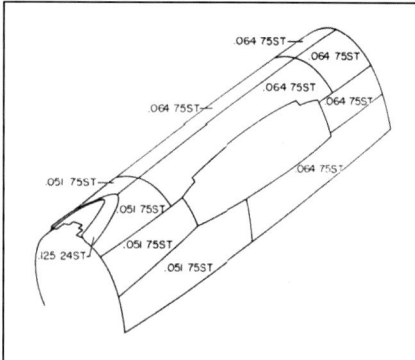 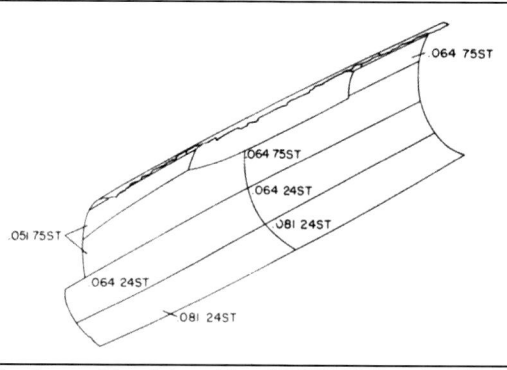 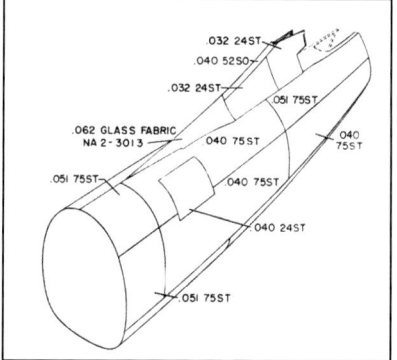

Drawing 24: Shows the B-45A and B-45C upper fuselage and bomb bay skins. (NMUSAF) Drawing 25: Shows the B-45A and B-45C starboard side and bomb bay door skins (NMUSAF) Drawing 26: Shows the B-45A and B-45C aft fuselage skins. (NMUSAF)

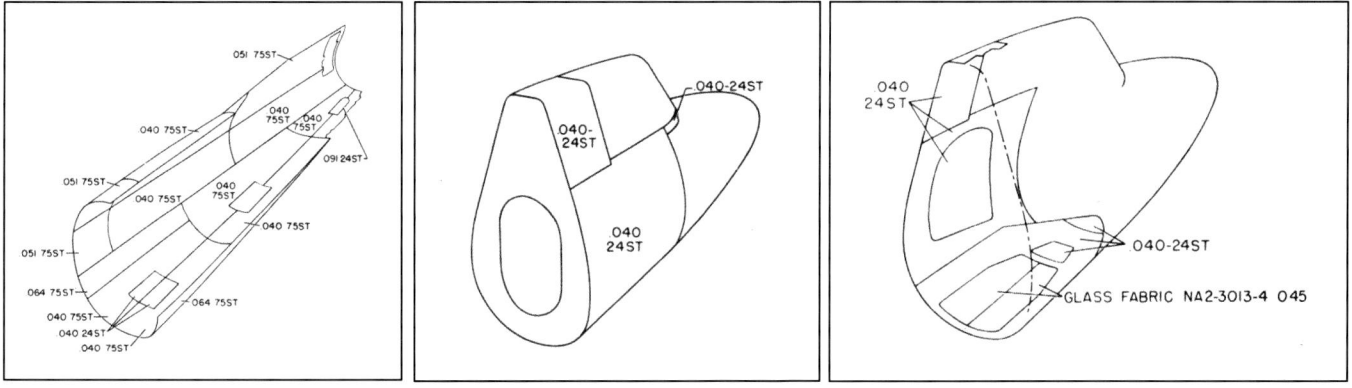

Drawing 27: Shows the interior B-45A and B-45C starboard, lower aft fuselage, and gunner's station skins. (NMUSAF) **Drawing 28:** Shows the exterior B-45A and B-45C production gunner's station and tail cone skins. (NMUSAF) **Drawing 29:** Shows the production gunner's station starboard and lower surface skins. (NMUSAF)

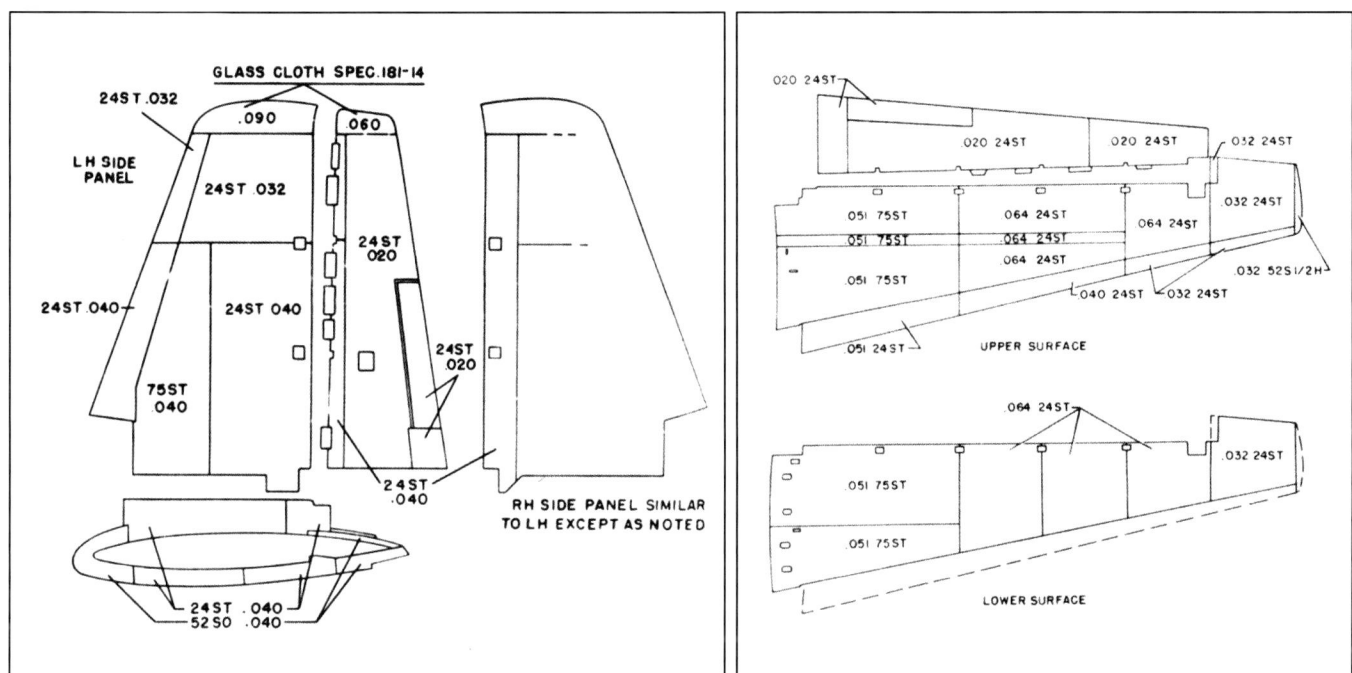

Drawing 30: Shows the B-45A and B-45C vertical fin and rudder skins. (NMUSAF) **Drawing 31:** Shows the B-45A and B-45C horizontal stabilizer and elevator skins. (NMUSAF)

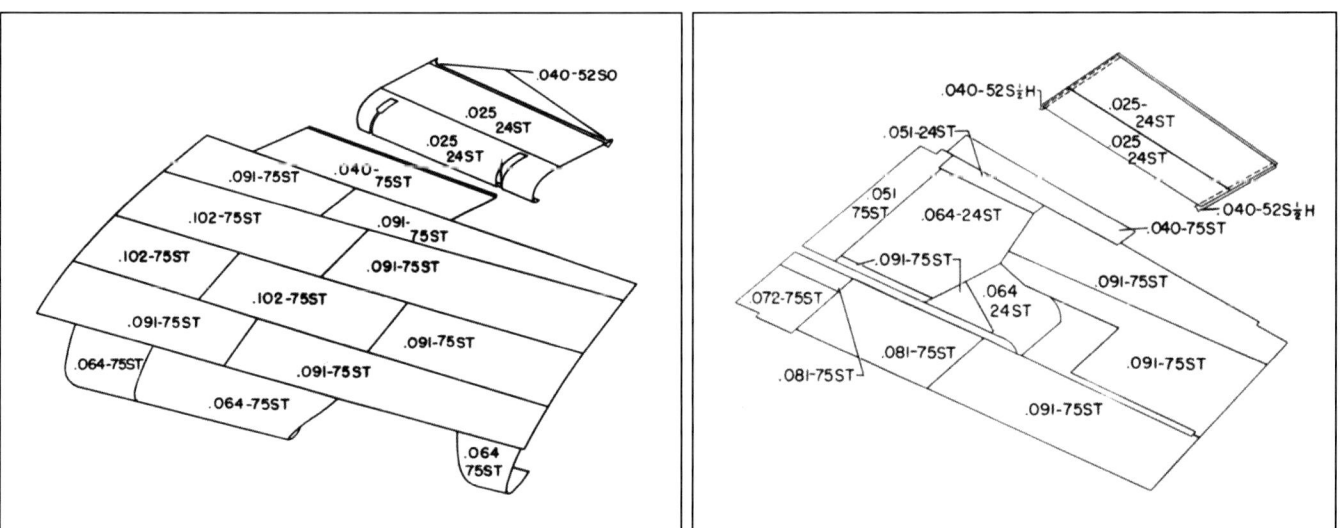

Drawing 32: Shows the B-45A and B-45C upper center wing section skins. (NMUSAF) **Drawing 33:** Shows the B-45A and B-45C lower center wing section skins. (NMUSAF)

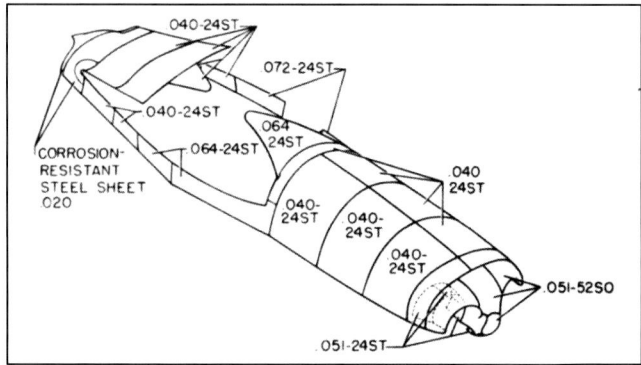
Drawing 34: Shows the B-45A and B-45C upper engine nacelle skins. (NMUSAF)

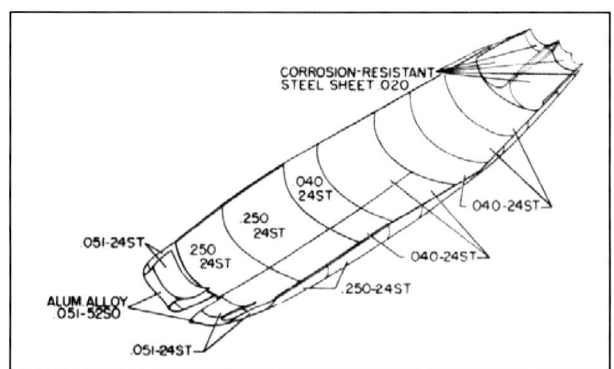
Drawing 35: Shows the B-45A and B-45C lower engine nacelle skins. (NMUSAF)

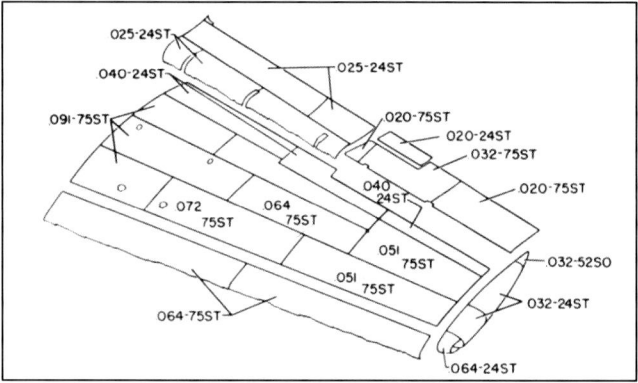
Drawing 36: Shows the B-45A and B-45C upper outboard wing skins. (NMUSAF)

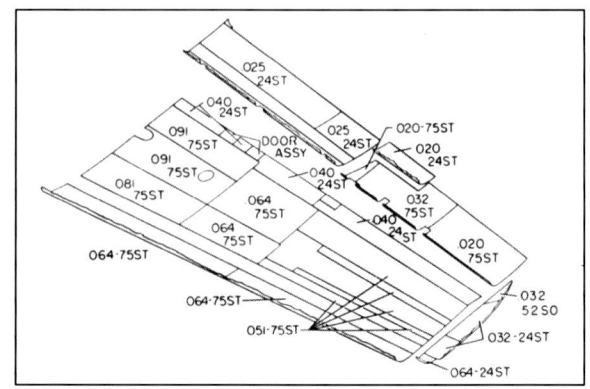
Drawing 37: Shows the B-45A-1, -5, and B-45C lower outboard wing skins. (NMUSAF)

Forward Fuselage Illustrations of the B-45A-1....

The Bomb/Nav Station in the B-45A-1 was equipped with a substitute bombing system and navigation gear, the AN/APS-23. Three or four aircraft assigned to the 47th Bombardment Group at Barksdale AFB, LA, with this equipment, provided a limited opportunity for crews to train in the basics of high speed, high altitude radar bombardment. It was not until mid-1951 that operational B-45A-5s had the AN/ASPQ-24 bomb/nav Radar, the E-4 Autopilot, and A-1 Fir Control System installed making the aircraft fully combat capable. Prior to these installations, it was possible to see straight through the nose enclosure, however, once the full range of bomb/nav gear was in place, the compartment became over crowded. The aircraft were transferred to the United Kingdom and the 47th Bombardment Wing became fully operational at RAF Sculthorpe in support of United States Air Forces in Europe (USAFE) and the US commitment to the North Atlantic Treaty Organization.

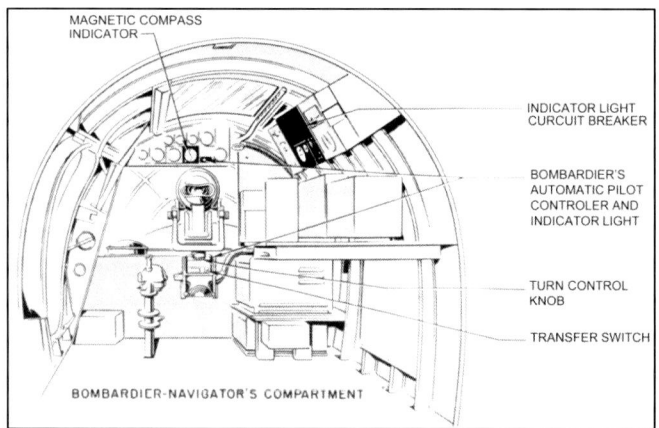
Drawing 38: This is the bomb/nav station looking through the forward transparency. (NMUSAF) **Drawing 39:** (Right) The drawing shows the aft view of the bomb/nav station highlighting the seat which rotated counter-clockwise. (NMUSAF)

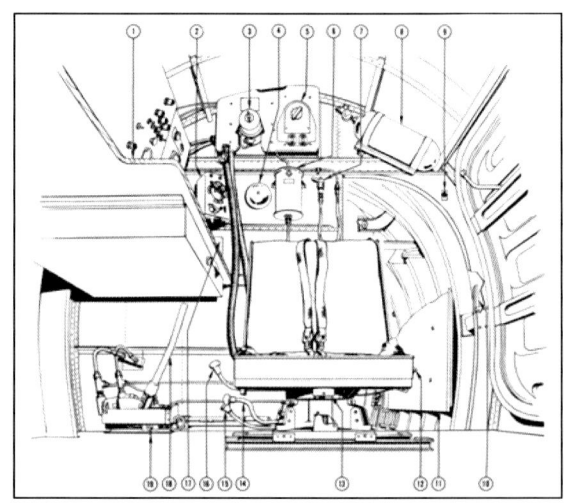

1. Chart Table
2. Camera Intervalometer Control
3. Oxygen Regulator
4. Alarm Bell
5. Heated Clothing Control Panel
6. Nose Gear Emergency Hydraulic Reservoir
7. Refiller Valve – Portable Oxygen Bottle
8. Portable Oxygen Bottle
9. Passageway Light Switch
10. Main Entrance Hatch Control
11. Ditching Panel
12. Shoulder Harness Lock Control
13. Seat Swivel Control
14. Seat Vertical Control
15. Seat Lateral Control
16. Emergency Escape Control
17. Landing Gear Emergency Operation Instruction Placard
18. Hand-pump – Landing Gear Emergency Release
19. Manual Check Valve – Emergency Landing Gear Lowering System

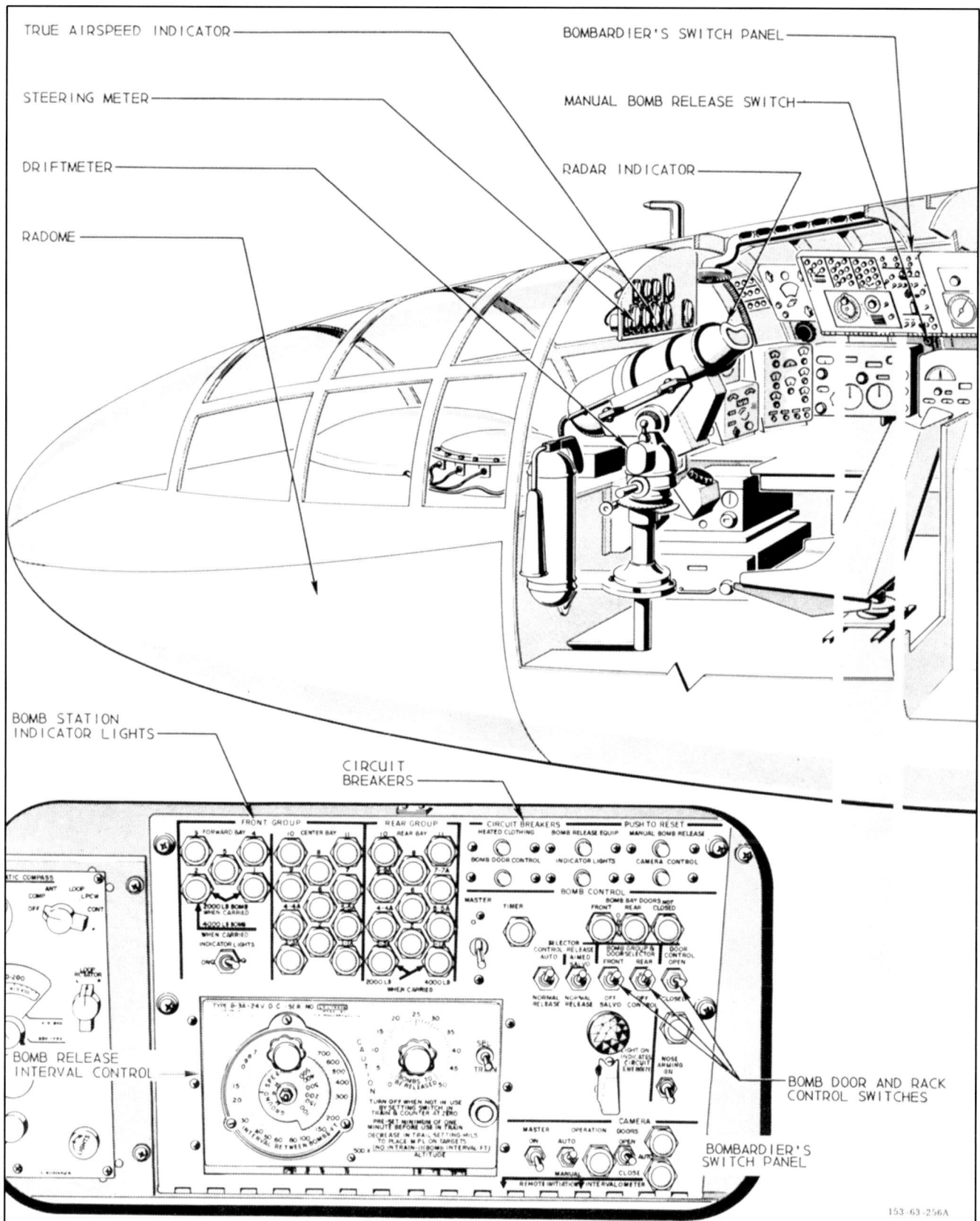

Drawing 40: As built in 1947 and 1948, few B-45A-1 and B-45A-5s were equipped with an early radar used to aim conventional bombing runs. The A-1 Fire Control System was not installed before 1952 therefore airspeed and crosswinds were based on dead reckoning. The bombardier would set the release parameters on the bomb/nav switch panel to include the time between each release and the desired rate of explosions in the target area. When the bomb/nav was ready to take over the release he would notify the pilot and fly the aircraft via the autopilot in reference to his instruments and the radar display. At the desired release point, the bomb doors would open and release the weapons in a pre-determined order based on the type of target and to maintain its center of gravity within safe parameters. The bomb station Indicator Panel would light as each bomb station released. As the illustration shows, the early radar and associated equipment took up a very small amount of room, which provided a good view out of the windscreen. There was no visual bombsight, such as the Norden, installed in any of the B-45A-1s or B-45A-5s. (NMUSAF)

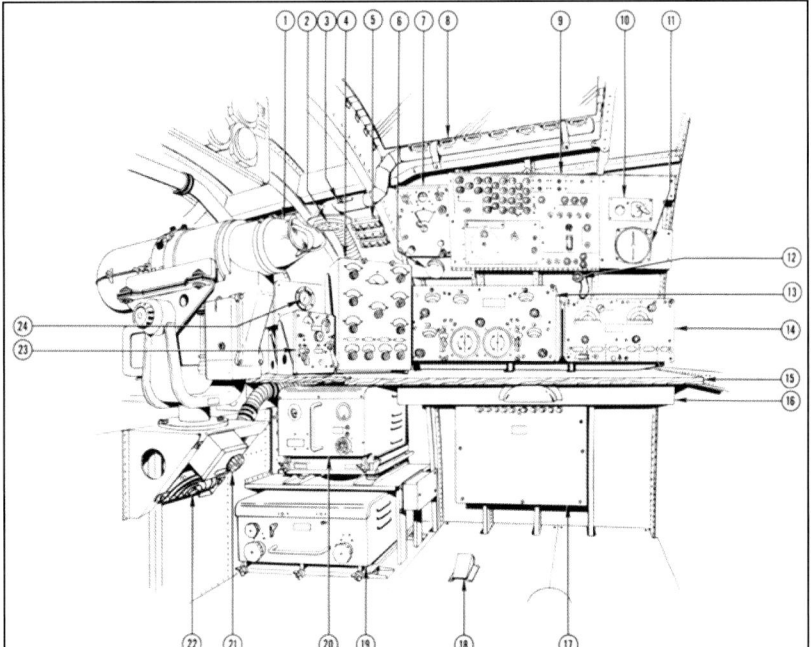

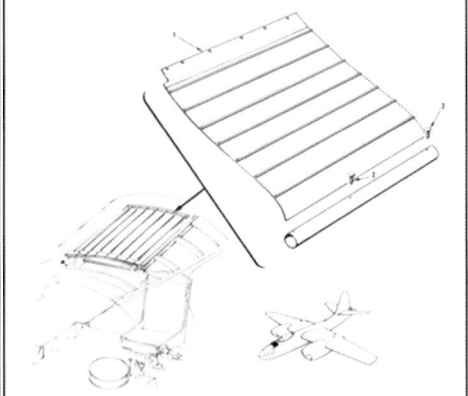

Drawing 41: Shows the control panel for the AN/APS-23 Bomb/Nav radar, a substitute for the AN/APQ-24 in the B-45A-1. (NMUSAF)

1. ID-218/APS-23 Indicator
2. Heat and Vent Outlet
3. Astrodome Defrost Shut-off Valve
4. C-413/1PS-23 Control Unit
5. Spare Lamps
6. Defrost and Ventilating Air Control
7. AN/ARN-6 Radio Compass Control Panel
8. Astrodome Defroster Outlets
9. Bomb Control Panel
10. Interphone Panel
11. Radio Compass Indicator
12. Bomb Release Button
13. CP-21/APA-44 Ballistics Computer
14. CP-22/APA-44 Mileage Computer
15. Chart Table
16. Chart Table Drawer
17. J-166/APA-44 Junction Box
18. Microphone Switch
19. AM-116/APA-44 Junction Box
20. PP-185/APA-44 Power Supply Unit
21. C-416/APS-23 Control Unit
22. Heat and Vent Floor Outlet
23. C-293/APA-44 Tracking Control
24. ID-168/APA-44 Airspeed Indicator

Drawing 42: View looking aft and down into the Bomb/Nav Station shows the astrodome flexible sunshade that was installed over the Bomb/Navstation on the B-45A-1, B-45A-5, B-45C and the RB-45C. The shade allowed the Bomb/Nav to view the radarscope in broad daylight without reflections. It was retracted by the spring-loaded tube to the port side. (NMUSAF)

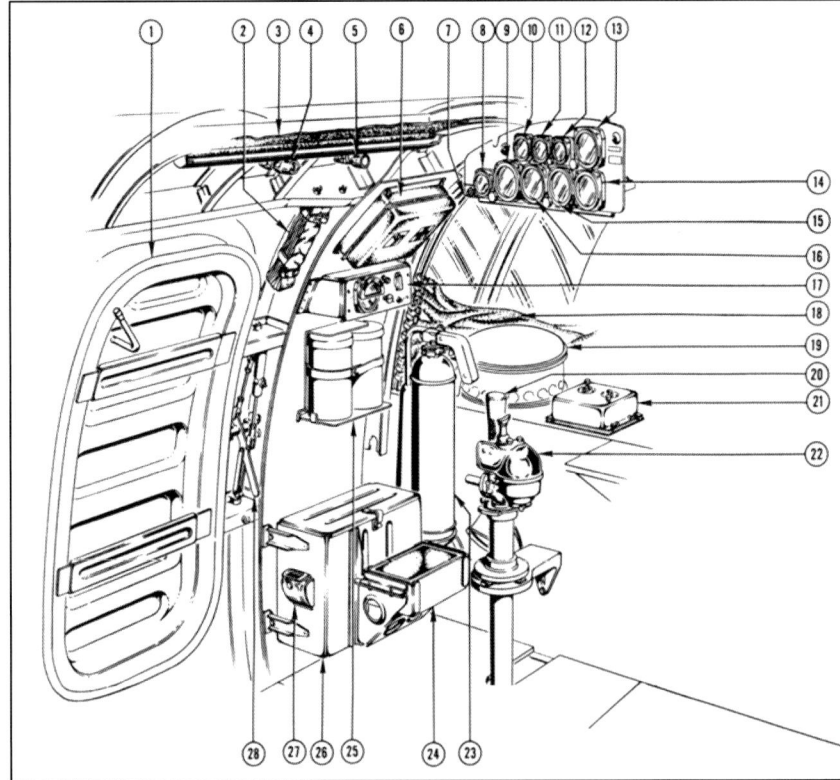

1. Main Entrance Hatch
2. Parachute Static Line
3. Astrodome Close-out Curtain
4. Compartment Light
5. Fluorescent Light
6. First-aid Kit
7. Driftmeter Power Switch
8. Free Air Temperature Indicator
9. Altimeter
10. Oxygen Flow Indicator
11. Oxygen Pressure Indicator
12. Clock
13. Cabin Pressure Altimeter
14. Radar Steering Meter
15. Magnetic Compass
16. True Airspeed Indicator
17. Radar Pressurization Control Panel
18. Radome Close-out Curtain\
19. RT-124/APS-23 Receiver-Transmitter
20. Hand Axe
21. Camera Control Box
22. Driftmeter
23. Fire Extinguisher
24. Sextant Stowage Box
25. Disposal Containers
26. Map and Data Case
27. Ash Tray
28. Main Entrance Hatch Emergency Release

Drawing 43: View to port showing the forward crew access hatch. (NMUSAF)

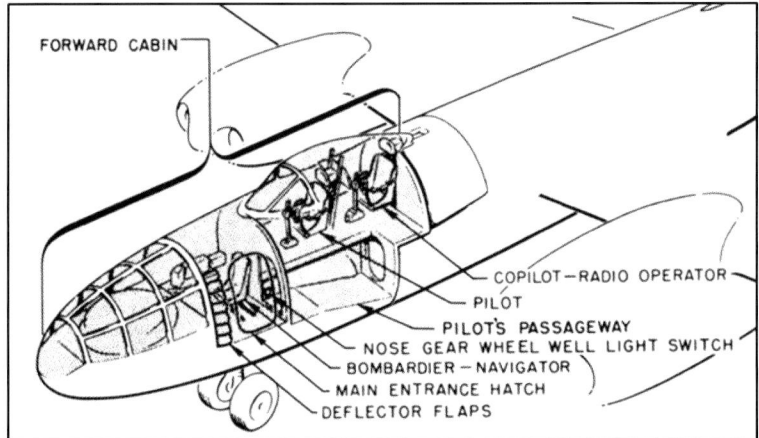

Drawing 44: This illustration shows the layout of the pilot's cockpit immediately behind the bomb/nav station. Just in front of the cockpit is the main crew access door which leads directly to the bomb/nav seat. Turning right after entry, the pilot and co-pilot step onto a level platform, then onto a downward slopping passageway. This narrow pathway gave them access to climb up and into the cockpit and strap into their ejection seats. The passageway continued downward beneath the co-pilot's communications equipment table toward a pressure bulkhead door that entered into the forward bomb bay. In front of that door was a harness that gave one individual a security harness for ride-along maintenance or other special mission needs. Supplemental instruments were generally mounted on the starboard side of the ejection seats up to the canopy, while communications and autopilot formation sticks were on the port side above the passageway. Entry into the bomb bay was possible inflight if the flight deck was de-pressurized.

The following images and drawings depict the B-45A-1 cockpit. It was remarkably roomy in design, but equipped with an opening canopy for ejection only, not ground operations. The interior temperatures, depending on the location, were consistently well above 100°F. While aircraft did have an engine driven air-conditioning system, its use was restricted during ground operations due to J35 overheating until the aircraft was airborne and climbing to altitude. The aircraft was equipped with two NAA-designed and built ejection seats. These were powered by large black powder charges which provided near 100% of their power instantaneously and were known for inducing serious back and spinal injuries to the pilot and co-pilot. They were also recommended to be used only above 1,000-feet ground-level. Ejection was difficult at best as the crew had to immediately lower their heads as far as possible while the canopy was jettisoned then raise the head to brace it against the headrest prior to ejection. All crew station compartments were pressurized. The pilot had full control of all systems in the cockpit such as electrical power, hydraulics, and fuel management and a full set of instruments on a large, easily-to-read panel immediately to his front. The copilot had only a small set of the "basic six" flight instruments, a stick with a control wheel, throttles, and rudder pedals. His primary function was communications, and giving the pilot short breaks from flying the airplane. He had neither the instruments nor the visibility for instrument flight or landings. Both cockpit crewmembers had autopilot formation control sticks to their left.

Photo 16: This image shows an early B-45A-1 pilot's instrument panel. (Steve Ginter Collection)

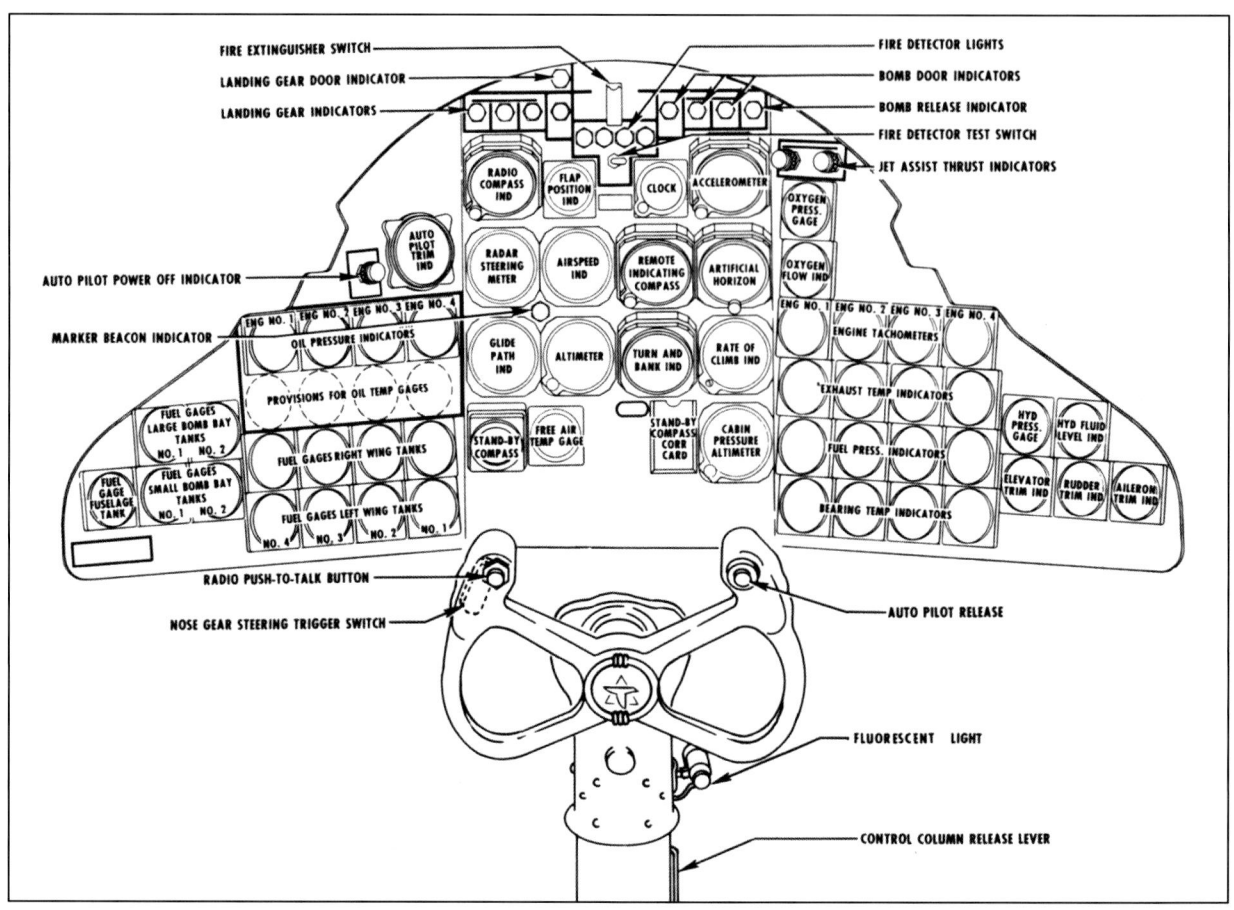

Drawing 45: The pilot's main instrument panel for the B-45A-1. (NMUSAF)

Photo 17: The pilot's electrical controls and throttle pedestal, on the starboard side of the pilot's seat in the B-45A-1. (Steve Ginter Collection)

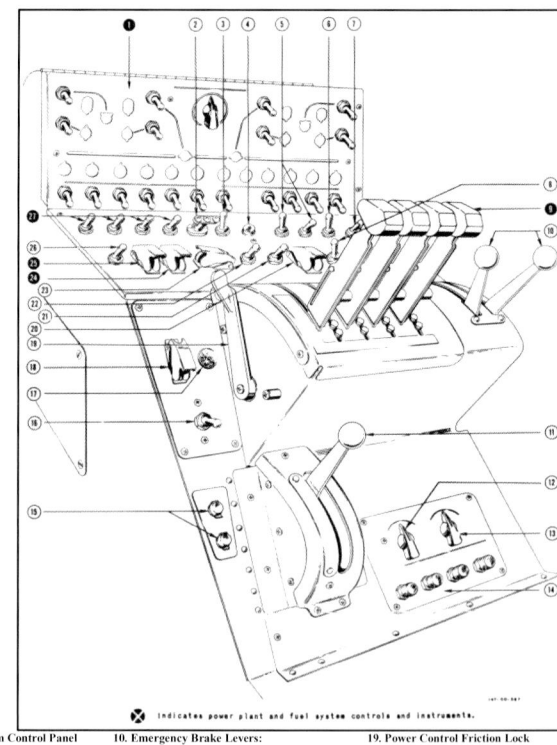

1. Fuel System Control Panel
2. Landing Light Switch
3. Fuselage Lights Switch
4. Fuselage Lights Keying Switch
5. Navigation Lights Switches
6. Formation Lights Switch
7. Taxi Light Switch
8. Bomb Door Switch
9. Power Controls
10. Emergency Brake Levers:
11. Wing Flap Control
12. Wing Anti-Icing Rheostat
13. Tail Anti-Icing Rheostat
14. Surface Control Boost Test Lights
15. Spare Lamps – Fuel System Control Panel
16. Bomb and Fuel Tank Salvo Selector Switch
17. Salvo Circuit Indicator Light
18. Bomb & Fuel Tank Salvo Control Switch
19. Power Control Friction Lock
20. Emergency Alarm Switch
21. Horn Cut-Off Switch – Cabin Pressure
22. Horn Cut-Off Switch – Landing Gear
23. Surface Control Boost Switch
24. ATO Units Release Switch
25. ATO Units Ignite Switch
26. Battery Switch
27. Starter Switches

Drawing 46: This is the B-45A-1 pilot's main electrical control and throttle pedestal with identification key. (NMUSAF)

The pilot's monopoly on flight controls and instruments was a holdover from World War II light and medium bombers— the A-20, A-26, and to some extent the B-25 and Martin B-26—where the pilot assumed the greatest part of the cockpit tasks. This would change only slightly in the B-45A-5. Such layouts could become a safety issue in major emergencies for the crew. Also note that many controls were in awkward locations and positions and would require focusing solely inside the cockpit or be difficult to reach in higher "G" situations or if crew injuries were involved.

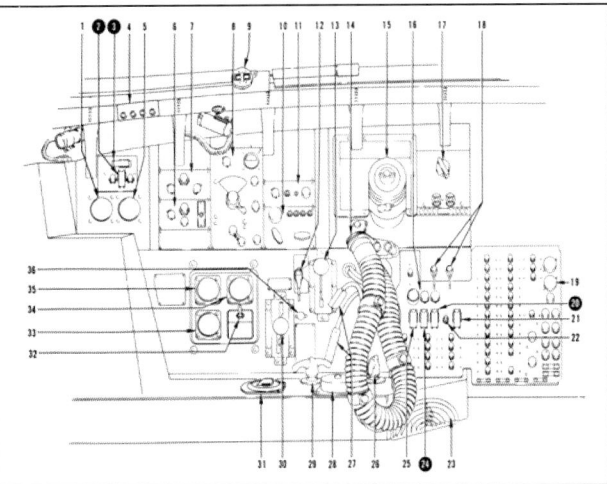

1. Oxygen Pressure Gage
2. JATO Units Ignite Switch
3. JATO Units Release Switch
4. Surface Control Boost Test Lights
5. Oxygen Flow Indicator
6. VHF Radio Control Panel
7. Liaison Transmitter Remote Control Panel
8. Radio Compass Control Panel
9. Ventilating Air Outlet
10. Interphone Control Panel
11. Instrument Approach Control Panel
12. Hydraulic Main System Pressure Valve
13. Surface Control Lock
14. Oxygen Hose
15. Pressure Breathing Oxygen Regulator
16. Bomb and Tank Salvo Indicator Lights
17. Heated Clothing Rheostat
18. Emergency Hydraulic Selector Valves
19. Circuit Breaker and Fuse Panel
20. Large Tank Salvo Switch
21. Emergency Alarm Switch
22. Bomb Door Switch
23. Ventilating Air Outlet
24. Bomb and Small Tank Salvo Switch
25. Bomb Salvo Switch
26. Life Raft Release
27. Radio Headset and Microphone Cord
28. Cabin Air Emergency Valve
29. Auto-pilot Release
30. Landing Gear Selector Control
31. Cabin Temperature Rheostat
32. Elevator Trim Tab Control Switch
33. Elevator Trim Indicator
34. Hydraulic System Pressure Gage
35. Hydraulic Reservoir Fluid Level Gage
36. Landing Gear Control Lock Emergency Release

Photo 18: This is the B-45A-1 pilot's aft starboard side control and circuit breaker panel. (Steve Ginter Collection) Drawing 47: This is the B-45A-1 pilot's aft starboard side control and circuit breaker panel with identification key. (NMUSAF)

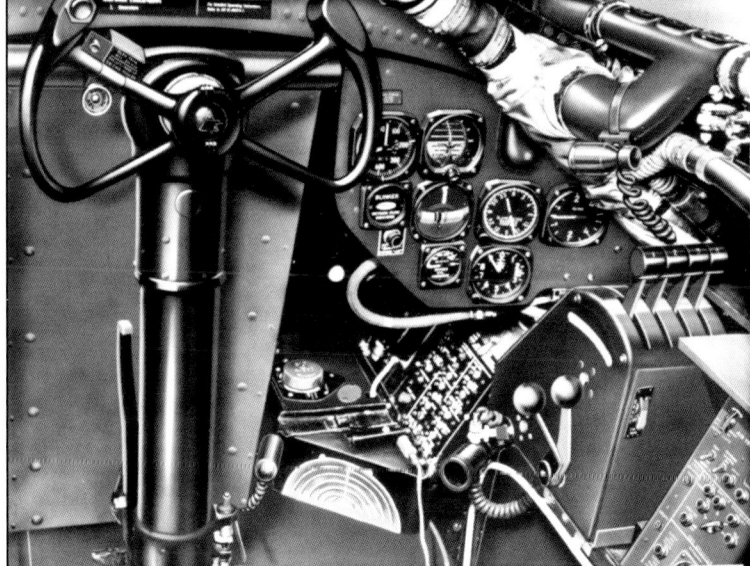
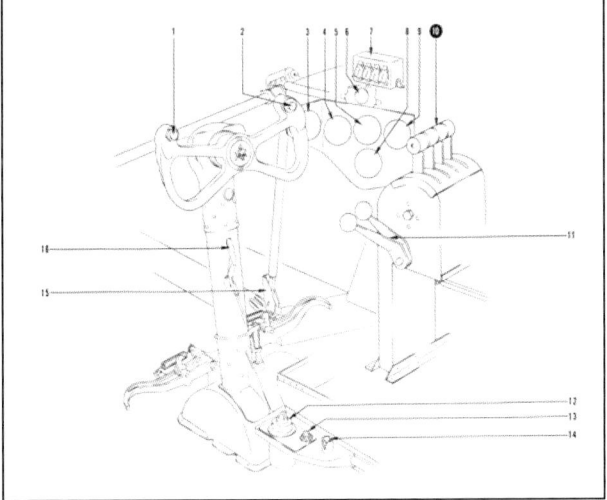

1. Radio Rush-to-Talk Button
2. Automatic Pilot Release
3. Turn-and-Bank Indicator
4. Airspeed Indicator
5. Altimeter
6. Ventilating Air Control
7. Surface Control Boost
8. Remote-indicating Compass
9. Rate-of-Climb Indicator
10. Power Controls
11. Emergency Brake Levers
12. Automatic Pilot Controller
13. Radio Push-to-Talk Button
14. Automatic Pilot Clutch Switch
15. Canopy Emergency Release Test Switches
16. Control Column Lock Release

Photo 19: This drawing shows the B-45A-1 co-pilot's control column and compact flight instrument panel. The co-pilot's flight instruments included only the "basic six" allowing him to fly in visual conditions and not make instrument approaches. (Steve Ginter Collection) Drawing 48: This shows the B-45A-1 co-pilot's control column and compact flight instrument panel with identification key. (NMUSAF)

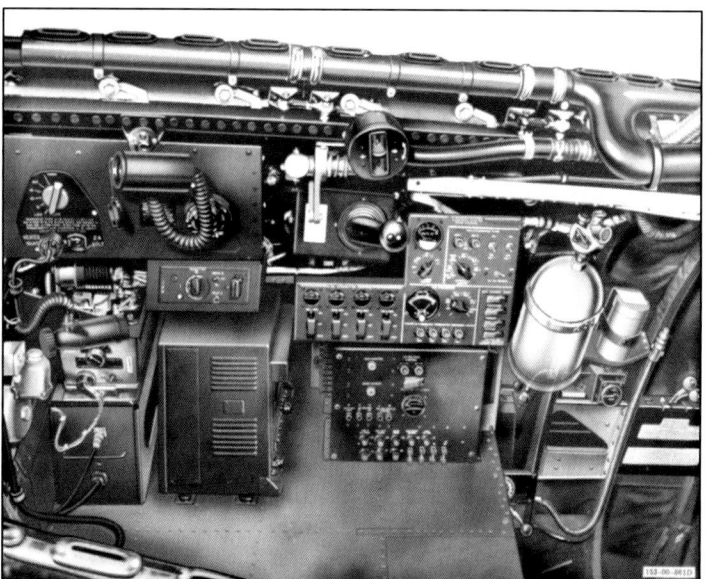
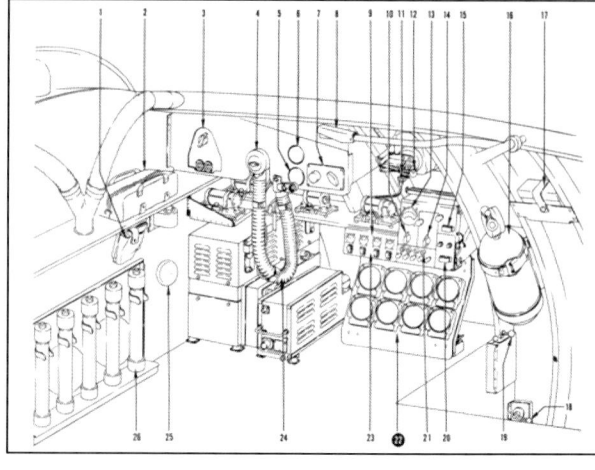

1. Pyrotechnic Pistol
2. Pyrotechnic Cartridge Case
3. Heated Clothing Rheostat
4. Pressure Breathing Oxygen Regulator
5. Oxygen Pressure Gage
6. Oxygen Flow Indicator
7. Interphone Control Panel
8. Canopy Emergency Release
9. Ammeters
10. Voltmeter Selector
11. Portable Oxygen Cylinder Filler Valve
12. Pyrotechnic Pistol Mount
13. Voltmeter
14. Alternator Switch
15. Three-Phase Inverter Switch
16. Portable Oxygen Bottle
17. Signal Lamp Spare Filter Case
18. Hydraulic Reservoir Filling Pump
19. Bomb Door Manual Control
20. Single-Phase Inverter Switch
21. Generator Overvoltage Indicators
22. Bearing Temperature Indicators
23. Generator Switches
24. Oxygen Hose
25. Emergency Alarm Bell
26. Drop Message Containers

Photo 20: This is the B-45A-1 co-pilot's starboard side radio and fuel system panel. (Steve Ginter Collection) Drawing 49 (Right): This is the B-45A-1 co-pilot's starboard side radio and fuel sytem panel with identification key. (NMUSAF via the Author)

Drawing 50 and 51: These show the B-45A-1 and B-45A-5 pilot's and co-pilot's formation control sticks. These were very similar to those installed in B-17s and B-24s produced during World War II. They allowed very fine adjustment to the autopilot while in close formation with other aircraft. (NMUSAF)

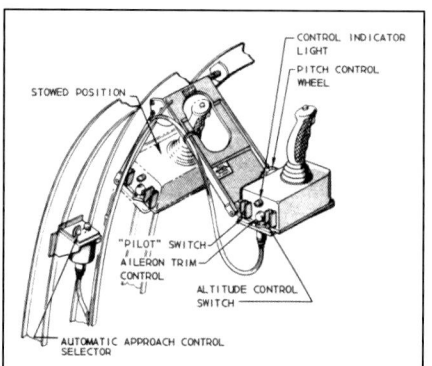
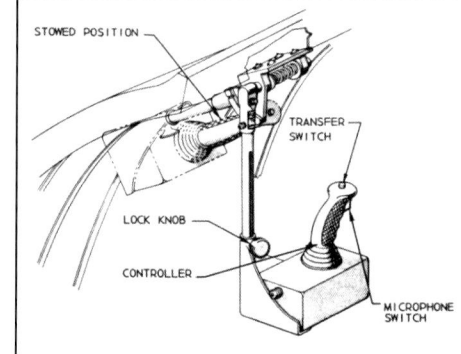

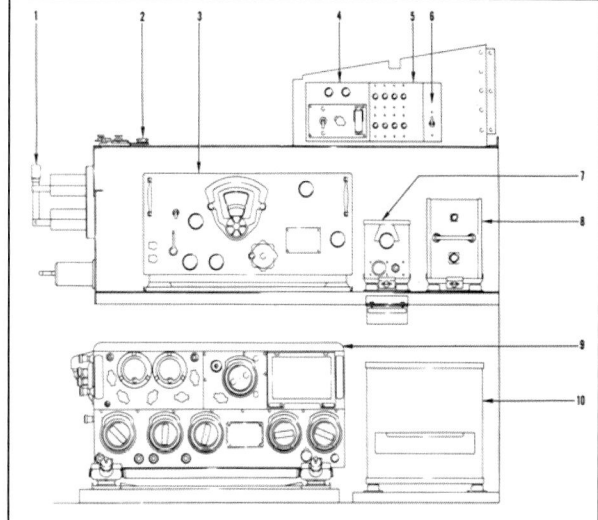

1. Capacitor Switch
2. Transmitting Key
3. An/ARC-8 Radio Receiver
4. IFF Radio Control Panel
5. Radio Circuit Breakers
6. AN/ART-13 Monitor Switch
7. Interphone Amplifier
8. AN/ARN-5A Liaison Receiver
9. AN/ARC-8 Liaison Transmitter
10. AN/ARM-6 Radio Compass Receiver

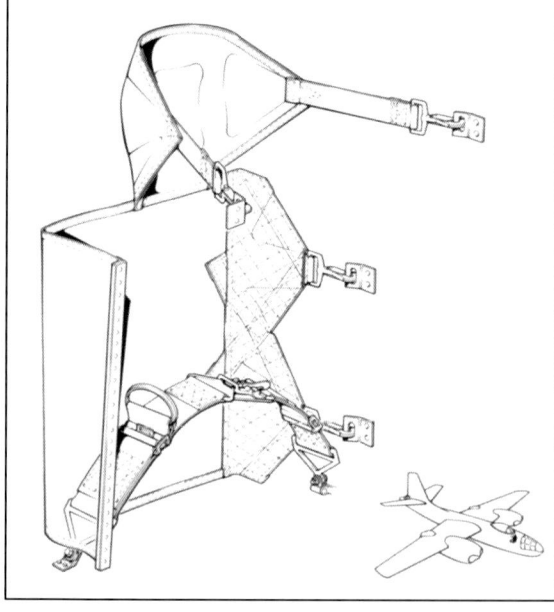

Drawing 52: These units were part of the B-45A-1 and B-45A-5 co-pilot's radio equipment installed on the starboard side of the station just below the canopy. (NMUSAF) Drawing 53: This was the B-45A-1 and B-45A-5 canvas crew seat used to carry an extra crew member during special missions or maintenance flights. The harness was mounted below the co-pilots portside radio and fuel panel with the harness facing forward in front of the bomb bay bulkhead at the aft end of the crew passageway below the cockpit floor. (NMUSAF)

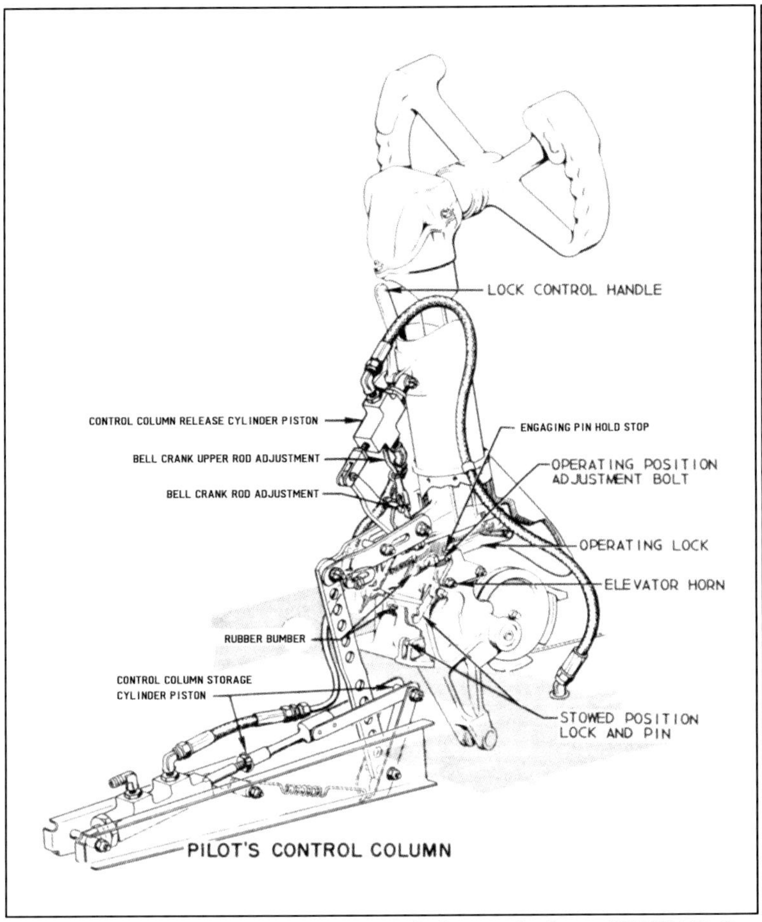

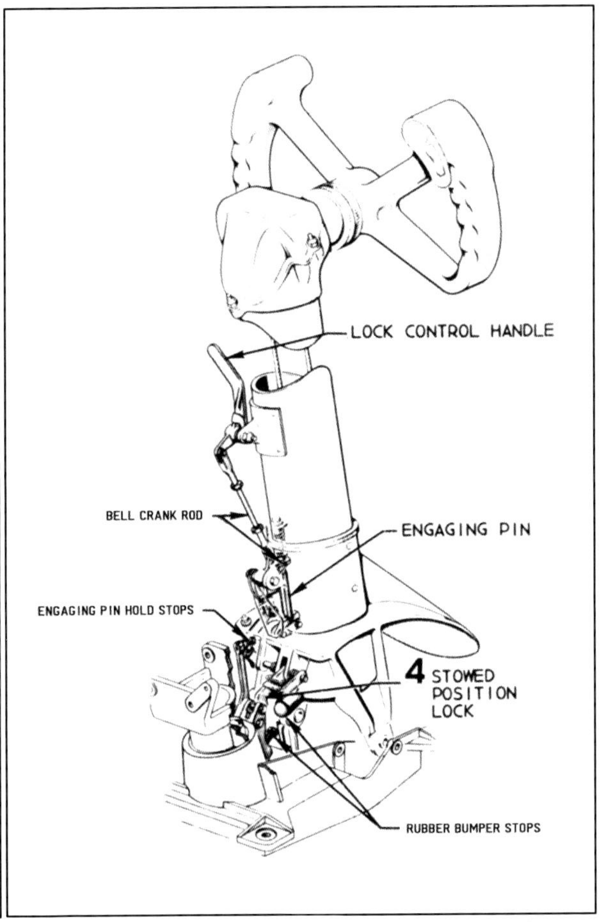

Drawing 54: This shows the B-45A-1 and -5 pilot's control column. (NMUSAF) Drawing 55: This shows the B-45A-1 and -5 co-pilot's control column. (NMUSAF)

North American designed and produced the ejection seats used in all B-45 models. This version was the first production series and was powered by a single black-powder charge. The full force of the charge was nearly immediate and often caused severe back and spinal injuries to crewmembers who had to eject. The ejection seat was new and still a work in progress in 1948 and 1949. Note the amount of movement required by the pilot and co-pilot to be ready to fire the seats in **Drawing 57**, especially lowering their heads and bodies to insure they were missed by the jettisoned canopy.

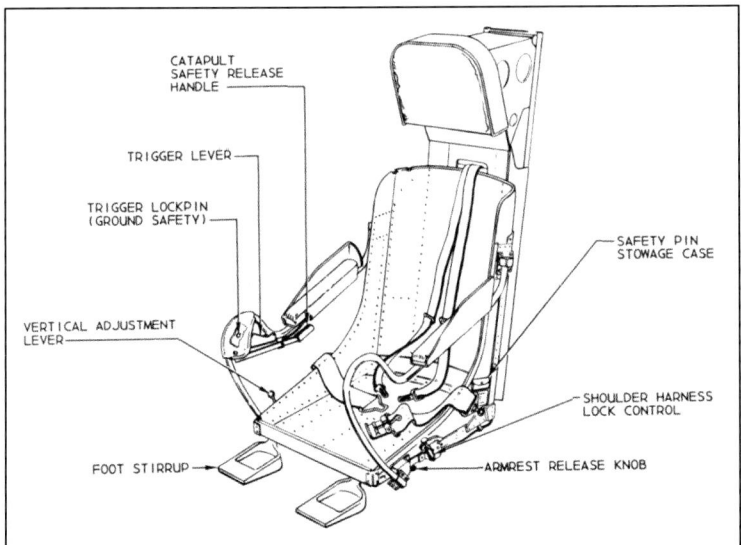
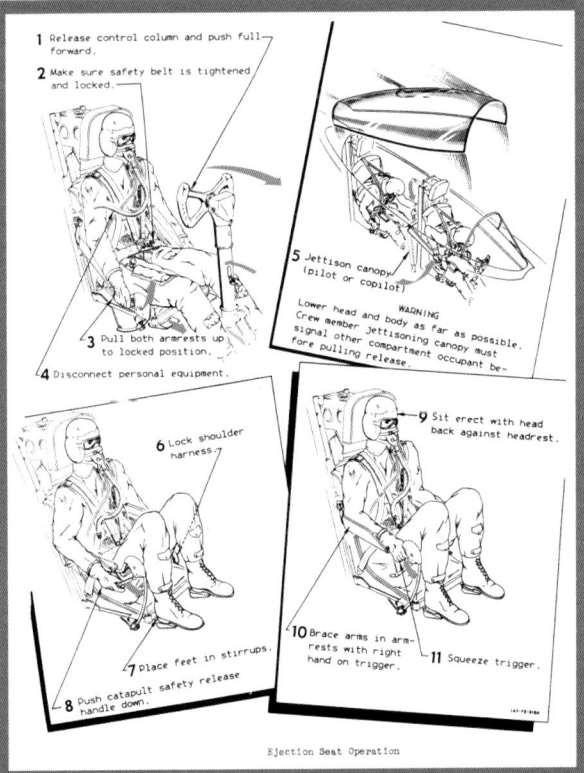

Drawing 56: This drawing shows the North American designed and manufactured ejection seat used in the B-45A-1 and B-45-5. (NMUSAF) Drawing 57: This is the pilot's flight manual illustration depicting the 11 separate and sequential steps required for a safe ejection from the B-45A-1 (NMUSAFM)

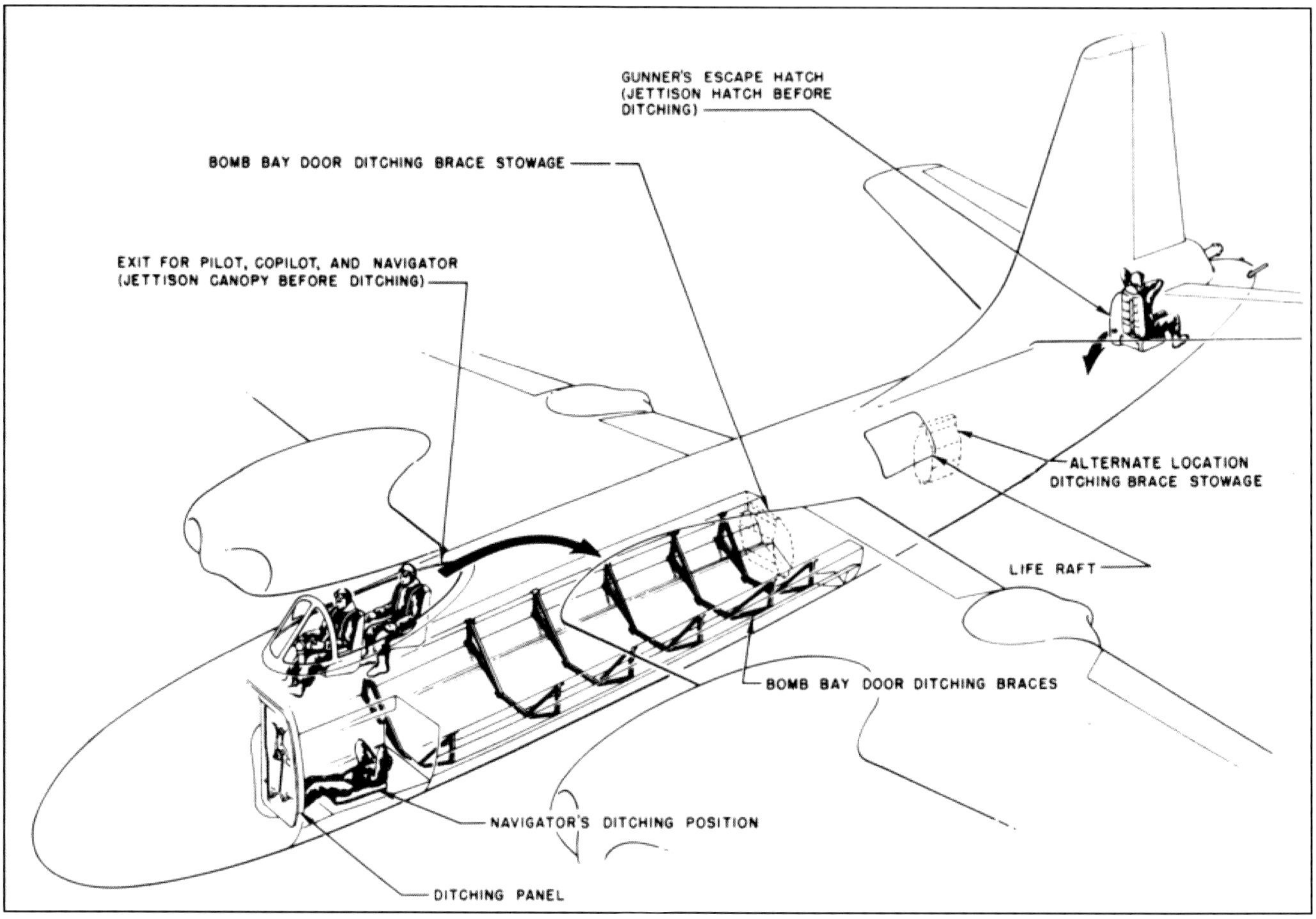

Drawing 58: The B-45 shared a safety feature with the Consolidated B-24 for ditching. A series of oak braces were installed along both sides of the bomb bay doors to help brace them from water impact and help in keeping the lower fuselage intact until the crew could evacuate the aircraft. (NMUSAF)

Bomb Bay Interior Illustrations for the B-45A-1 and -5 as built....

The B-45A was never recognized as a potential atomic bomber before the XB-45 flew in 1947. The aircraft was simply too narrow and the bomb bay too short to accommodate any of the existing atomic weapons or those known to be planned. Therefore it was designed and planned to be a short range light bomber capable of carrying the majority of existing World War II weapons up to and including the 22,000 pound Tall Boy bomb developed for the RAF. The aircraft had two separate bomb bays divided by the wing spar with an unobstructed area in the lower bomb bay beneath the wing. Carrying two 318-gallon fixed and two jettisonable 872-gallon fuel tanks in each bay, the B-45A-1 had a total fuel capacity of 5,800 gallons, all internal. Each bomb bay was enclosed by two sets of bomb bay doors for a total of four doors over each half-bay. The bays could be opened individually.

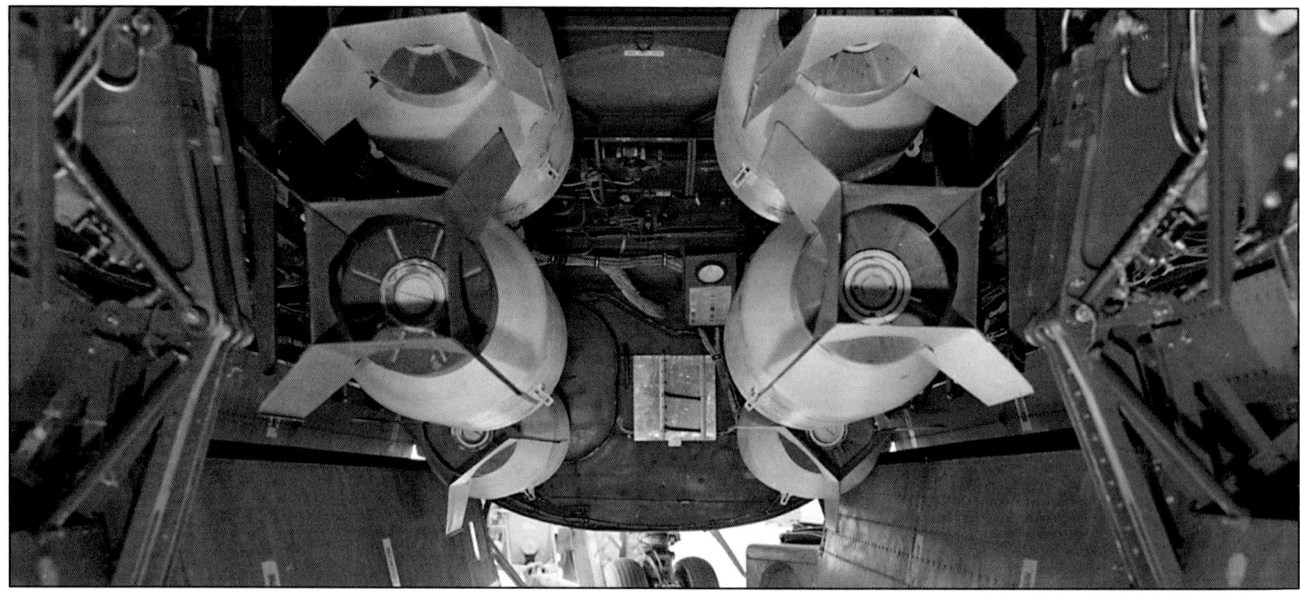

Photo 21: B-45A-5 7026 was photographed on the ramp at Egiln AFB in 1951. (Craig Kaston via Steve Ginter)

Photo 22: B-45A-5 7025, assigned to the 3200th Proof Test Group was photographed above the Gulf of Mexico just north of Eglin AFB, FL in the fall of 1949. (Steve Ginter Collection)

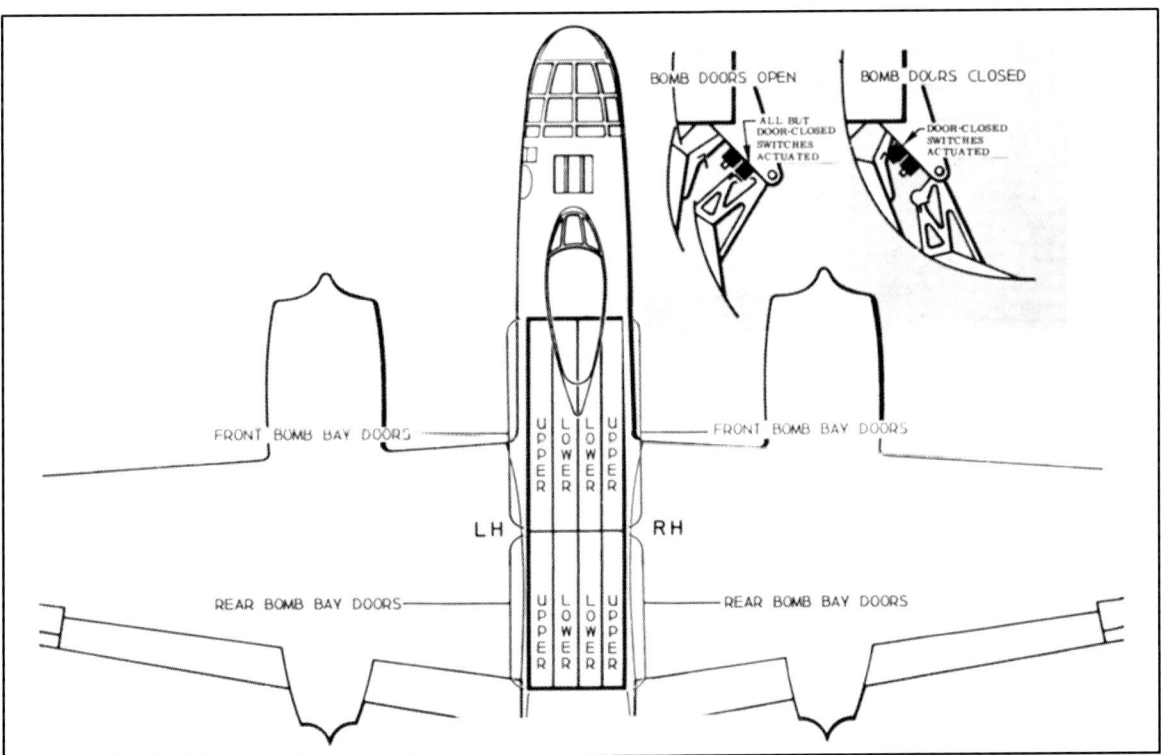

Drawing 59: The B-45 family featured a very unusual style of bomb bay doors as shown in the drawing above. Two sets of doors were located on the belly, one set covering the forward and center group of bomb racks, and one covering the rear set of bomb racks. Each set of doors consisted of four separate panels, two covering the upper portion of each half of the bomb bay, and the other two panels covering the bottom of each half of the bay. The narrow width of each set of panels allowed the belly itself to be fairly close to the ramp surface for bomb loading. The forward and rear sets could be opened separately and each set could opened or closed in 1-second. A view of each set looking directly aft is shown on the right side of this drawing above. (NMUSAF)

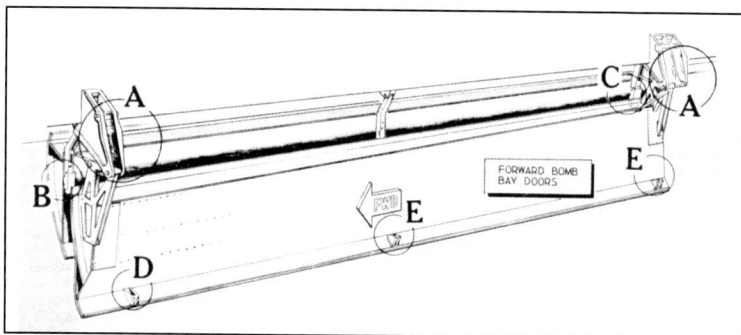 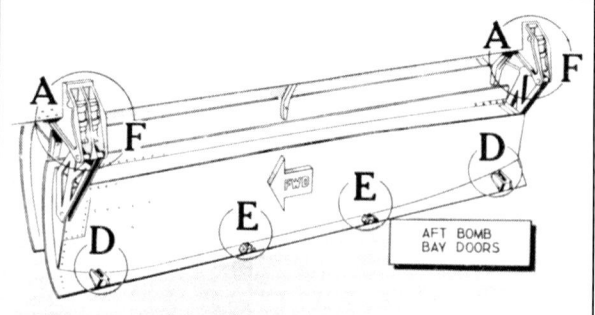

Drawings 60 and 61: Above are drawings of the starboard side forward and aft bomb bay doors, looking aft. Items within the A and F circles were hydraulically powered actuators and hinges, while Items B through E were alignment fixtures. (NMUSAF)

The B-45's weapons would be those developed and used to fight World War II. The aircraft's technical data specific 46 different types of ordinance including General Purpose, Incindiary and Incindiary Cluster, Fragmnetation, Depth Charge, Chemical, Parachute Flares, Photo Flash, Cluster, Armor Peircing, and Semi-Armor Peircing munitions weighing from 90 to 22,000 pounds each. Those few aircraft with a functional conventional bombing system were accurate and capable of delivering these weapons from an altitude of 40,000-feet to under 250- feet in radar-guided attacks, all at over 500 miles per hour. They could deliver entire multiple weapon loads in one location or across a lengthly area depending on the target, and could attcak in level or shallow dive-bombing profiles. The aircraft could attack in single or multiple aircraft formations as required. One of the short comings of the B-45 was that it could not be re-armed, or "turned" rapidly for repeated attacks by the same aircarft. Weapons had to be loaded and armed indvidually, and even with multiple crews, it took time. In general, a weapon had to be brought to the aircraft via a fork lift and placed adjacent to the portion of the bomb bay where its rack was located. Depending on the size of the weapon, it might be possible to slip the weapon beneath the bomb doors below the bomb racks. However if it was too large, it would have to loaded onto a low-level towed trailer, then manuvered into the bomb bay area from the forward or aft ends.

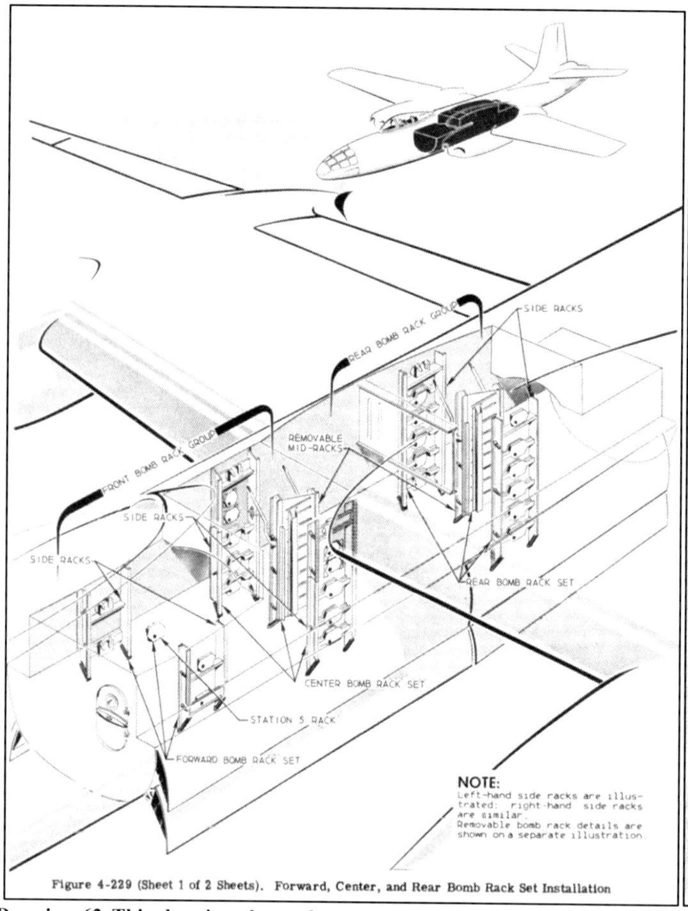

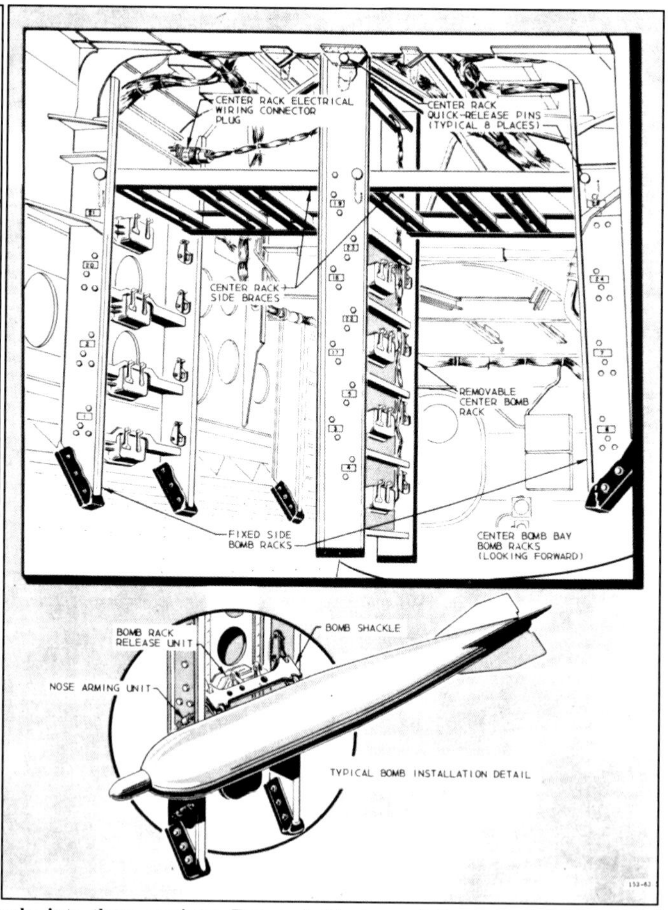

Drawing 62 This drawing shows the arrangement of the aircraft's bomb racks into three sections, Forward, Center and Rear. Each held multiple, individual mumitions units. (NMUSAF) **Drawing 63:** This drawing shows the details of the racks which allowed the weapons to be secured to the racks and released under the control of the bomb/nav. Each weapon had its own station and shackle with electrical connections to operate the arming and release controls on each bomb shackle. (NMUSAF)

Next it had to be placed on a handling cart or cradle, safety chains installed, and moved directly beneath its rack. The weapon and its cart would be placed on an electrically-powered hoist. Four cables were then run from the hoist through the handling cart and upward to the top of the bomb bay and hooked into specific mounting holes in the bomb bay ceiling. The weapon was electrically hoisted to its particular bomb station in order, and its shackle mated to the weapon station's bomb rack release. Then safety pins would be installed into the bomb rack releases. One or two bomb fuzes were installed as required by the weapon or the mission. With a fuze or fuzes installed, safety wires were attached to the bomb rack release, routed through one or both bomb lugs, and then forward and/or aft to the blades of each fuze. The wires kept the fuze blades from rotating in the bomb bay, "safeing" the weapon. Then the bomb bay could be pre-flighted along with the rest of the aircraft. The saftey pins in the bomb rack release would be removed before take-off allowing the loaded bombs to be released on command. The fuze safety wires remained in place until the bomb was released and the wires were pulled out of the fuze blades being held by the releases on the bomb racks. This allowed the fuze blades to turn and arm the weapon after a specified number of revolutuions.

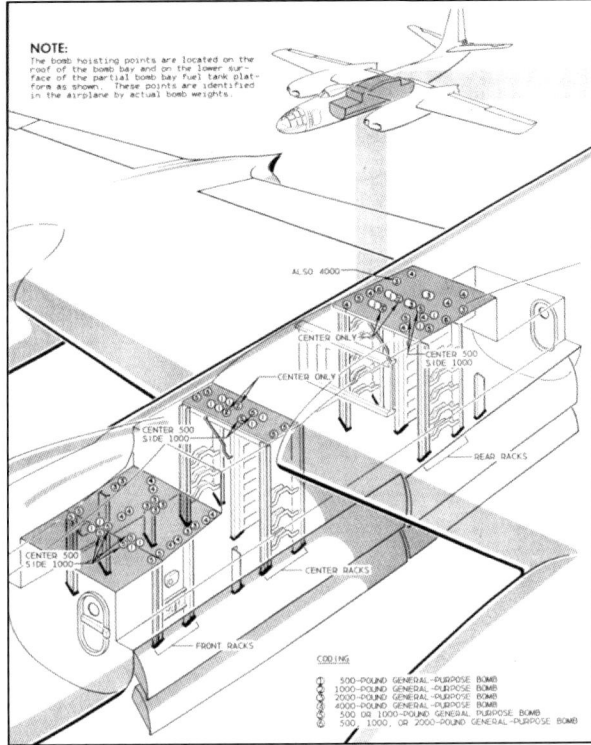

Drawing 64: to the left: This drawing shows the hook holes on the ceiling of the bomb bay and aft partial fuel tank cover where the four cables used to lift each weapon were secured to the hoist and cart. (NMUSAF)

Drawing 65 below: Image A shows a general purpose bomb loaded on its hoist cart and the electrical hoist located underneath the load. Image B shows a bomb hoisted out of its cart, safety chains and shackle installed, and on its way upward in the bomb bay via the two metal cables to its station on the bomb rack. (NMUSAF)

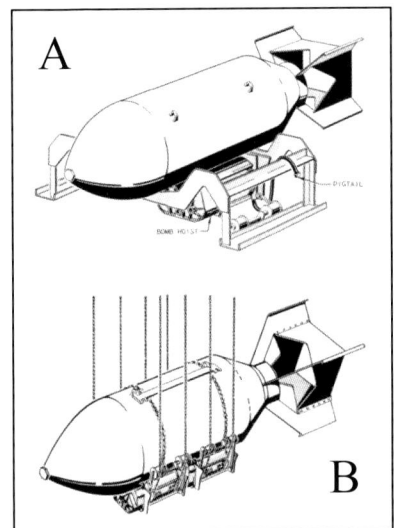

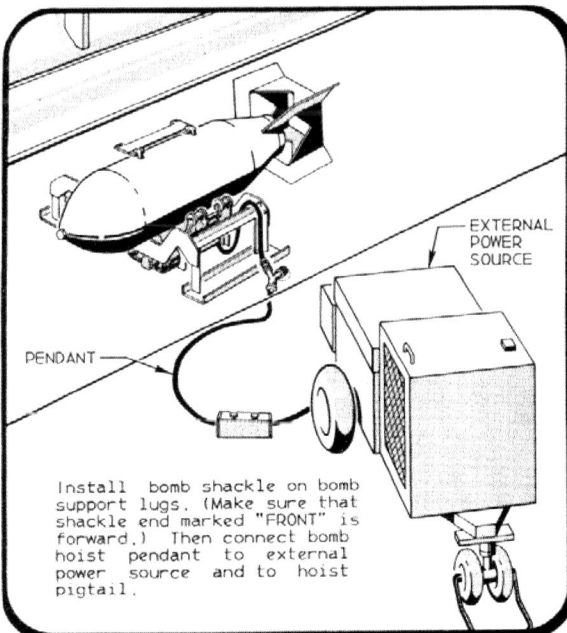

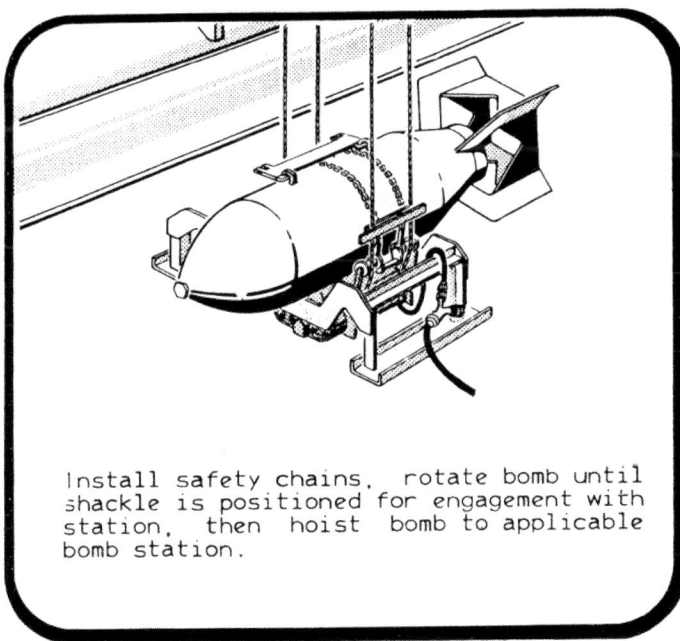

Drawing 66: This drawing shows a general purpose weapon in place on its handling cart, below its designated bomb rack with its electrically-powered hoist in place. The auxiliary power cart is in place to power the hoist via the "pendant' or cable. (NMUSAF) **Drawing 67:** The hoist has its four metal cables extended and secured to the ceiling of the bomb bay. The weapon's bomb shackle has been installed onto the bomb's lugs and it is ready to be hoisted to its bomb rack station. (NMUSAF)

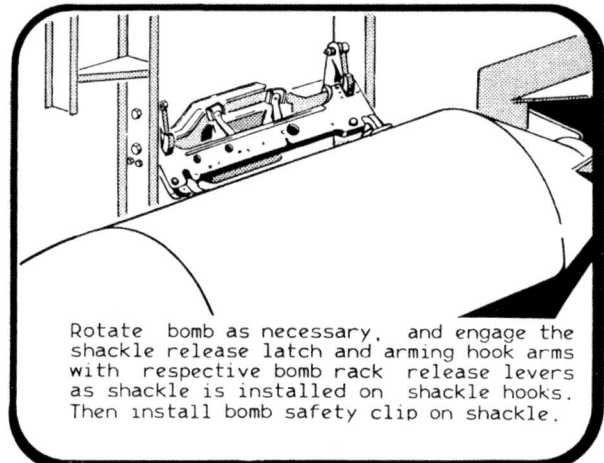
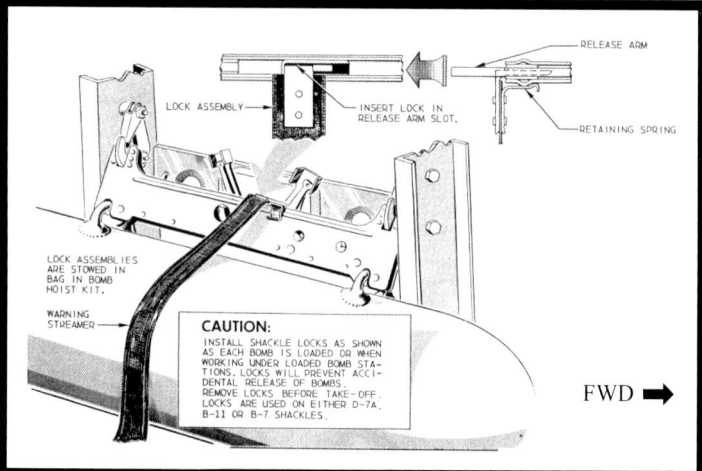

Drawing 68: The weapon has now been raised to its station level on its bomb rack and has been rotated to bring its shackle into alignment to be connected with the bomb rack's release unit. The safety chains have been removed and the hoisting cables have been unfastened from the bomb bay ceiling and lowered to the ramp for the next weapon. The next steps will be fusing and safeing weapons. (NMUSAF) Drawing 69: The safety pin has been inserted per the drawing first. Assuming this weapon will only have a single fuse in the nose, a heavy copper wire would be run from the aft bomb lug which is also attached to the bomb shackle, then forward and through one of the holes in the release unit attached to the bomb rack. Then the wire would be routed down and forward through the front bomb lug and through the forward fuse blades and trimmed just a few inches ahead of the blades. (NMUSAF)

The Tall Boy and Grand Slam....

The largest weapons the B-45A-1 and -5 could carry were the British designed and manufactured 12,000 pound Tall Boy and 22,000 pound Grand Slam. The range penalty was high, but there were still targets throughout Europe that would need their kind of destructive impact. In the RAF the weapons were carried by the Avro Lancaster against transportation targets and submarine pens in Northern Germany. The transportation targets were generally tunnels and the supporting foundations for bridges, which, although heavily built with deeply buried foundations, were destroyed by the earthquake-like destructive imapcts of the weapons. The RAF soon learned not to attempt to hit the bridge, but to aim the weapon to either side to allow the concussion to severely crack and shatter the bridge and drop it by breaking it into major peices. The Grand Slam was also built to penetrate deeply built concrete facilities whch made it a successful weapon against the few surviving submarine pens in northern Germnay at the end of the War.

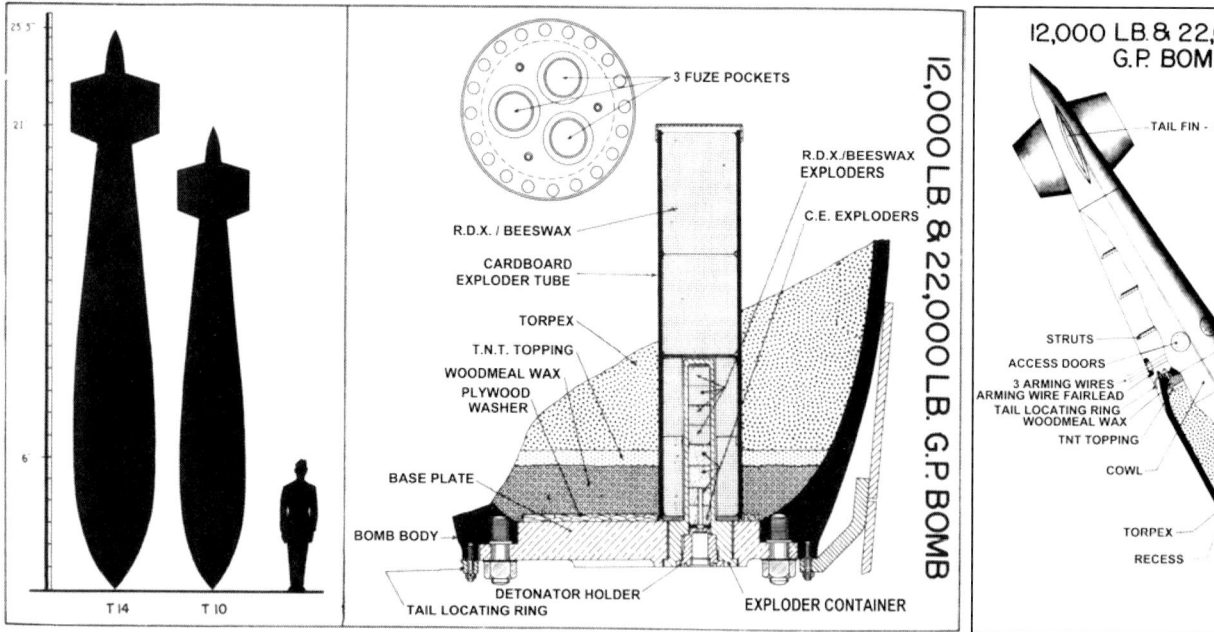

Drawing 70: These are the dimensions of the two weapons and what they would have looked like against a crew member. The explosive of each weapon was primarily Torpex with a T.N.T. surface-coating on the entire weapon The weapon's three fuzes were reached through a hatch on the side that gave access to the fuze compartment on the aft end of the explosive chamber in the mid-portion of the weapon. Drawing 71: Note that the four tail fins were twisted 5° to the right to give the weapon a counterclockwise rotation to improve accuracy. (NMUSAF)

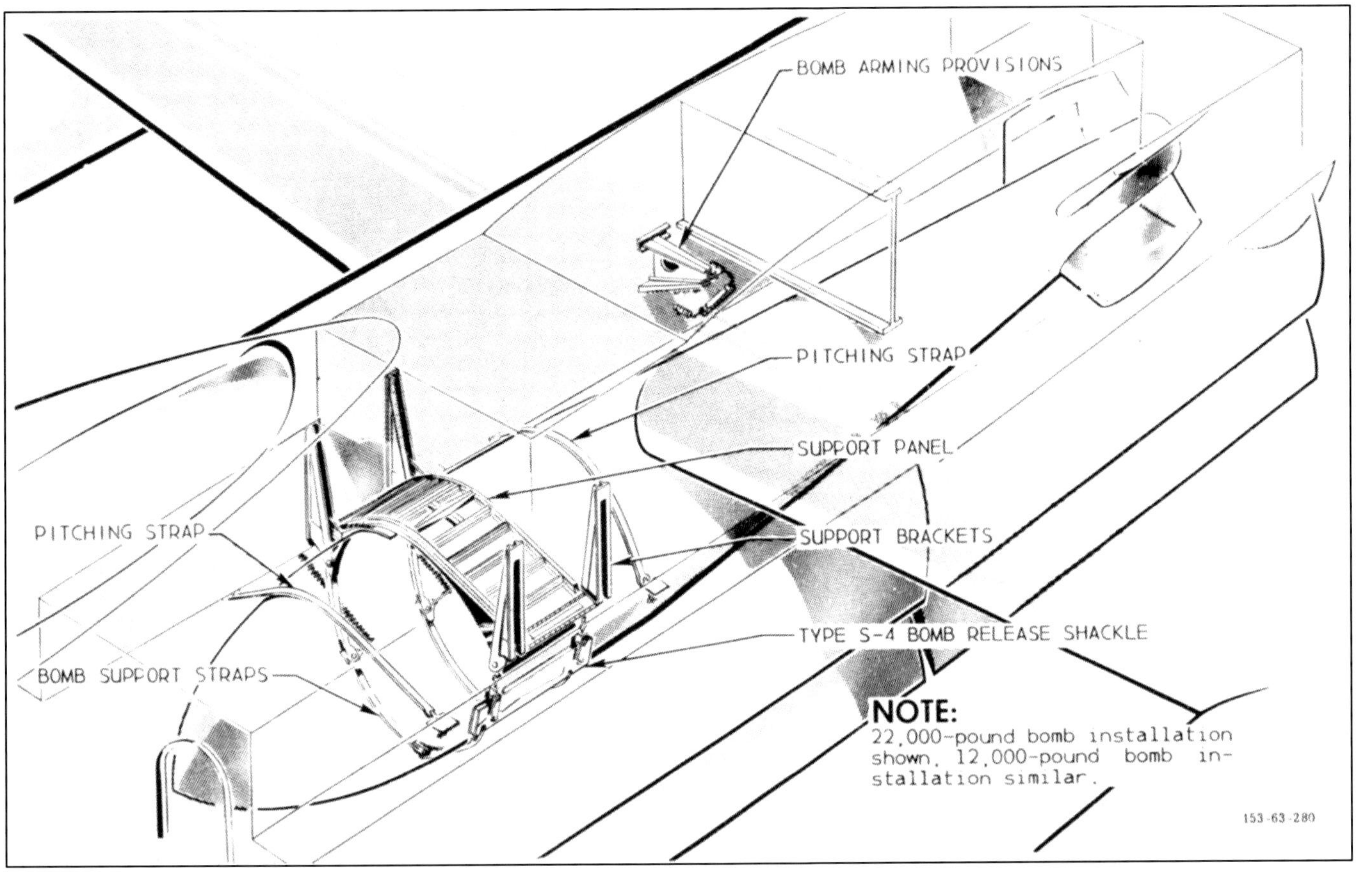

Drawing 72: This drawing gives several details of the installation of the Grand Slam that filled both bomb bays of the B-45A. (NMUSAF)

Drawing 73: This drawing shows a representation of what the Grand Slam would have looked like fully loaded in a B-45, and how much it totally filled the depth of the bomb bay. The racks were installed in special fittings on the bomb bay wall structure designed for extremely heavy loads. No other weapons or bomb bay fuel tanks could be carried if the aircraft were loaded with a Grand Slam. (NMUSAF)

Weapons Technical Data and Hoist Gear....

The aircraft's Operations Manual contained a long list of illustrations called Bomb Release Sequence Diagrams (**Drawing 73**) for the bomb loading crews to use to load different types or numbers of weapons. The illustrations identified the weapon and how many were to be loaded. This Diagram shows guidance for loading 27 500-lb. General Purpose bombs. A view that listed all of the bomb racks by group was given with guidance on which racks and stations were to be used to maintain the proper CG. Certain stations were identified for use with heavy weapons such as 2,000-lb. and 4,000-lb. bombs. At the lower left is a drawing of the bomb/nav's Bomb Station Indicator Light Panel. This panel had indicator lights for each weapons station that lit when that station released its weapon.

Next to **Drawing 74** is **Drawing 75,** the Bomb Hoisting Kit which shows details of the hoisting gear used in the B-45 Bomb Bay. All of the equipment was housed in a Hoisting Equipment Kit, and every aircraft had two assigned. At the 12 o'clock position were drawings of the cable assemblies, two capable of handling up to 500-lb. loads and two capable of handling 2,000-lb. loads. These cables were the ones that fitted into the holes in the ceiling of the bomb bays. Note that they had simple hooks at each end, no bolts were threaded into the openings in the ceiling panels. At the two o'clock position were hoist cables fitted with hoist sling assemblies that kept the cables at a standard separation distance from each other. The cables were also fitted with hoist sling hanger assemblies, or loops that the hooks on the end of the hoist cables could fit into.

At the three o'clock location on **Drawing 75** was the cluster bomb adapter cradle used to protect fragmentation munitions during loading. This was a metal cover that helped prevent the cluster weapons from hanging on any portion of the bomb rack, release assembly, bomb station or arming levers. The cover had four metal loops that allowed the hoisting cables to be hooked on. At the four o'clock position was the canvas bag that held the safety pins used on the bomb shackles. At the five o'clock location was the bomb hoist, powered by an auxiliary power cart. The kit contained cables, plugs and switches. On the side of the hoist were specific locations for fitting hoist couplings depending on the specific weapon being loaded.

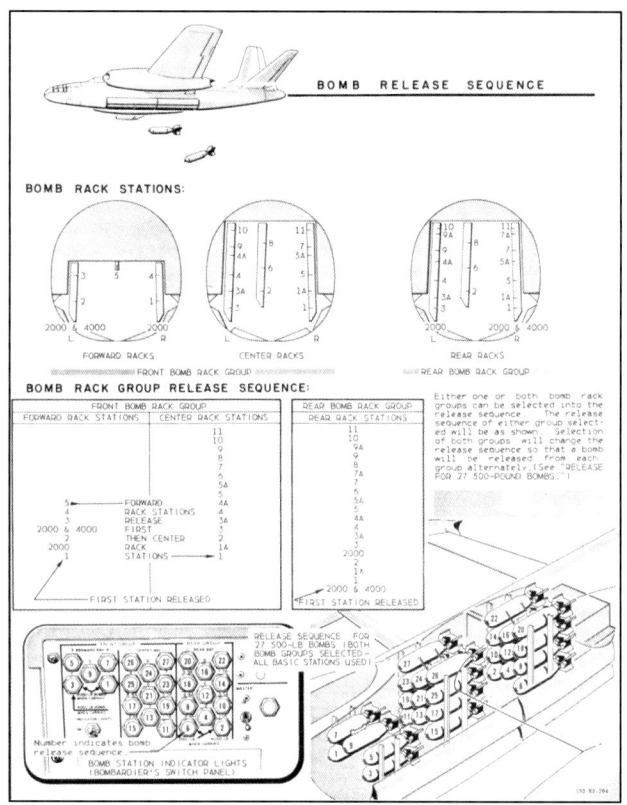 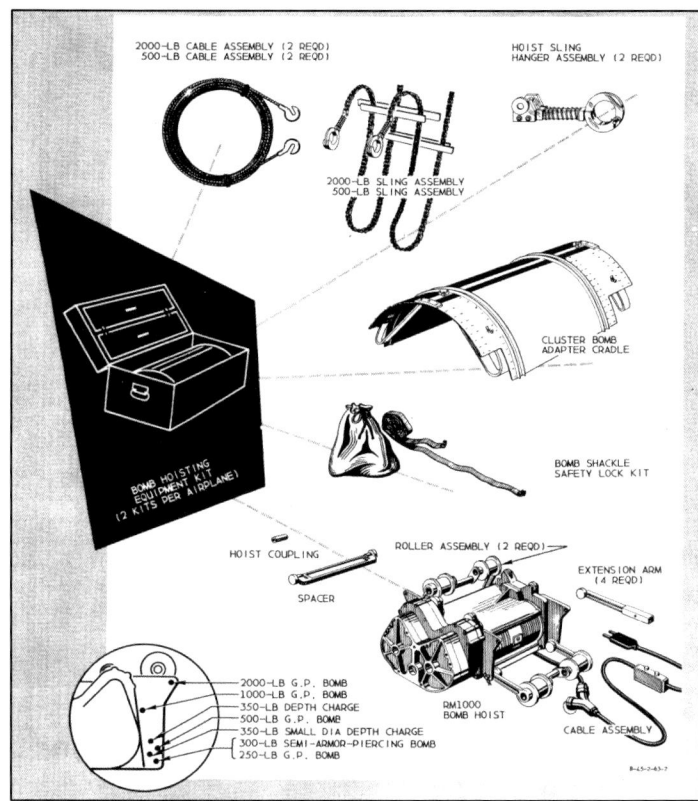

Drawing 74: This drawing shows the Bomb Release Sequence Diagram for 500-lb. General Purpose bombs. (NMUSAF) Drawing 75: This drawing shows the contents of the Bomb Hoisting Kit, of which two were assigned to each aircraft. (NMUSAF)

Fuel Systems....

The wings of all B-45As could not hold weapons stores but they were fitted to hold jet fuel in four tanks in each wing, one permanent tank in the fuselage just behind the cockpit, two jettisonable tanks and two stationary, but removable tanks in the bomb bays. Each tank in the wing was made up of a specific number of fuel cells, or metal structures that fitted within the wing's structure surrounded by the fuel tank enclosure. The cells of each tank were plumbed together to automatically send or receive fuel as each tank as directed to manage its fuel load. Because the total number of cells was driven by the wing's structure, combining the cells through plumbing, pumps, and electrical controls into a lesser number of tanks made the task of managing fuel throughout the aircraft easier.

For example, look at the port wing of the aircraft in **Drawing 76**. The farthest outboard tank is Tank 4 Left (4L) with a capacity of 376 gallons. This tank is made up of four cells. Tanks 3L and 2L have two cells each, while Tank 1L consists of three cells and a total capacity of 509 gallons.

To transfer fuel from Tank 4L to Tank 1L, it takes only two switches, one for Tank 4L to pump to Tank 1L, and 1 for Tank 1L to receive from Tank 4L. Controlling two switches is far easier than controlling the seven switches involved if you had to control individual cells for this transfer.

All wing tanks, and the upper bomb bay tanks were self-sealing. This meant that each cell or tank had one or more rubber bladders inside the cavity that would seal a hole made by an entering shell or shrapnel and stop the loss of fuel or fumes from that location. The "FT" tank in the fuselage was a single cell self-sealing tank permanently built into the fuselage. The J1 and J2 tanks were single-cell, non-self-sealing jettisonable tanks which hung from two sets of stations in the forward and aft bomb bays each. Above J1 and J2 tanks were two self-sealing bomb bay tanks (BT), removable but not jettisonable, one each hung in the forward and aft bomb bay. Fuel in these BT and JT sets was managed as a single combined unit in the forward or aft bomb bay. The fuel management system in the B-45A was controlled by the pilot. Refueling the B-45 was accomplished by filling each tank individually by hose from above the wing and fuselage.

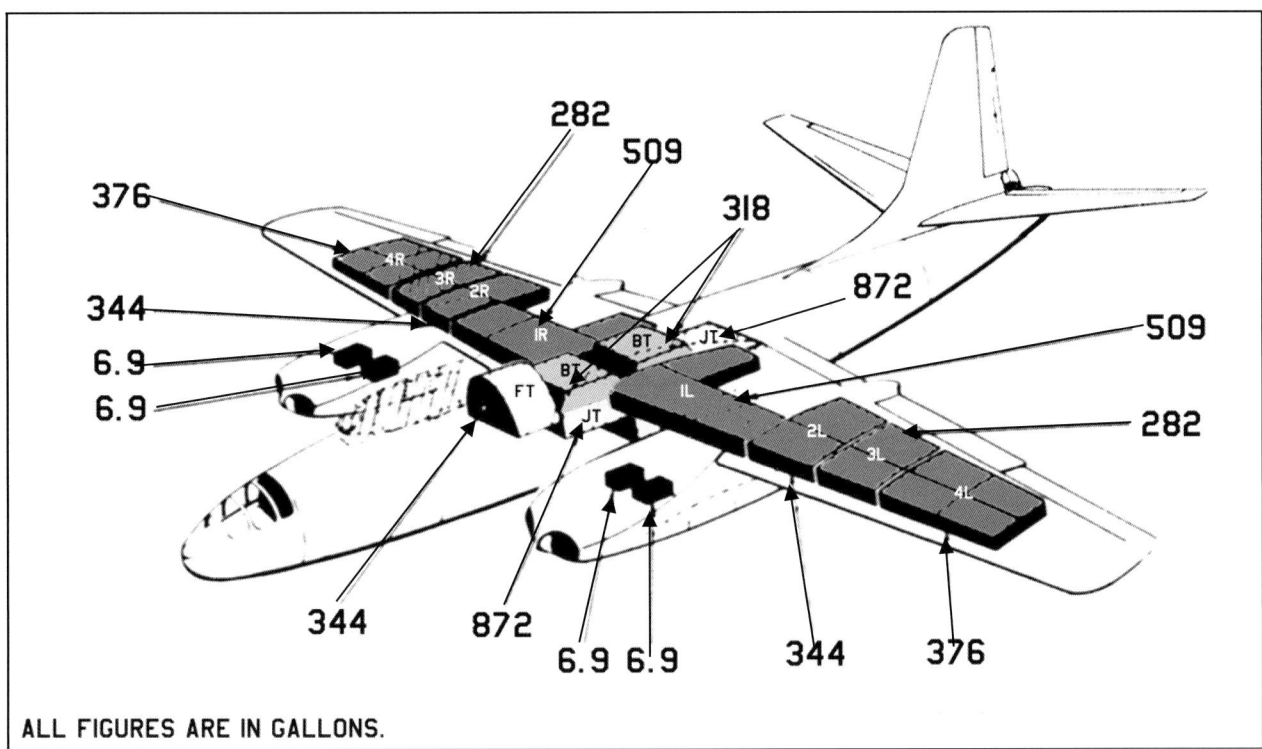

Drawing 76: This drawing shows the location, capacity and names of all fuel tanks and the number of cells in each wing tank in the B-45A-1 and A-5. Capacities are in gallons and are listed in bold three digit numbers. FT was the 344-gallon single cell fuselage tank; tanks J1 and J2 were two 872-gallon jettisonable but non-sealable single cell lower bomb bay tanks; tanks B1 and B2 were two self-sealing removable, but non-jettisonable single cell 318-gallon upper bomb bay tanks. If both an upper bomb bay tank and jettisonable lower tank were installed, the fuel was fed from each pair of tanks through the upper bomb bay tank as a single forward or aft 1190-gallon tank. If no upper bomb bay tanks were fitted the forward and aft jettisonable tanks were individually fed to the fuel system as individual fore and aft 872-gallon tanks. If no lower jettisonable tanks were fitted, but upper bomb bay tanks were, they were individually fed to the fuel system as individual fore and aft upper 318-gallon tanks. Tanks 1 through 4L were the left wing tanks and 1 through 4R were the right wing tanks, the number of cells in each tank indicated by the number of rectangles in each tank. A gallon of JP-4 weighed approximately 6.6 lbs., therefore the B-45A-1 and -5 carried aircraft carried 37,923.6 lbs. of fuel. The four numbers leading to black boxes on the nacelles are the 6.9-gallon capacities of the oil tanks in each nacelle, one dedicated to each engine. As built, B-45A-1s and -5s did not carry drop tanks. (NMUSAF)

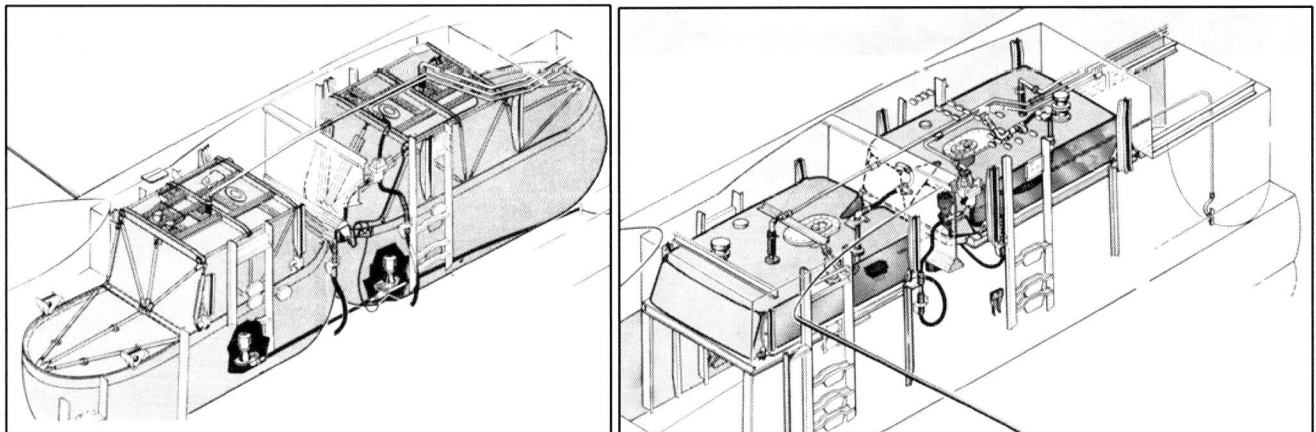

Drawing 77: This drawing shows the two lower bomb bay jettisonable 872 gallon non-self-sealing tanks used on B-45A-1 and -5s. (NMUSAF) **Drawing 78:** This drawing shows the two upper bomb bay 318 gallon removable, but non- jettisonable self-sealing tanks. (NMUSAF)

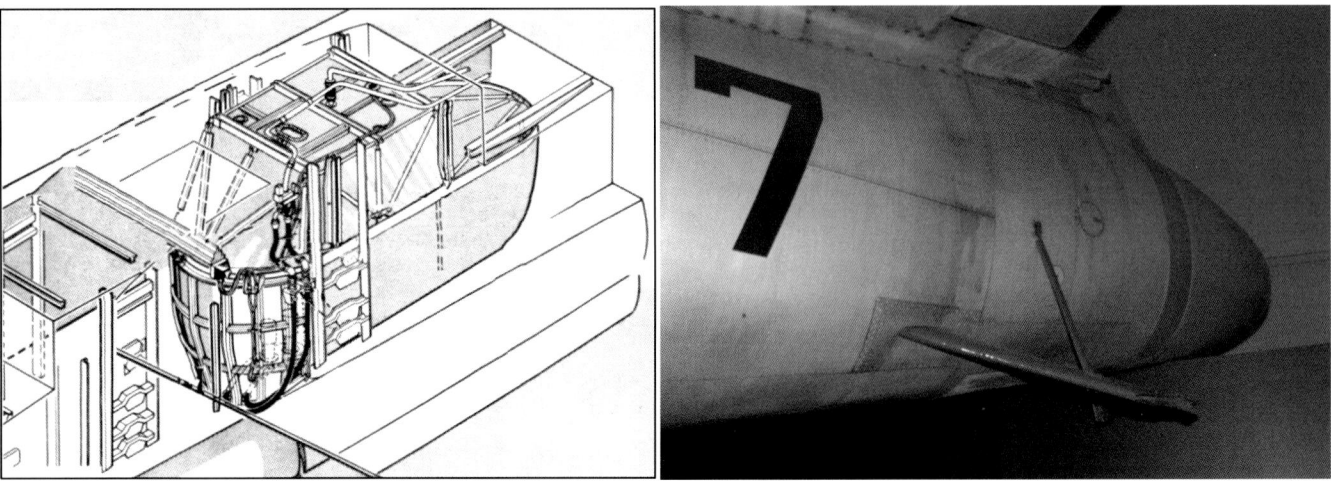

Drawing 79: This drawing shows the single lower bomb bay 872 gallon self-sealing jettisonable bomb bay tank. (NMUSAF) Photo 23: Shows the B-45's single emergency fuel dump. (Courtesy Chuck Holte)

Engine Nacelles....

The four J35s or J47s were housed in a single nacelle containing two engines on the leading edge of each wing. On the outboard side of the port nacelle was Engine No. 1; on the inboard side, Engine No. 2. On the starboard nacelle, Engine No. 3 was on the inboard side; on the outboard side was Engine No. 4. Each nacelle was nearly identical and it was virtually impossible to tell which GE engine was on the inside. In fact the only way to tell the difference between the B-45A-1 and the B-45A-5 visually on the ramp was to look at the jet nozzles at the aft end of the nacelle. The J47 exhausts were angled 5° downward at the very end of their duct. In **Photo 24**, notice that the nozzle is not centered in the nacelle opening. The lower portion of the nozzle is much closer to the bottom of the duct indicating the nacelle is carrying a J47 and must be a B-45A-5. It is obvious that the nacelles drew heavily from those on the Arado Ar 234C and featured some of the same advantages such as having the compressor, combustion, and turbine sections forward of the wing. This aided in a small way the possibility of reducing significant damage to the wing should one of these sections came apart catastrophically in flight. Such major failures were still quite common in these early years of jet prolusion in military aircraft and were the most significant causes of the lengthy delays in getting both the J35 and J47 into service in a timely manner. Further it added flexibility to loading areas forward, above, and aft of the wing in the fuselage while maintaining the aircraft's CG or center-of-gravity.

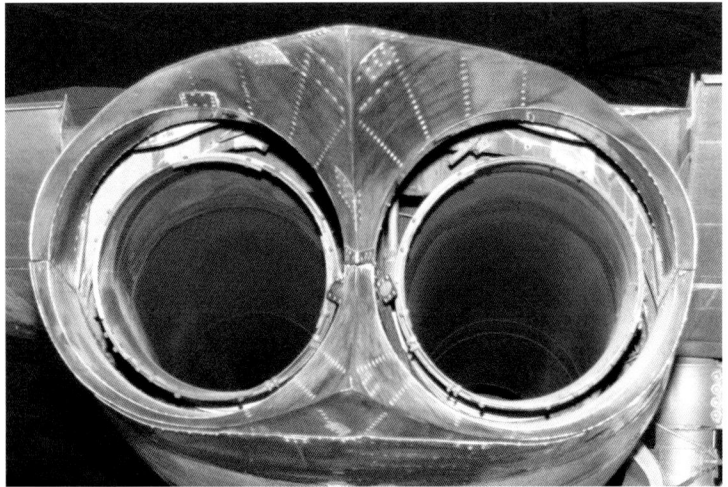

Photo 24: B-45A-5 nozzles. (Courtesy Chuck Holte)

By mid-September 1948 the 22nd B-45A-1 had been accepted by the USAF and put into service. And with the J35s, there was still quite a distance to having a consistently dependable engine. The service and GE foresaw this eventuality as early as October 1945 and GE convinced the USAAF that their new J47 would be the solution to the significant performance and dependability problems of the J35. GE's 1946 offer to the USAAF was to spend three to four million dollars of GE funding on a facility that could handle the production of the new engines that would support the B-45, F-84, F-86, B-47, YB-49, and B-36 jet engine pods, if the USAF guaranteed GE such a contract before the engine was even developed. Air Materiel Command and the Air Staff remembered quite well the impact of forcing the production of the J35 to Allison and were aware that if GE did not survive as a turbine engine producer and developer, the US would be severely behind in turbine technology throughout the world.

And GE made it clear that if it were not able to make such a deal it would indeed leave the turbine technology business immediately. The first Contract between the USAAF and GE was signed on in May 1946 for development and test and production for J47s. Over the next four years there were numerous changes, reductions, additions, and modifications to these contracts. But by May 1948 the USAF had accepted 75 production engines and another 1,694 J47s were on order.

The J47 suffered several serious issues such as turbine blades cracking and fracturing, compressor disks failing and malfunctioning engine controls which had grounded aircraft equipped with the J47s several times. The USAF pressed hard for answers and solutions including bringing Allison back on board to build a GE product once more. The most significant solution was made by GE Managing Engineer Harold D. Kelsey. In previous years, the GE technical development center at Lynn River, MA had grown to work both R&D and production. The combination had not worked and Kelsey ordered a massive reorganization of GE Engines. All R&D would take place only at the Lynn River Works and all production would be relocated at the facility located north of Cincinnati, OH, the Lockland Plant.

This facility was the largest production facility in the US and had produced Wright Aeronautical Engines during the War. Both GE and the USAF funded the $15,000,000.00 bill to get the plant in shape as soon as possible. Much of that funding went into tooling and machine tools and jet engine test cells. One of the first priorities faced by GE was to get the water-injection system for the J47-15s (known to GE as the TG-190C) to be used on B-45C and RB-45Cs, functional as soon as possible so these airframes could be finished on the assembly line. That the USAF approved of Mr. Kelsey's leadership was seen in a non-committal briefing to GE at Wright-Patterson AFB that indicated the requirement for J47s exceeded 12,000 engines by March of 1953.

The J47 was designed to the same basic dimensions as, and to be interchangeable with the J35-3. This was to make eventual replacement of J35s in service by J47 far less expensive. The J47 started with an eleven-disc compression section, but this was increased to a twelve-disc compressor section that was 3 inches smaller in diameter early in its development. It held eight combustion chambers, a single-stage turbine, an exhaust cone, and an accessory section on the nose which held a starter, main and emergency fuel pumps, and oil pump. The weight was to be 2,300 pounds; maximum thrust was 5,220 lbs.

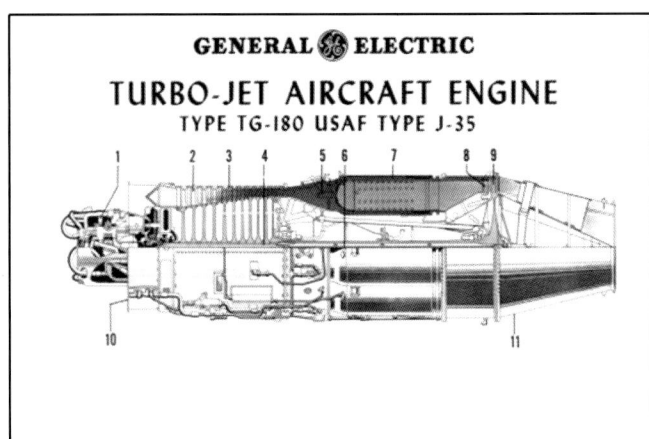

Drawing 80: This drawing shows the internal configuration of the J35 Turbojet engine and a list of the key operational components. (Author's Collection)

Photo 25: This is an image of a cut-away J35 engine on display at the National Museum of Naval Aviation. (Courtesy Stephen Kahl)

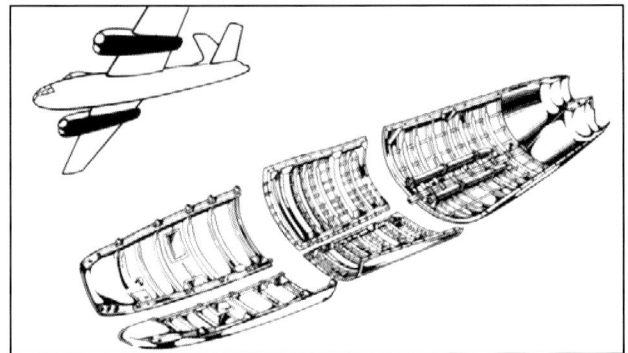 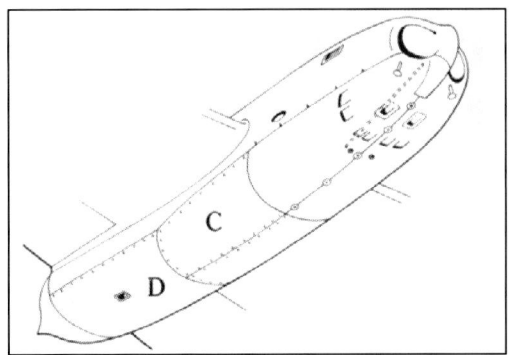

Drawing 81: This drawing shows the very heavy structure inside the five sections of lower cowling on each B-45 nacelle. (NMUSAF) Drawing 82: All five loweowling sections are shown closed in the drawing. (NMUSAF)

 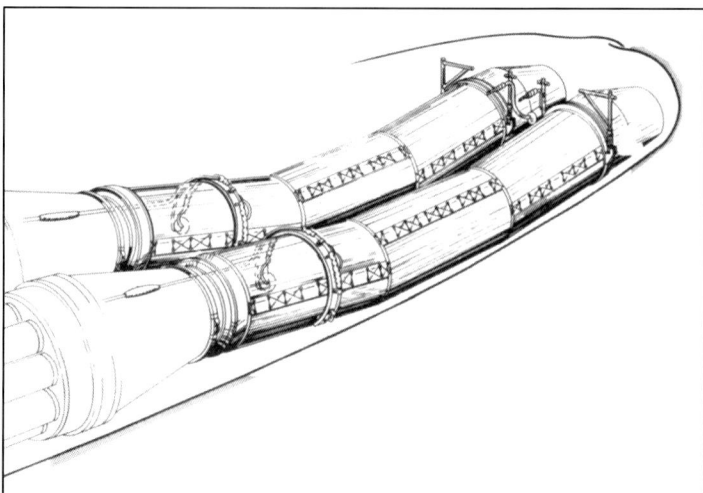

Photo 26: This image shows some of the directional shifts in the exhaust ducts for the J35s and J47s. Any change in direction of the exhaust, no matter how small, caused a slight decrease in thrust, likely caused the lost about 3 percent in power. (The Boeing Airplane Company) Drawing 83: This drawing shows all four of the thrust directional changes in the ducts. (NMUSAF)

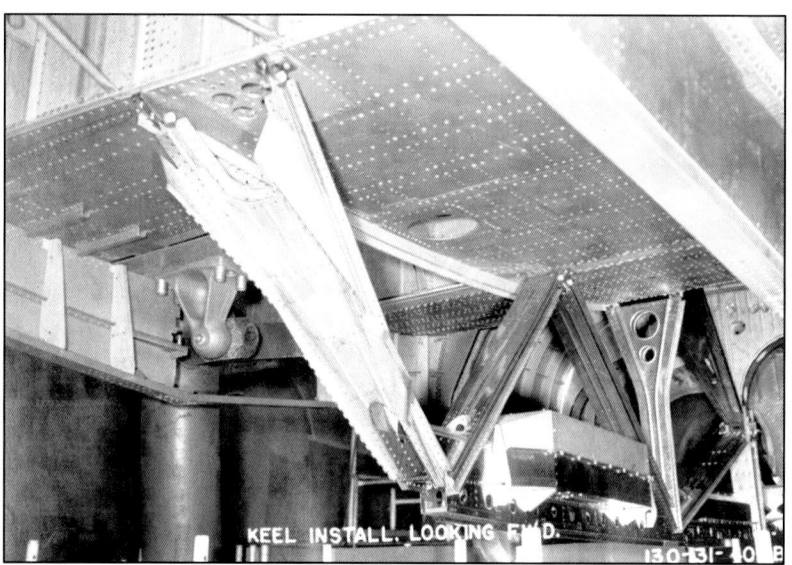 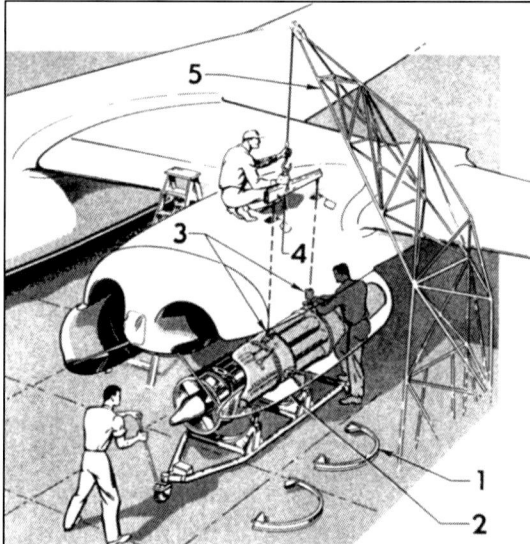

Photo 27: This is the keel structure that supported both engines in each nacelle. The diagonal triangular structure on the left connected the keel to the aft spar, while the tri-angular topped structure to the right connected the keel to the forward spar. (Alan Griffith) Drawing 84: This drawing shows some of the critical ground support equipment that was not available to most B-45 units upon their formation. The engine hoist and cradle were some of the most important units because of the amount of time their absence cost (NMUSAF)

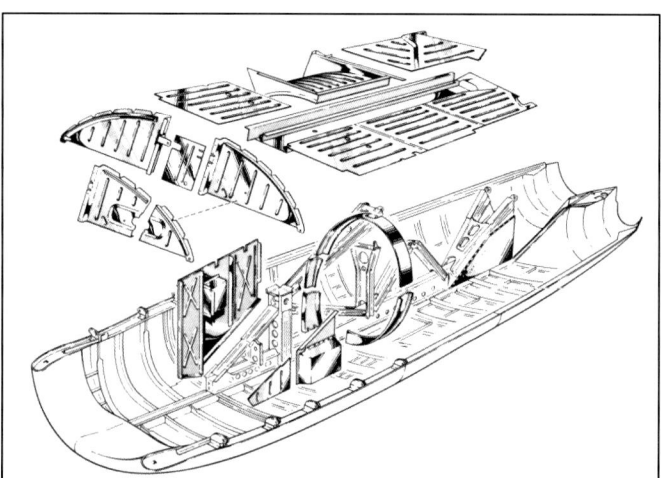 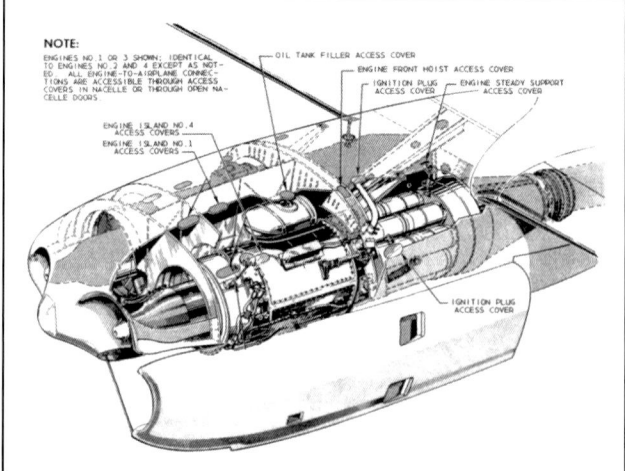

Drawing 85: This drawing shows the structure built into both nacelles to support the two engines they housed. The engines bolted onto the triangular beam assembly in the keel of the nacelle, which in turn was mounted into the front spar by the diagonal beam from the center arch and by the vertical beam with the triangular apex. The triangular structure aft of the structure mounting to the front spare was the nacelle mount to the aft spar. (NMUSAF) Drawing 86: This drawing shows the diagonal mounting beam from the keel of the nacelle to the front spar and the No. 1 engine mounted in place. (NMUSAF)

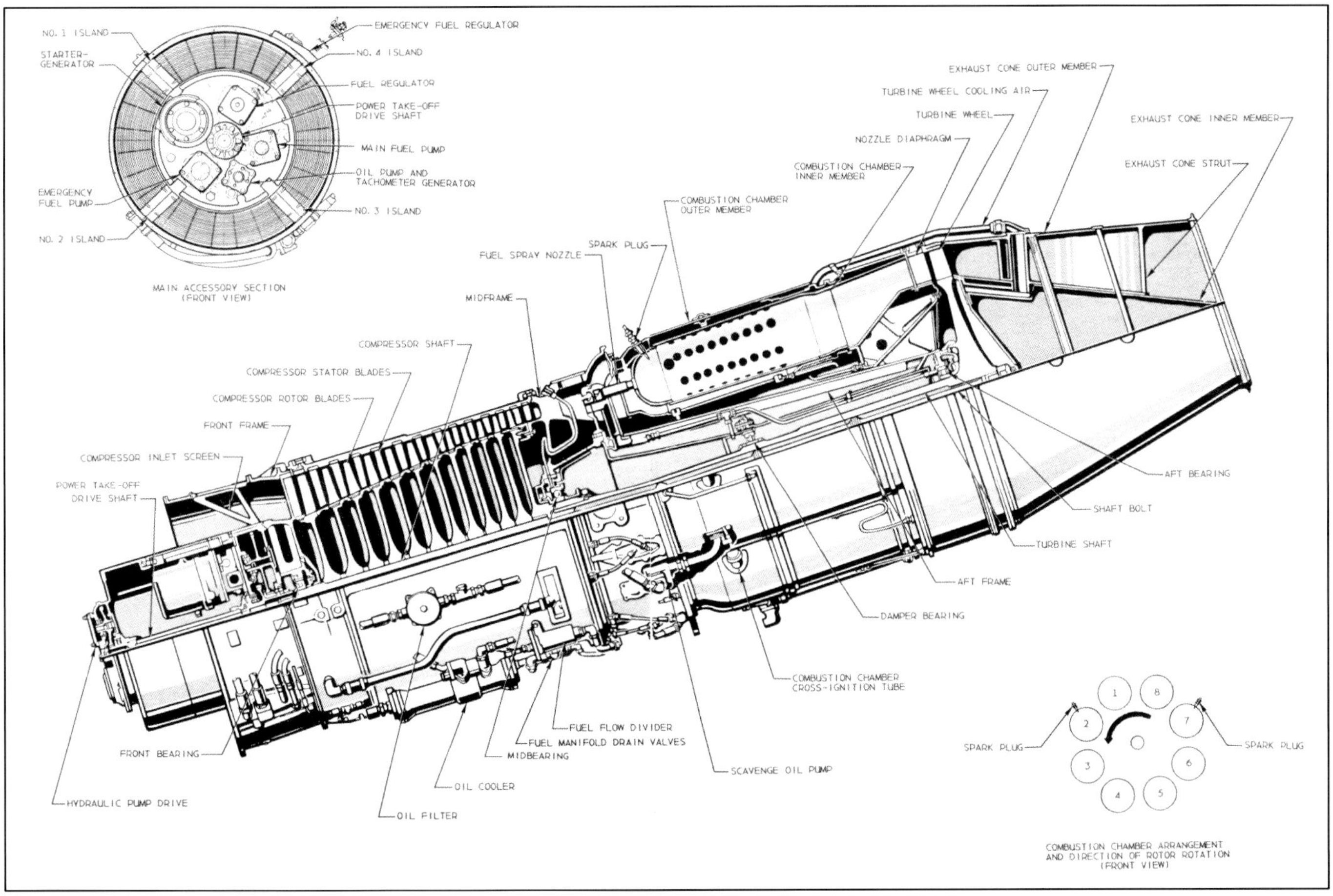

Drawing 87: This is a maintenance technical order drawing of the J47-9 as installed in the B-45A-5. (NMUSAF)

Canopy Reinforcements....

Shortly after the B-45 entered service, a problem that was cropping up on newly manufactured Boeing B-50Ds appeared on the B-45A. Repeated pressurization cycles on high altitude missions and the following depressurization for descent and landing caused increasingly frequent and more wide-spread cracks in the Bomb/ Nav station transparency and flight deck canopy. A Time Compliance Technical Order (or TCTO) was issued in late 1948 or early 1949 to apply bands of fiberglass tape and resin over the metal framework in the Bomb/Nav nose transparency and the cockpit canopy. (TCTOs had a specific, usually short-term time required to complete and inspect the requirements of the order.) The strips were approximately two inches wide and layered in place until the reinforcements were 1/4 inch thick on the inside of the surfaces and 3/16 of an inch on the outside. All seams had triangular joints laid over them as well. Additional re-enforcements could be added as required and frequently additional reinforcements were placed between the metal framing supports on the Bomb/Nav nose. In 1955 another TCTO was issued to remove all transparency material over the bomb/nav station and re-cover the station with a factory produced all-metal cover.

Photos 28: This image shows fiberglass reinforcements on the canopy and nose. (Steve Ginter) Photo 29: Note the triangular bases of the vertical stripes. (David Menard) Photo 30: A B-45A over the UK clearly shows the patterns of fiberglass reinforcements on the nose and canopy. (The BBA Collection)

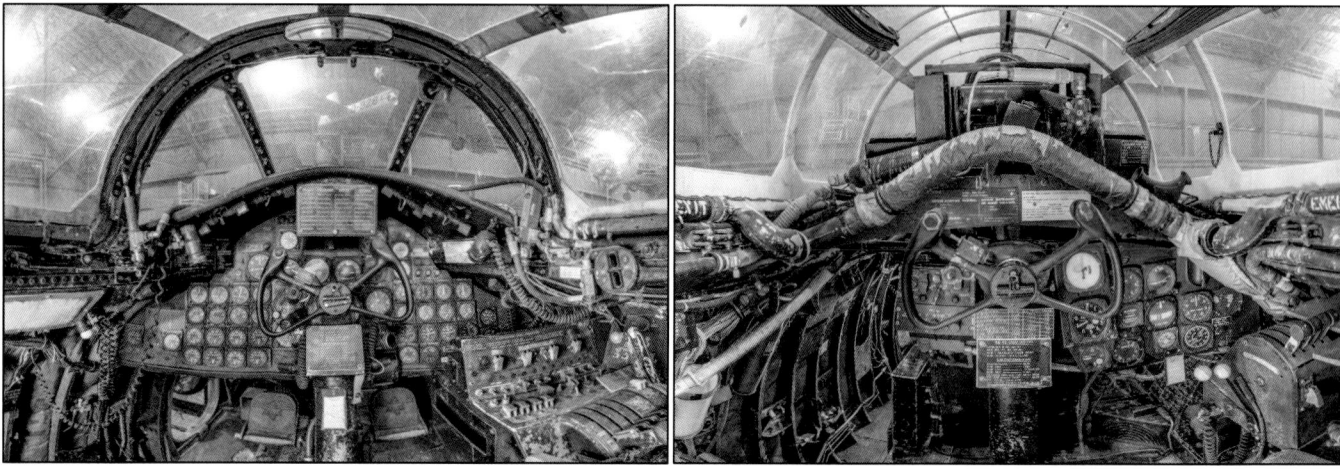

Photos 31: NMUSAF's B-45C pilot's panel with fiberglass reinforcements. (NMUSAF) Photo 32: NMUSAF's B-45C co-pilot's panel with fiberglass reinforcements. (NMUSAF)

Landing Gear....

The landing gear on the B-45A were hydraulically retractable shock struts made by the Cleveland Pneumatic Tool Co. The main wheels were Goodyear 12-ply 30-inch diameter and 7.7-inch wide tires with inner tubes. The struts and wheels were completely enclosed by doors when retracted.

Photo 33: This shows the nose landing gear looking aft and showing the starboard side. The taxi light was mounted on the nose gear strut, while the landing light was mounted in the center fairing of the port-side nacelle. Courtesy Chuck Holte) Photo 34: The port side of the nose gear wheel and tire. (Courtesy Chuck Holte) Photo 35: The starboard side of the nose strut looking forward and to port. (Courtesy Chuck Holte) Drawing 88: The nose gear strut, retraction actuators. (NMUSAF)

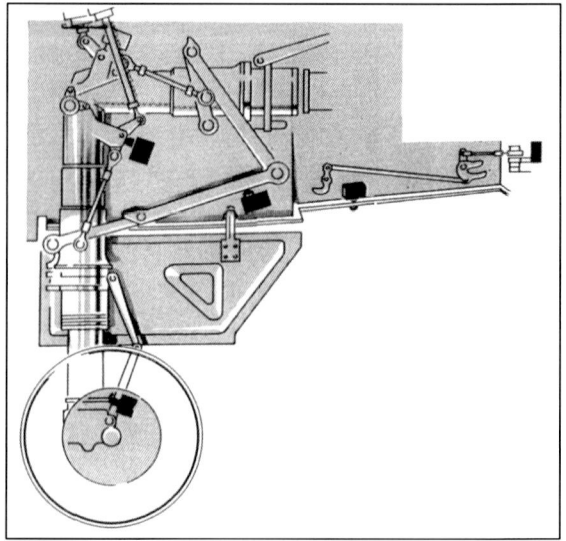

(NMUSAF) Drawing 89: Shows the structure of the tail skid looking to starboard. (NMUSAF) Photo 36; Shows the tail skid looking forward. (Steve Ginter) Photo 37: Shows the aft fuselage entry door looking forward. (SDAM via Craig Kaston)

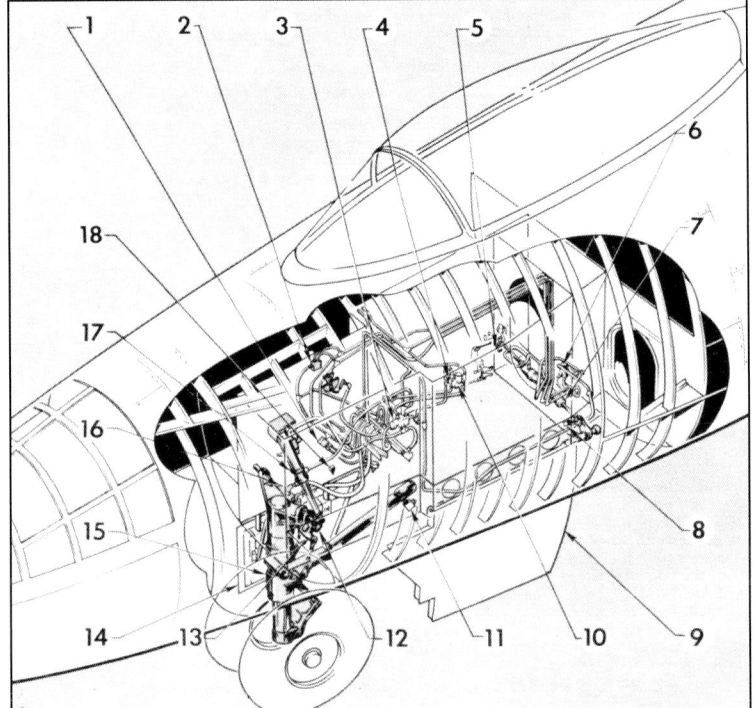

Drawing 90: This drawing shows the many details located in the nose wheel well which are listed below. (NMUSAF)

1. Hydraulic Thermal Relief Valve
2. Hydraulic Thermal Relief Valve
3. Nose Gear Hydraulic Control Valve
4. Nose Gear Steering Pressure Reducing Valve
5. Wheel Door Lock Position Indicating Switch
6. Wheel Door Hydraulic Control Valve
7. Nose Wheel Door Actuating Cylinder
8. Nose Gear Sequence Switches
9. Nose Wheel Door
10. Nose Gear Steering Pressure Shut-off Valve
11. Gear Down Position Indicating Switch
12. Nose Wheel Down-lock Indicating Switch
13. Wheel Door Sequence Switch
14. Nose Gear Strut Door
15. Nose Gear Strut Assembly
16. Nose Wheel Hydraulic Steering Unit
17. Nose Gear Hydraulic Actuating Cylinder
18. Wheel Door Sequence Switch

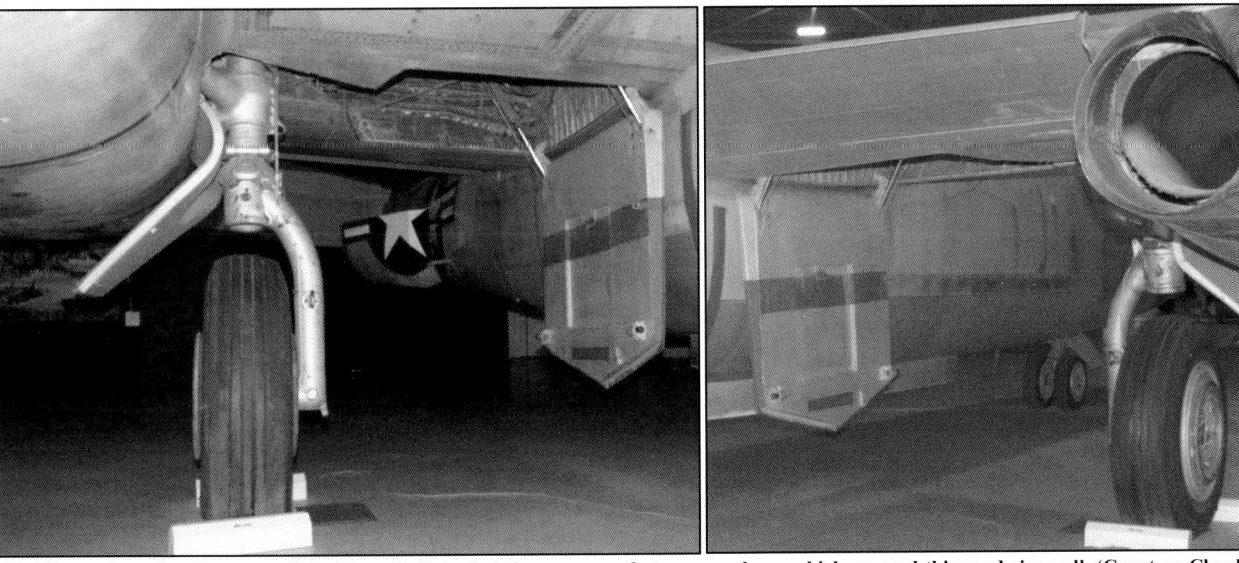

Photo 38: The starboard main gear-well looking aft. Note that there were only two gear doors which covered this good-size well. (Courtesy Chuck Holte)
Photo 39: This is the starboard gear-well looking forward. (Courtesy Chuck Holte)

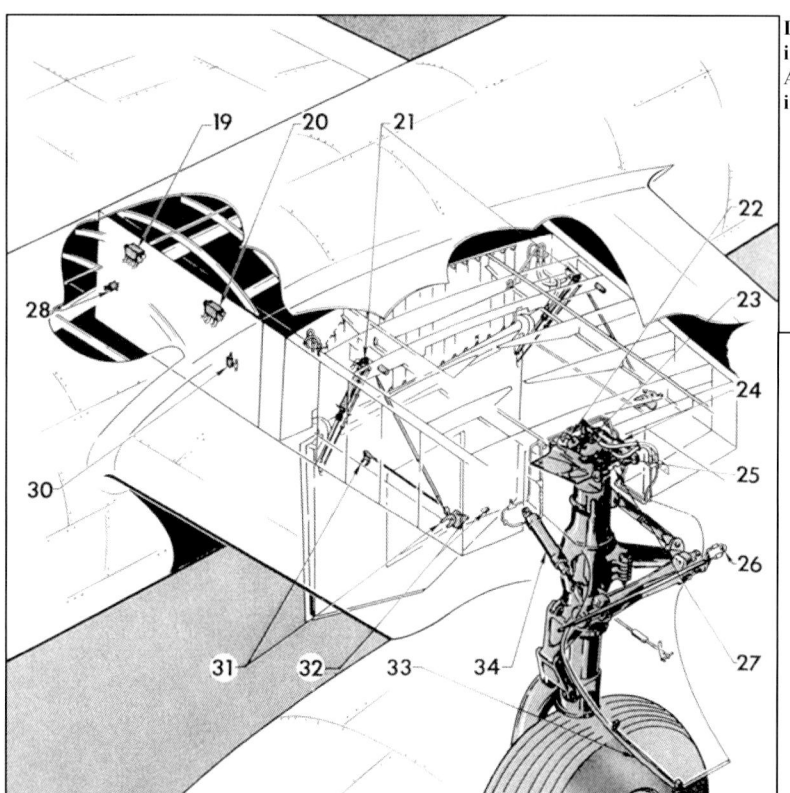

Drawing 91: This drawing illustrates the many details housed in the main landing gear wells of the B-45 landing gear system. A key for all numbered items (19 through 34) in the drawing is in the box below. (NMUSAF)

Main Landing Gear System Key

19. Main Gear Control Valve
20. Main Gear Door Control Valve
21. Main Gear Door Actuating Cylinders
22. Main Gear Up-Lock
23. Door-Closed Sequence Switch (Operated Gear-Up)
24. Main Gear Up and Down-Lock Actuating Cylinder
25. Door-Closed Sequence Switch (Operated Gear Down)
26. Shock Strut Fairing Attaching Pin Mechanism
27. Main Gear Down-Lock
28. Main Gear and Door Dump Valve
29. Main Gear Thermal Relief Valve
30. Main Wheel Door Thermal Relief Valve
31. Main Wheel Door Locks
32. Door Closed & Locked Position Indicating Switches
33. Shock Strut Fairing
34. Main Gear Actuating Cylinder

Main Gear Door and Strut

Photo 40: The starboard main gear door. These doors could be opened for use as airbrakes without the main gear being extended. They were the only airbrakes on the aircraft which made the B-45 difficult to slow rapidly. (Photo courtesy of Chuck Holte) Photo 41: This is the starboard main gear strut looking outboard to the starboard nacelle. The heavy metal loops on the lower strut were used to secure tie-down chains. (Courtesy Chuck Holte)

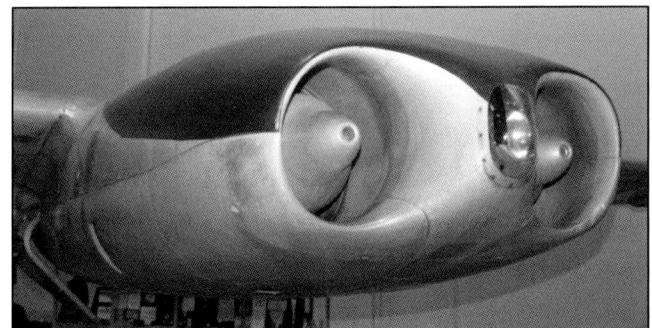

Photo 42: This image shows the port nacelle intakes looking aft. The sole landing light is mounted in the center of the nacelle air intake fairing. The shape of each intake showed that the compressor air intakes are narrower in diameter in the middle of the opening right at the tip of the bell-shaped accessory section cover. Then, as the intake extends aft aft, the opening expands to a larger width for the rest of the compressor intake duct. This, in effect, creates a venturi at the front of the intake which increases the air pressure, air speed of the intake air, and increased volume of the air that enters the compressor section. The result was that the engine was made more efficient and powerful by a structural design rather than any equipment or other energy-absorbing means. The bell-shaped cover protected the engine starter, the main and emergency fuel pumps, and an oil pump. (Courtesy Chuck Holte)

Photo 43: This is the port main gear door and inboard and aft portion of the main wheel well. (Courtesy Chuck Holte) Photo 44: This is the port-side main gear strut with detail of the gear mount just inboard of the nacelle. (Courtesy Chuck Holte)

Photo 45: Looking outboard in the starboard main gear well, shows the main strut and its mount. (Courtesy Chuck Holte) Photo 46: This image looks to the outbaord and aft walls of the starboard wheel well. The bottles held hydraulic fluid supporting the brake system. (Courtesy Chuck Holte)

Photo 47 and 48: These show the aft wall of the starboard wheel well and the inboard wall of the starboard wheel well above the main gear door. (Courtesy Chuck Holte)

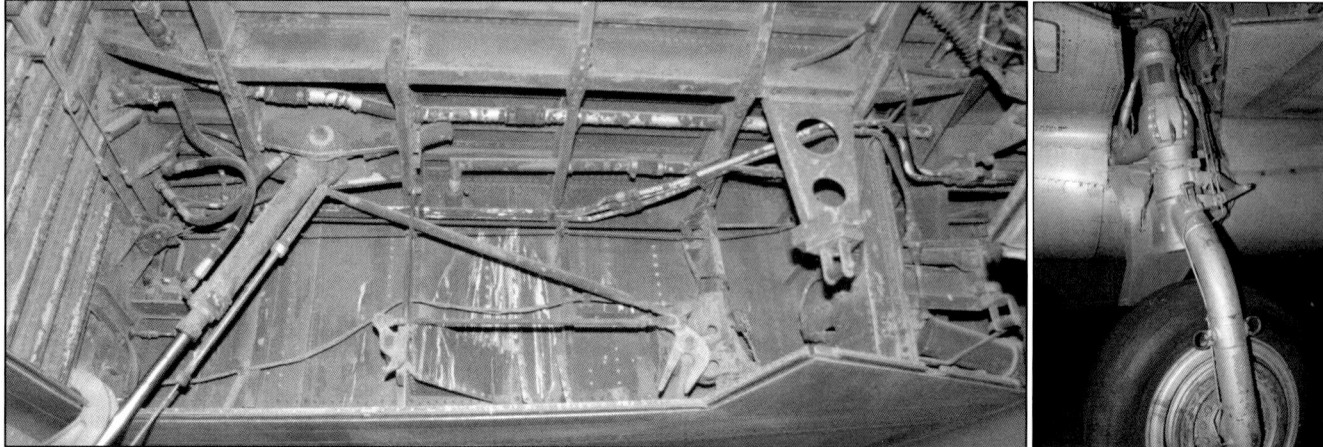

Photo 49: This image shows the forward wall of the starboard wheel well. The hooks hold the main gear door closed once retracted. (Courtesy Chuck Holte) Photo 50: This image shows the port main gear strut and the forward leaning angle of both struts. It also shows detail of the strut mount in the wing structure, and the gear door between the nacelle and the main gear. The details and layouts of both main gear wells and the details within are nearly identical on each side of the aircraft. (Courtesy Chuck Holte)

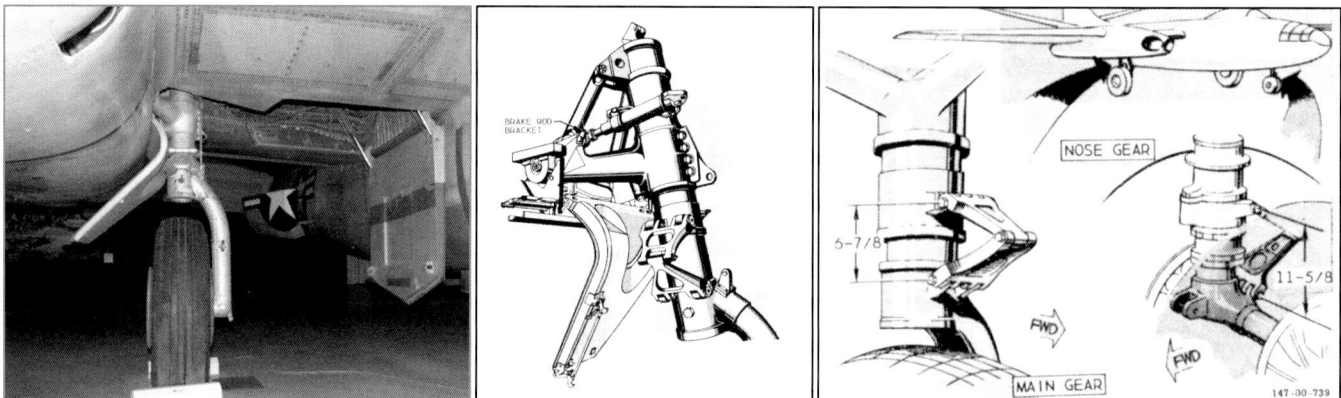

Photo 51: This photo looks head-on and aft to the starboard main strut showing the mount detail and nacelle-main gear door. (Courtesy Chuck Holte) Drawing 92: This is detail of the main strut mount and nacelle gear door. (NMUSAF) Drawing 93: This drawing shows the proper oleo expansion heights for all struts regardless of aircraft gross weight. (NMUSAF)

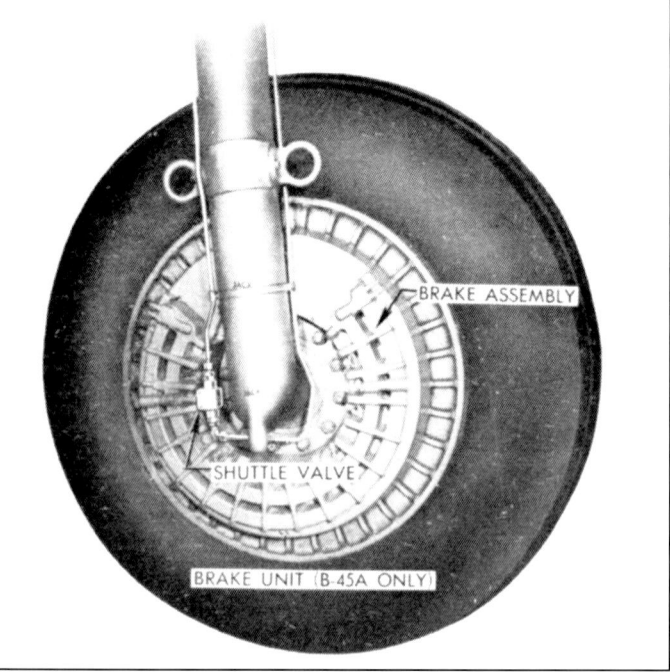

Photo 52: This is the outboard side of the B-45A main wheel. (Steve Ginter) Photo 53: This is the inboard side of the B-45A main wheel showing the brakes. (NMUSAF)

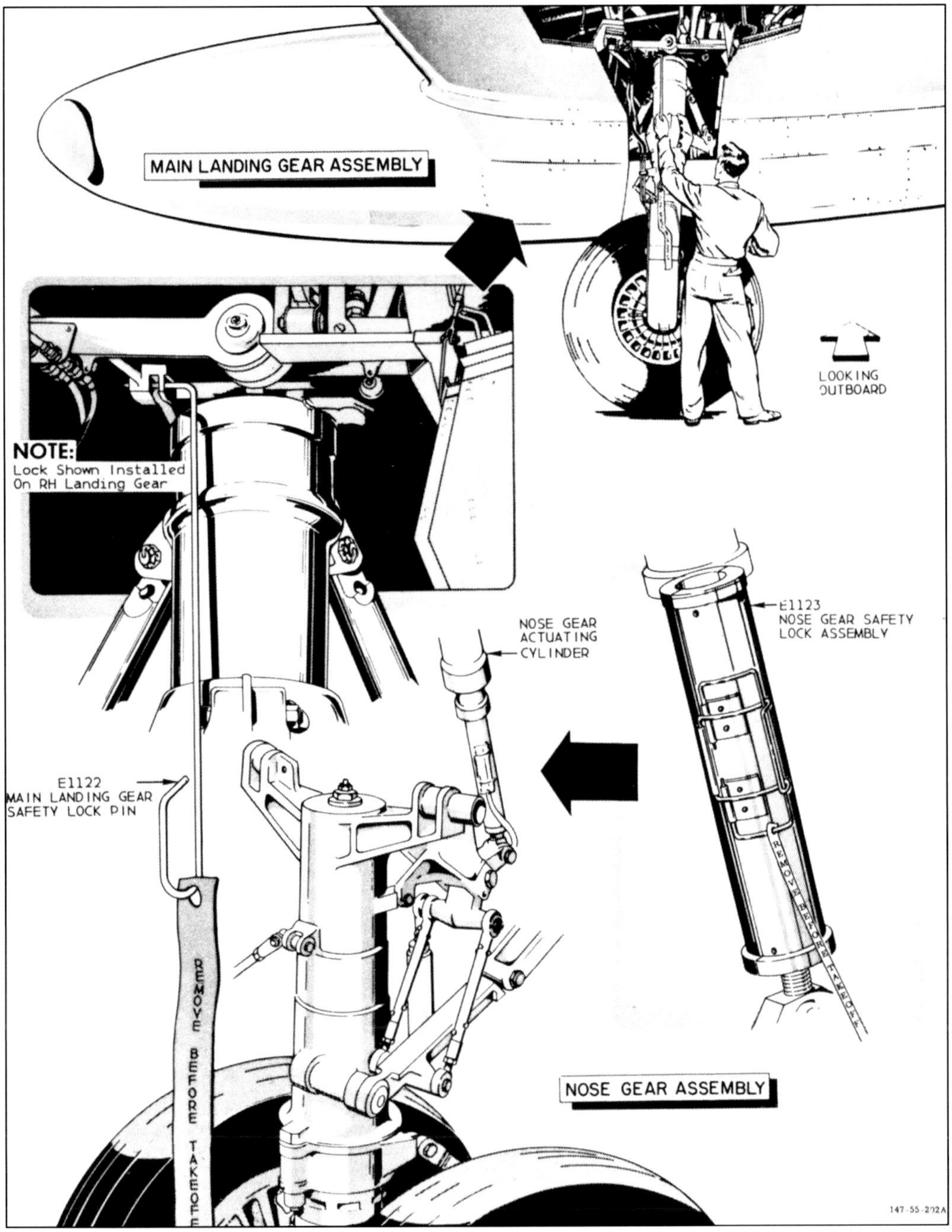

Drawing 94: This drawing details the location of the landing gear safety pins and their "Remove Before Flight" tags. (NMUSAF)

Photo 54: Even with the gear down, the B-45A was still an attractive and graceful aircraft. (NARA II via Tommy Thomason)

Flaps....

Photo 55: The B-45A had four sections of hydrauic powered wing flaps. Two were between each nacelle and the fuselage, and two were outboard between the nacelle and the alerons. (Courtesy Chuck Holte) Photo 56: This image shows the starboard nacelle looking forward at the starboard flaps, one on each side of the nacelle. The nacelles were streamlined to slope behind the aft wing and flaps, therefore a vertical sheet metal partition (enclosed within the black/white circle) was mounted on the inboard upper surface of the outboard flap to smooth the airflow over that section of the flaps. (Courtesy Chuck Holte)

The Gunner's Station....

All B-45A-1s and A-5s were built without production turrets. Gunners did sometimes ride in their stations to monitor formation discipline for aircraft commanders, but they had no weapons. To say the least, space was at a premium even without turrets installed as photo 23 shows.

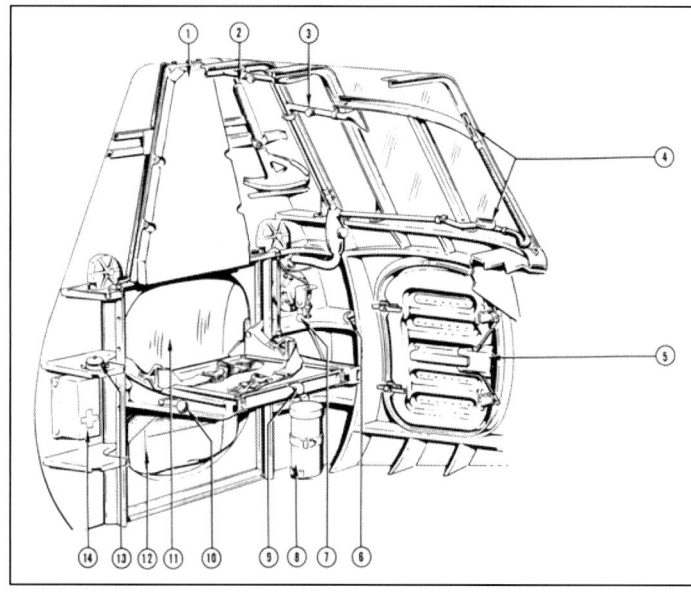

1. Back Rest
2. Hand Axe
3. Air Shut-off Valve
4. Defroster Outlets
5. Escape Hatch Release Handle
6. Ditching Belt Attaching Hook
7. Cabin Air Emergency Dump Valve
8. Disposal Container
9. Seat Horizontal Adjusting Handle
10. Seal Vertical Adjusting Handle
11. Cabin Access Door
12. Ditching Bell
13. Alarm Bell
14. First-aid Kit

1. Gun Sight
2. Oxygen Flow Indicator
3. Cabin Pressure Switch
4. Cabin Pressure Altimeter
5. Oxygen Pressure Indicator
6. Cabin Light
7. Spare Lamps
8. Heat and Vent Outlet
9. Radio Jacks
10. Ash Tray
11. Interphone Panel
12. Ditching Belt Attaching Hook
13. Refiller Valve – Portable Oxygen Bottle
14. Disposal Container
15. Heated Clothing Control Panel
16. Emergency Escape Switch
17. Circuit Breakers
18. Cabin Temperature Rheostat
19. Door Zipper Heater Indicator Light
20. Heat and Vent Outlet
21. Oxygen Regulator
22. Turret Controller
23. Portable Oxygen Bottle
24. Turret Control Panel

Drawing 95: This drawing shows the starboard side of the Gunner's Station in the B-45A-1. The retractable airbrakes at the front of the hatch were to the left of this drawing at the same level. (NMUSAF) Drawing 96: This drawing shows the port side of the gunner's station. This drawing originally contained elements of the planned Emerson E-1 turret in it, but they were removed from the illustration and the key because no Emerson E-1 through 3 turret was ever installed on B-45A-1s or A-5s. (NMUSAF)

Photo 57: This shows an exterior image of the Gunner's Station on a late B-45A-1. Note how little room there was for the gunner to scan from side to side and how the position lights in the horizontal stabilizer (one orange and one white on each side) could and did cause a lot of glare for the gunner during night missions. The Gunner's Station ended at a bulkhead where the skin seam separated the aft fuselage and the tail cone in this image. (NMUSAF)

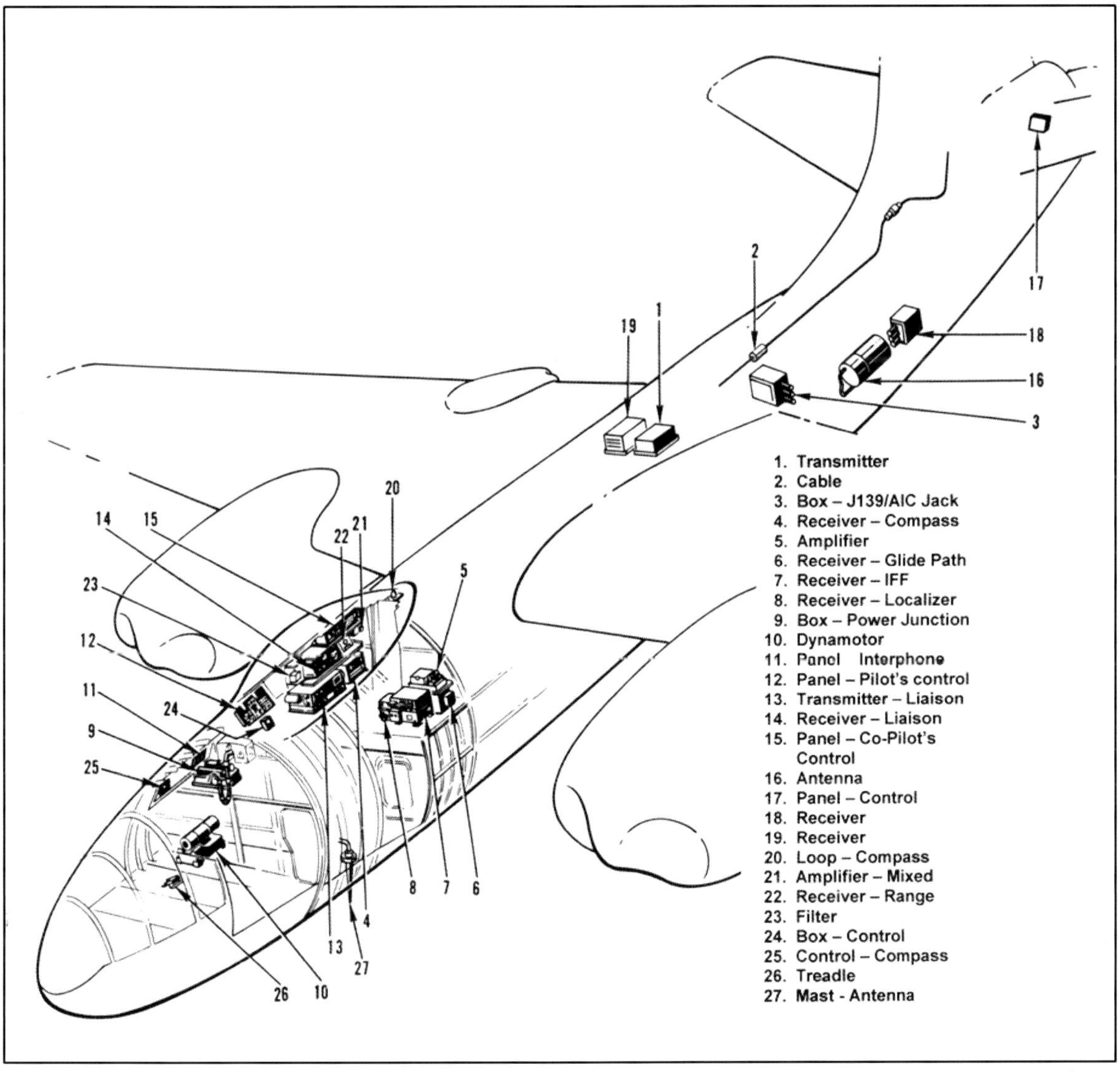

1. Transmitter
2. Cable
3. Box – J139/AIC Jack
4. Receiver – Compass
5. Amplifier
6. Receiver – Glide Path
7. Receiver – IFF
8. Receiver – Localizer
9. Box – Power Junction
10. Dynamotor
11. Panel   Interphone
12. Panel – Pilot's control
13. Transmitter – Liaison
14. Receiver – Liaison
15. Panel – Co-Pilot's Control
16. Antenna
17. Panel – Control
18. Receiver
19. Receiver
20. Loop – Compass
21. Amplifier – Mixed
22. Receiver – Range
23. Filter
24. Box – Control
25. Control – Compass
26. Treadle
27. Mast - Antenna

Drawing 97: This drawing shows B-45A-1 radio equipment locations in the B-45A-1 fuselage with identification key. (NMUSAF)

Final Notes on Production....

With the B-45A-5 production having started at the Long Beach Plant, production rates did increase but there were still problems. GE had yet to get the compressor and turbine problems resolved on the J47 and it took time to get the production facility at the Lockland Plant in Ohio fully operational. The non-delivery of the AN/APQ-24 radar, the A-1 Fire Control, and the E4 Autopilot meant that the B-45A-5s were far from combat capable. Hence, Pentagon doubts about the longevity of the B-45A lingered. It appeared that by 1950 the B-45As might all become to target tow aircraft or worse yet, slated to become range targets. While one Bomb Group had become fully equipped with non-combat ready B-45As, the Pentagon and the Tactical Air Command decided to put the non-assigned B-45A-5s into storage and cocoon them at the Long Beach Plant for indefinite storage. However, this is far from the end of the story.
(Craig Kaston via Steve Ginter)

Photo 58: 32 By the end of January 1949, the USAF had contracted with North American Aviation for a total of 22 B-45A-1s, 74 B-45A-5s, 10 B-45C-1s and 33 RB-45C-1 conversions. By the end of December 1949, the USAF had contracted with NAA to bail 32 B-45A-5s for an unlimited period of outdoor cocooned storage at the Long Beach Production facility to begin on 28 February 1950. The purpose was to put the aircraft into a safe storage environment until HQ USAF determined what the ultimate fate of the B-45 family was going to be. Options being considered by the Air Staff ranged from a fleet-wide conversion into target tow aircraft to their use as ground-placed targets for air-to-ground weapons delivery training at a variety of live-fire ranges around the United States. Three of the stored aircraft are in this photograph taken in late 1950. (Craig Kaston via Steve Ginter)

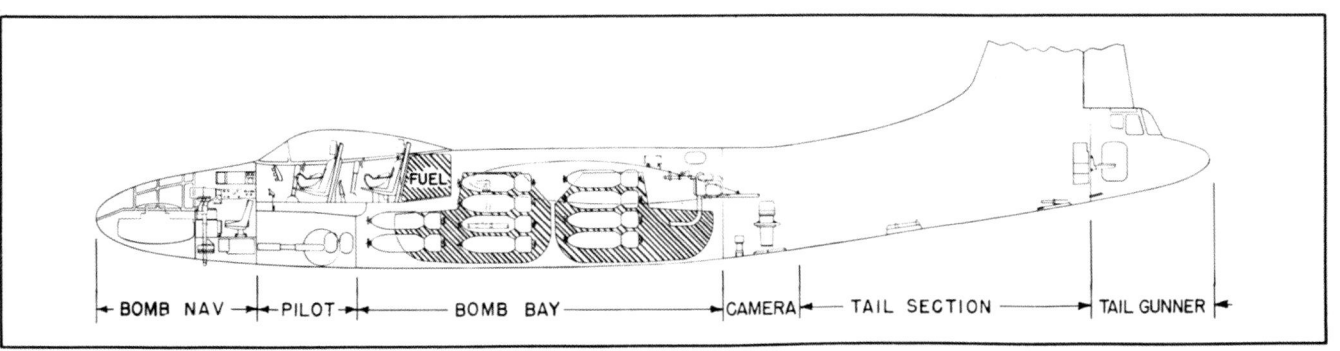

Drawing 98: B-45A-1 and A-5 Inboard Profile as Built. (NMUSAF)

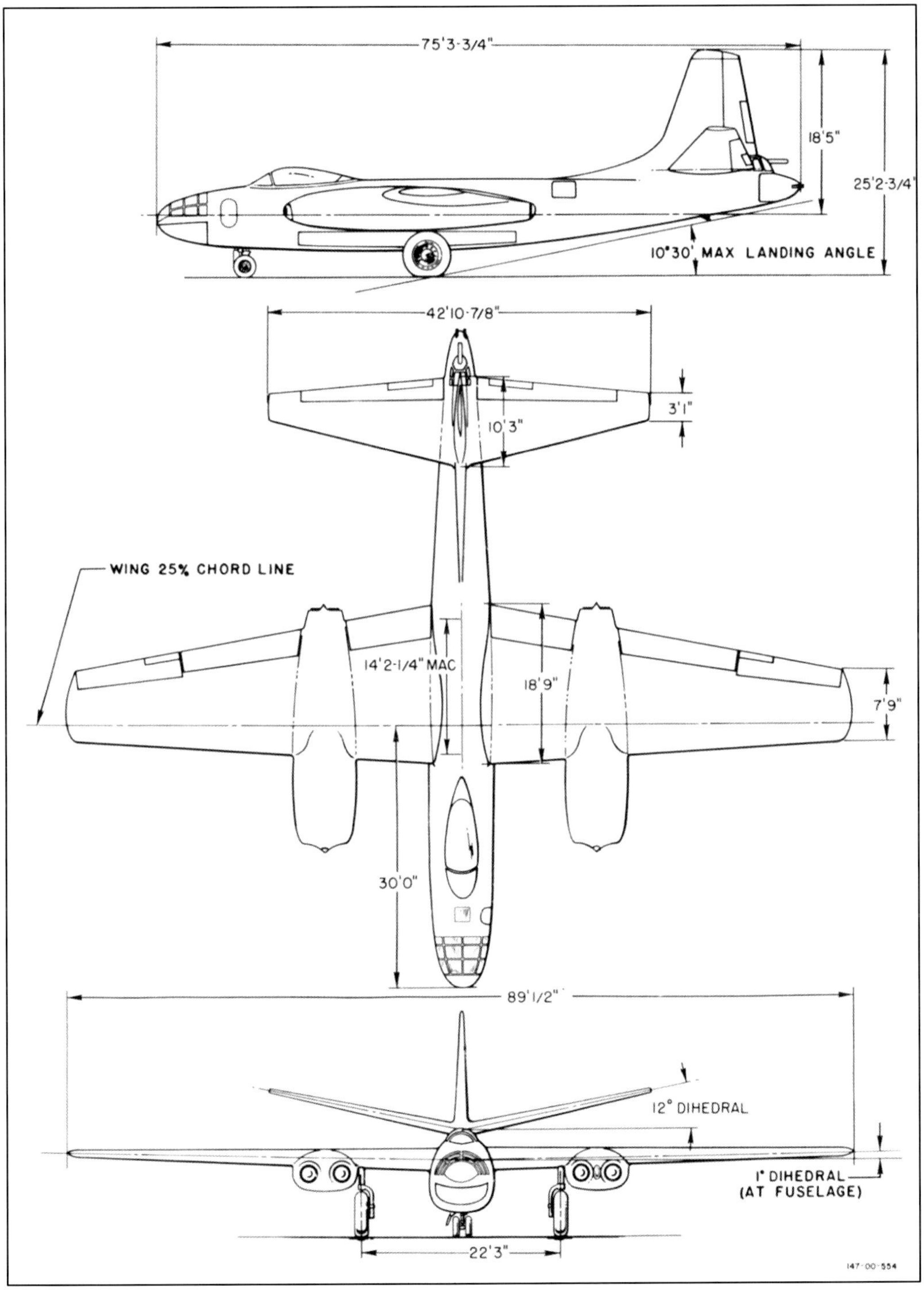

Official USAF Three View Drawing of the B-45A

While production of the B-45A-1 and A-5 was underway, North American submitted two new proposals based on the NA-147 to the Air Staff in 1948. Drawing 99: This drawing shows their 1948 proposal for the "B-45B, a larger, long-range, inflight-refuelable, high altitude, conventional air-to-air guided-missile-armed interceptor using ground-based-radars for guidance to a directed point where the missiles would be launched at a tracked target. Drawing 100: This shows a B-45A-based ground-controlled "drone" conventional-missile-armed all-weather-interceptor. Lack of sufficiently sophisticated electronics for guidance of both the aircraft and missiles meant neither aircraft were pursued. (Both Steve Ginter)

Photo 59: B-45A-5 7078 undergoes pre-delivery engine run-ups at the Long Beach plant prior to being delivered to and accepted by the USAF. The sprayed over-all aluminum lacquer was a rare finish on B-45s. (Sheet metal covered forward compartment) (Steve Ginter Collection)

Photo 60: A late production B-45A-5, 7063, is shown just after take-off. (NMUSAF)

CHAPTER 4 – THE B-45C DESCRIPTION AND SERVICE

DESCRIPTION....

Production....

By the end of 19April 1951 the USAF had accepted 22 B-45A-1s and 74 B-45A-5s. The first of 32 B-45A-5s began its storage at the Long Beach Plant on 28 February 1950 and the last came out of storage and entered active USAF service on 23 April 1951. The last B-45A-5 was accepted by the service from the North American Plant was on 27 February 1950. North American began design work on the new model on 22 September 1947; two months after the USAF had approved production for 51 B-45Cs for a second light bomber group by General Spatz. However by 12 January 1949, the 51 B-45C order was cancelled and the order book was modified under Contract AC-18000 to read 10 B-45Cs for Research and Development. The first North American Model NA-153 or B-45C-1 was accepted by the USAF at the Long Beach Plant on 4 May 1949. The last B-45C was accepted by the USAF at Long Beach on 14 April 1950, less than two months after the last B-45A-5 was accepted.

Photo 1: The first B-45C 8001 climbs over the southern California desert and mountains on a test flight out of Muroc AFB. (Courtesy of The Boeing Aircraft Company)

Structural Differences....

The B-45C was an improved B-45A model with the same engines as the B-45A-5 and intended initially to be a high speed, high altitude early warning, Anti-Aircraft Artillery (AAA) radar, and air to air interceptor target aircraft. But its future was radically changed into a state-of-the-art technical research and development aircraft. Its performance was that the AMC and the ARDC pressed the Air Staff to have all ten airframes transferred to those commands as soon as possible.

The B-45Cs featured the installation of a 1,125 gallon jettisonable fuel tank on each wing tip thus increasing the radius main difference was the installation of two 1,125 gallon jettisonable fuel tanks, one on each wing tip. The tanks in combination increased the mission radius of the C-model to over 1,000 miles from the B-45A-5 model's 850 miles. The B-45C also was equipped with a single point refueling system which reduced the refueling time by almost 70%. Single point refueling allowed the ground crew to use one filling receptacle to load JP-4 into all or any combination of the aircraft's tanks. The remainder of the structure and skin of the B-45C was the same except for the use of some stronger materials in the construction. The B-45C-1 with the addition fuel capacity and those heavier materials had an empty weight of 1,942 pounds heavier than the B-45A-5.

Exterior Differences....

Photo 2: One interesting feature of the tip was that the forward fins were significantly canted in such a way that they drove the nose of the tip tank away from the leading edge of the wing. Photo 3: Along with Photo 3 this image highlights that the aft horizontal fins were slightly canted down to drive the nose of the tank up. (Jim Hawkins via Steve Ginter)

Fuel Capacity....

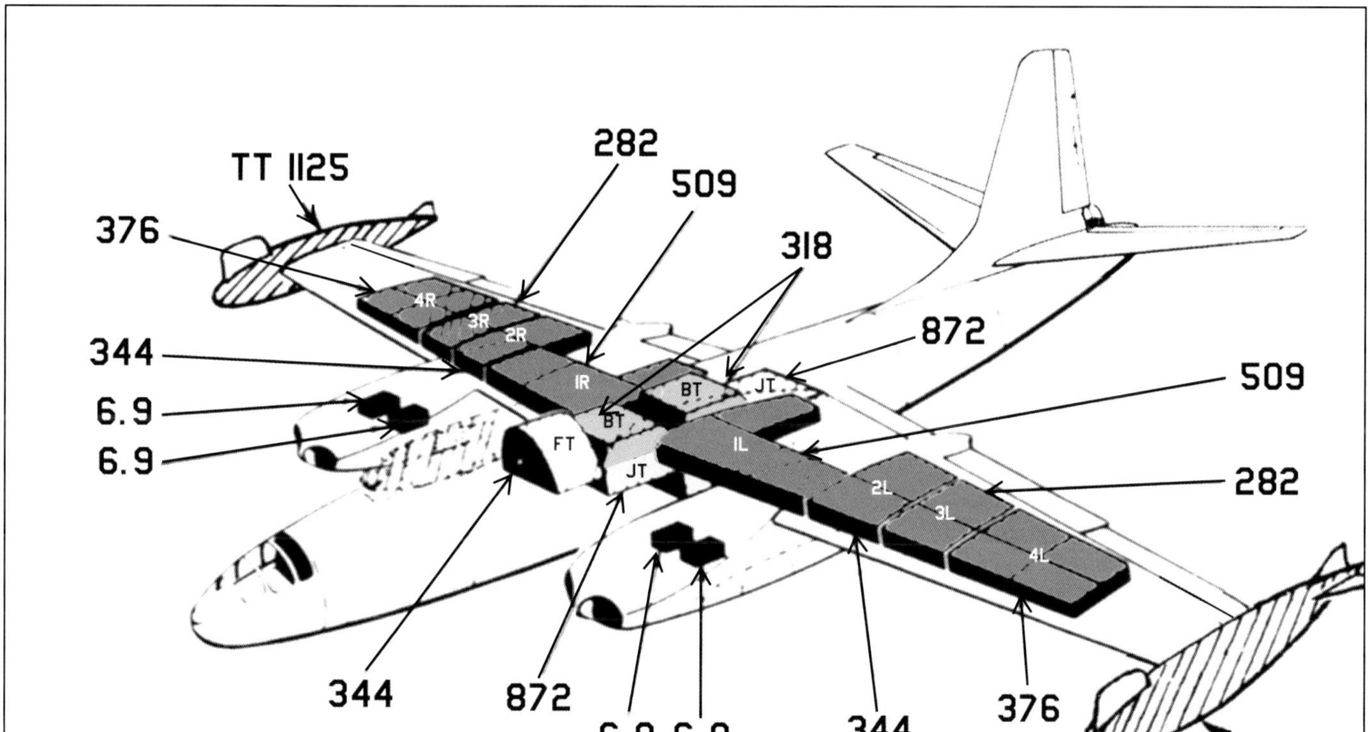

Drawing 1: This drawing shows the location, capacity and names of all fuel tanks and the number of cells in each wing tank in the B-45C-1. Capacities are in gallons and are listed in bold three digit numbers. FT was the 318-gallon single cell fuselage tank; tanks J1 and J2 were two 872 gallon jettisonable but non-sealable single cell lower bomb bay tanks; tanks B1 and B2 were two self-sealing removable, but non-jettisonable single cell 318-gallon upper bomb bay tanks. If both an upper bomb bay tank and jettisonable lower tank was installed, the fuel was fed from each pair of tanks through the upper bomb bay tank as a single forward or aft 1190 gallon tank. If no upper bomb bay tanks were fitted the forward and aft jettisonable tanks were individually fed to the fuel system as individual fore and aft 872-gallon tanks. If no lower jettisonable tanks were fitted, but upper bomb bay tanks were, they were individually fed to the fuel system as fore and aft upper 318-gallon tanks. Tanks 1 through 4L were the left-wing tanks and 1 through 4R were the right-wing tanks, the number of cells in each tank indicated by the number of rectangles in each tank. TT are jettisonable tip tanks carried only by the B-45C-1 and RB-45C-1. A gallon of JP-4 weighed approximately 6.84 lbs., therefore the B-45A-1 and -5 carried aircraft carried 37,923.6 lbs. of fuel, while the B-45C-1 carried 52,774 lbs. The B-45C-1s was not equipped for water/ alcohol injection. (NMUSAF)

The B-45C-1was equipped with a single point ground-refueling-system which allowed a single specified receptacle to route the intake of fuel to any or all the 16 internal or two tip tanks. In-flight-refueling was not fitted to the B-45C-1 due to lack of funds and a perceived lack of its requirement to support planned test programs. One other major difference between the B-45C and its offspring was that the B-45C J47-15 engines were not equipped for water/injection take-off thrust augmentation as were those on the RB-45C. The drawing below shows the inboard profile of the B-45C as basically a B-45A-5.

Photo 4: NAA official photo of the first B-45C-1 at Muroc AFB in May of 1949. (Gerald Balzer Collection via Tony Landis and HQ AFMC/HO)

Photo 5: ¾ port side view of 8001. The first B-45C-1 at Muroc AFB. (Gerald Balzer Collection via Tony Landis and HQ AFMC/HO)

Photo 6: Port side plan view of the first B-45C-1 8001 at Muroc in May 1949. (Gerald Balzer Collection via Tony Landis and HQ AFMC/HO) Photo 7: ¾ aft view of the port side of 8001 at Muroc AFB. (Gerald Balzer Collection via Tony Landis and HQ AFMC/HO)

Photo 8: Full aft view of 8001 at Muroc AFB. (Gerald Balzer Collection via Tony Landis and HQ AFMC/HO) Photo 9: ¾ aft view of the starboard side of 8001, the first B-45C-1. (Gerald Balzer Collection via Tony Landis and HQ AFMC/HO)

Photo 10: Starboard side plan view of the first B-45C-1 8001 at Muroc in May 1949. (Gerald Balzer Collection via Tony Landis and HQ AFMC/HO)
Photo 11: Starboard side ¾ front view of the first B-45C-1 8001 at Muroc in May 1949. (Courtesy of The Boeing Aircraft Company)

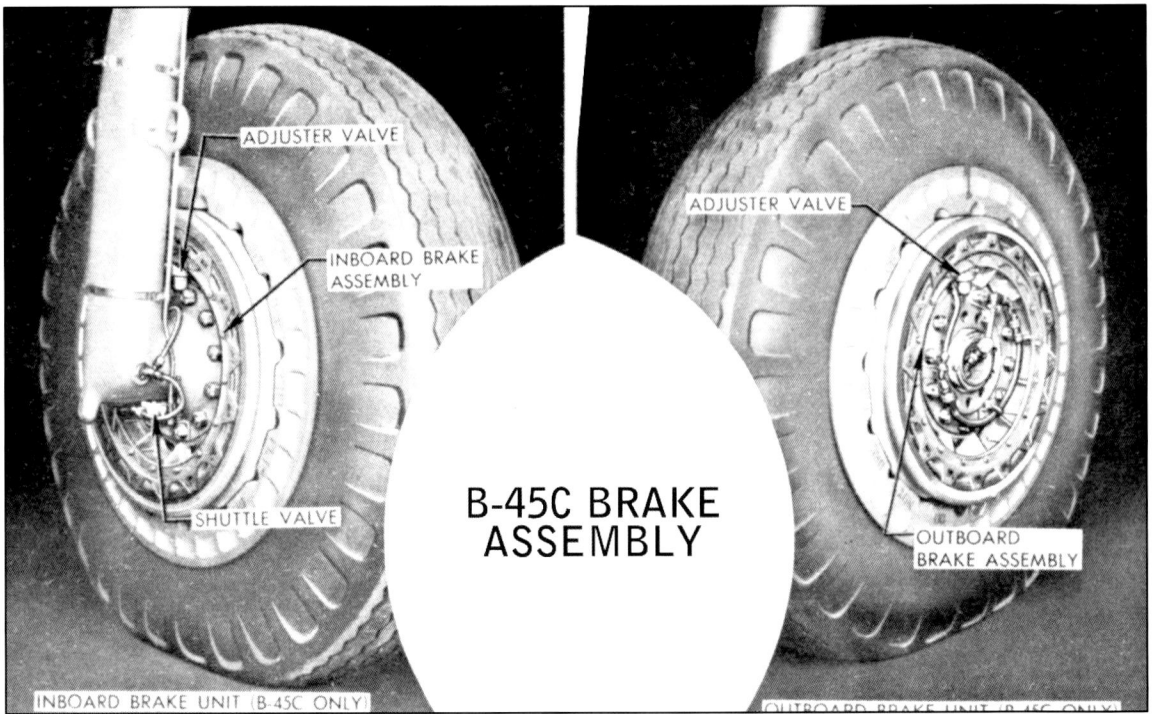

Photo 12: The main gear brakes on the B-45C-1 were still hydraulically powered units, but somewhat stronger to handle the heavier landing weights. Wheels and tires were the same size as the B-45A-5. (NMUSAF)

SERVICE, AN ATOMIC OPTION....

On 14 November 1950 NAA was directed to deliver three B-45C aircraft to the USAF Special Weapons Command at Kirtland AFB with a specific list of equipment installed including weapons racks for the upcoming Project BACKBREAKER. **8008** would be delivered in February 1951; **8005** would be delivered in May 1951; and B-45C **8001** was to be delivered in June of 1951 after the installation of a Norden visual bomb sight. These three aircraft were dedicated to the goal of equipping nearly all tactical USAF aircraft to carry and deliver atomic weapons.

This USAF atomic weapons program for fighter and fighter-bomber aircraft began nearly four years before these aircraft arrived in New Mexico. The SANDSTONE effort was primarily an emphasis of the Atomic Energy Commission rather than the services themselves. The AEC goal was to significantly reduce the amount of fissile material (especially plutonium) used in weapons while increasing their explosive power or yield. The required uranium and plutonium isotopes were in very short supply, and nowhere close to what would be needed to prevent an adversary from using atomic weapons against the US for fear of massive destruction to their own country. This concept would become formally known as Deterrence with the advent of hydrogen or thermo-nuclear weapons.

As covered in Chapter 1, Operation SANDSTONE had solved this fission materials problem and the series of smaller, more efficient weapons were nearing testing in the fall of 1951. One of the most important weapons for use by fighters, fighter-bombers and light bombers was the Mk 7. It carried a levitated core able to withstand the aerodynamic forces of being carried externally by tactical fighter bombers and the improved explosive sphere detonation system similar of the Mk 5. It was designed to be carried by high speed (but below Mach 1) fighters such as the F-84F and G, and the Navy's F2H-2 Banshee, and at high or low altitudes. The weapon, designed and built by Douglas Aircraft, featured a smaller diameter casing and more aerodynamically efficient body. All models of the weapon had an automatic IFI/IFE system controlled by the pilot, but the entire aft portion of the weapon had to be removed before the pit could be loaded onto the IFI/IFE gear. The tail section was then re-attached prior to take-off. The Mk 7 weighed 1,700 pounds and its yield could be set from 1 Kilotons or KT to 70 KT.

Testing and History Making....

The three B-45Cs sent to AF Special Weapons Command were assigned to the 4925th Test Group (Atomic) at KAFB The mission of the Group was to test live and unarmed test shapes and systems on ranges at the Nevada Test Site (NTS), northwest of Las Vegas, the Salton Sea Proving Ground in California southeast of Palm Springs, and the Pacific Proving Ground south southwest of Hawaii. One of the B-45s would test atomic weapon shackles and release gear, as well as electronic monitoring equipment for live weapons; one would serve as a chase aircraft and develop ground procedures for handling atomic weapons, and one would drop live weapons.

Interior Differences....

As built, all 10 B-45Cs carried the interior configuration of the B-45A-5 until they began their R&D roles. Some were equipped AN/APS-23 but none are known to have been equipped with AN/APQ-24 bomb/nav radar system or A-1 Fire Control system. When they became R&D platforms most of the similarities faded, save the cockpit. The B-45A-5 cockpit remained relatively unchanged except for the addition of new fuel system controls and gauges for the tip tanks. The bomb bay remained relatively unchanged throughout the service of 8004 through 8007 with the racks and stations being used to mount sensors and recorders. **8009** and **8010**'s bomb bays were highly modified with retractable engine mounts for in-flight testing. Interior and mission changes to **8001** (and to some degree **8002** and **8003**) were prompted by a fundamental change in mission for the B-45 and will be covered below.

Testing and Making History, B-45C 8001....

Photo 13: 8001 was imaged on the 4925th secure ramp at Kirtland sometime in mid 1951 before its history making drop of two Mk 7 devices over the NTS in 1951 and 1952. The presence of the fiberglass radome indicates that it has yet to receive the modification for dropping those weapons. By this time, in December 1950, the airframe had been re-designated an EB-45C. (Author's Collection)

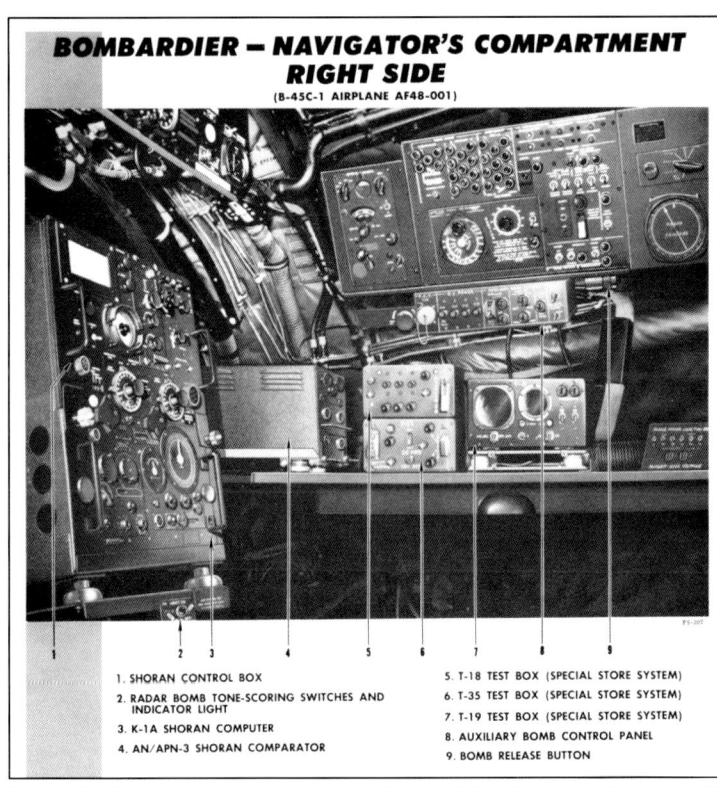
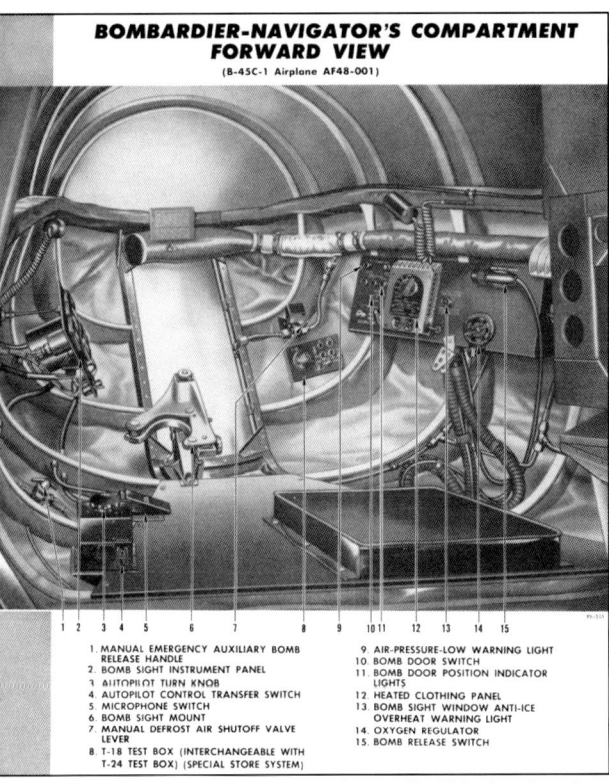

Photo 14: This image shows the starboard side of the Bomb/Nav Station on B-45C 8001. (NMUSAF via the Author) Photo 15: This drawing shows the main modifications to the forward section of the Bomb/Nav station in 8001. The Norden Bomb Site, a Model M-9C, would have been mounted at position 6 in flight. (NMUSAF)

The 4925th and the weapons staff at the NTS were not confident in the accuracy of bombing by the AN/APQ-23 or 24 radar for navigation and aiming. Therefore, the first modification to B-45C 8001 was to reconfigure the nose to install an optically clear panel for the use of a Norden Bombsight. The fiberglass radome was partially removed and the nose portion was replaced with a re-formed exterior metal section with internal re-inforcment and a dedicated mount for the Norden Bombsight. The partitian between the cabin and the antenna area was enlarged and the radar antenna removed. Neither the AN/APS-23 or 24 were installed for the test. Instead, a system called SHORAN, for Short Range Navigation System, consisting of its control panel, the K-1A SHORAN Computer, and the AN/APN-3 SHORAN Interogator-Responder was used.

Developed by RCA and the Wright Field Radio and Radar Laboratory, the SHORAN system was the most accurate short-range naviagtion method available at the time. Two radio signals were transmitted by the aircraft and intercepted and rebroadcast to the airplane by two ground stations with highly accurately determined positions. The AN/APN-3 compared the time of their round trips which indicated the distance to each station and thus the position of the aircraft from each station. SHORAN had a range limited only by the horizon, the farther and the higher the aircraft was determined the range. For example, at 40,000 feet, the range was about 280 miles. Three Special Weapon Test Boxes were installed to monitor and adjust any special weapon carried by the aircraft. In addition, the aircraft was equipped with a tone system that indicated the weapon's release to a ground observer station. With these modifications complete, B-45C **8001** was modified one last time by North American personnel to accept specially designed weapons racks to mount a variety of special weapons in the forward bomb bay. With this mod complete the aircraft was added to the test fleet of the 4925th at Kirtland in the late fall of 1951.

Its first missions involved dropping shapes designed to mimic the MK 7 Atomic weapon. The Mk 7 was destined to become one of the most numerous weapons fielded in the US tactical nuclear arsenal, with the last versions not retired until 1967. This was the weapon that made it possible for single engine fighter-bombers to become atomic capable. In late October 1951 **8001** was assigned to drop the first weapon during Operation BUSTER JANGLE over the Yucca Flats area of the NTS. On 5 November 1951 the aircraft made history by being the first turbojet-powered bomber to drop a live weapon when **8001** released Shot Easy, an experimental version of the Mk 7 in Area 7 of the NTS. The drop was made from 24,500 feet at 350 mph and yielded 31 KT. On 1 May 1952 the aircraft became the first jet bomber to drop a production version of a US atomic weapon during Operation SNAPPER-TUMBLER's Shot Dog on 1 May 1952 dropping a Mk 7 from 19,000 feet at 273 mph yielding 19 KT.

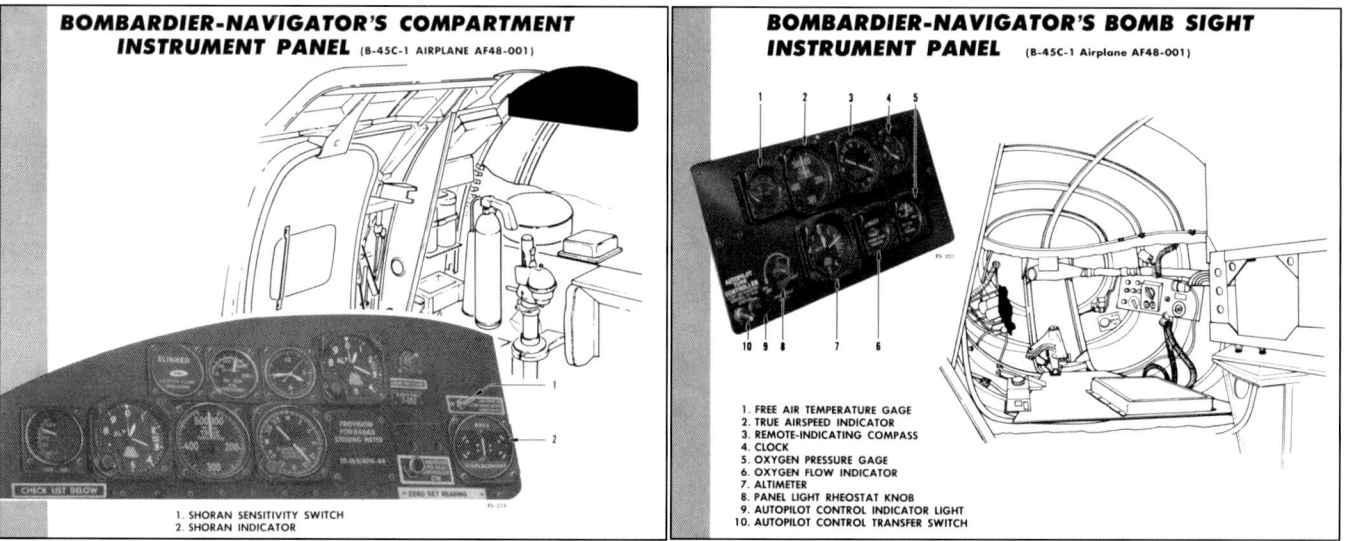

Photo 16: This drawing shows the flight instrument panel to the upper left of the Bomb/Nav. The main difference in 8001 was the presence of SHORAN instruments on the panel. (NMUSAF) Photo 17: This panel was located to the left of the Norden sight in the very forward part of the modified Bomb/Nav station. The panel duplicated many of the instruments on the previous panel so they could be referenced while using the Norden. (NMUSAF)

Photo 18: This practice loading and handling weapon on display at he National Museum of the United States Air Force matches the first live test devices and weapons dropped by 8001 during BUSTER-JANGLE for a yeild of 31 Kt and SNAPPER-TUMBLER for a yeild of 19Kt at the Nevada Test Site in December 1951 and March 1952. (Wikipedia CC)

Photo 19: B-45C 8001 appears as it was on Operation BUSTER JANGLE Shot Easy on 5 Novemebr 1951 and Operation SNAPPER-TUMBLER Shot Dog on 1 May 1952. This is when it made history as the first American jet bomber to drop a live Atomic weapon in November 1951 and the first to drop a production American Atomic Weapon on 1 May 1952. The patch on the nose is that of the 4925th Test Group (Atomic) (NARA II via Tommy Thomason)

Photo 20: 8001 sits at the United States Air Force Museum in March 1957, at the first location for the facility in Area A of Wright-Patterson AFB. This image clearly shows the modification to house the Norden Bomb Sight installation. (NMUSAF)

Special Weapons Center test bird 8001 remained at KAFB and assigned to the 4925th through March 1957 with two depot modification sessions at the San Bernidino Air Materiel Area, Norton AFB (NAFB), CA. On 27 March 1957 **8001** was assigned to the USAF Museum at WPAFB for static display. But it was Struck-off-Charge and disposed of shortly after arrival in March 1957 by senior base officials who directed that the Museum be "straigthened up" for being much too clutted and over-crowded.

B-45C 8002....

B-45C **8002** was delivered to and accepted by the USAF at Long Beach CA on 13 Jul 49. It was immediately flown to Wright-Patterson AFB, OH and assigned to the 2750th Air Base Wing as an EB-45C for flight test work until it was assigned to the ARDC, and the WADC at WPAFB on 22 Jun 51.

On 22 November 1951, the aircraft was bailed to the Allison Engine Company at Indianapolis, IN for modifications and flight test of the Allison J-71 engine. The J71 was a development of the GE J35 turbo-jet engines that the government had directed Allison to produce in order to solve serious problems at GE. The J71 powered the USN XP6M Sea Master flying boat, the McDonnell F3H-2 Demon, and the Douglas B-66 and RB-66 for the USAF. Unfortunately, it was outclassed by the GE J79 and the Pratt & Whitney J57 and the performance of the aircraft it equipped never lived up to promises. In fact, the US Navy chose a completely different engine, the J57, to replace the Westinghouse J40 engine in the early models of the A3D Skywarrier. The J71 engine was a success in the early versions of the Northrop SM-62 Snark Intercontinental Cruise Missile flown from Cape Canaveral, FL. **8002** was Struck-Off-Charge by the USAF at Tinker AFB on 30 April 1957.

Photo 21: 8002 was photographed on the ramp at Edwards AFB in January of 1952 carrying the Allison J71 that it was testing at the time. Note that it was also carrying a Bell M-7 defensive tail turret that would ultimately be fitted on three RB-45Cs of the 91st Strategic Reconnaissance Squadron, 6091st Reconnaissance Flight based at Yokota AB, Japan. Later all RB-45Cs of the 19th Tactical Reconnaissance Squadron assigned at RAF Sculthorpe in the UK carried this turret as well. (NMUSAF)

Photo 22: 8002 departs Edwards AFB in late 1951 of a flight test mission for the Allison J71. It was also carrying a chaff dispenser nozzle, painted red and located just below the forward portion of the National Insignia. This defensive equipment would later be carried by B-45A-5s. Note that the aircraft is equipped with two emergency fuel jettison tubes. (NMUSAF)

B-45C 8003....

B-45C **8003** was delivered and accepted by the USAF at Long Beach, CA on 16 September 1949. The aircraft crashed on take-off on 20 September 1949 when an engine disintegrated because of an over-speed caused by an emergency fuel control failure. The aircraft was destroyed in the resulting crash onto the Long Beach Golf Course. (The author was unable to find any factory images of 8003.)

B-45C 8004....

B-45C **8004** was delivered and accepted by the USAF on 26 September 1949 at Long Beach, CA. It was delivered to Eglin AFB, FL on 22 Nov 49, and assigned to the Air Proving Ground Command (APGC), 3200nd PTG on 25 November 1949. On 25 May 1950, the Tornado was assigned to the AFSWC, 4925th Support Wing at KAFB. In December 1950 it was bailed to the North American Plant at Long Beach for work on developing kits to convert USAF aircraft to carry special weapons. In October, the aircraft was re-designated as an EB-45C by the 4925th Special Weapons Group at KAFB. It returned to the San Bernardino Air Materiel Area (SBAMA) at Norton AFB, Ca in December 1951, August 1952, July 1953, and December 1954 for modifications and parts development for similar projects. After Depot–Level maintenance at SBAMA, **8004** was transferred to Griffiss AFB, NY, and assigned to the Rome Air Development Center as a B-45C from May 1956 until March 1958. The aircraft was turned in for reclamation at Griffiss AFB, NY in April 1958.

Photo 23: B-45C-1 8004 climbs immediately after take-off from an unknown location. (NARA II via Tommy Thomason)

B-45C 8005....

Photo 24: The image shows 8005 at high altitude on one of its varied test flights before its loss in April of 1953. (NMUSAF)

B-45C **8005** was completed by North American on 23 September 1949 and delivered to and accepted by the USAF at EAFB, the 2759th Experimental Wing, on 6 December 1949. It was bailed to NAA on 1 March 1951 through 16 May 1951 for Special Weapons Modifications. It was re-assigned to the AFSWC, 4925th Support Wing, 4925th Test Group (Atomic), 4925th Special Weapons Group as an EB-45C on 10 June 1951. On 7 April 1952 it was bailed to Northrop Aviation Company (NAC) at Los Angeles International Airport (IAP) in support of the Northrop SM-62 Snark Intercontinental Cruise Missile Program. It was relocated to the Northrop facility at Hawthorne Field by ARDC on 25 May 1952 for conversion to a drone control platform for the Snark. The aircraft was redesignated as an EDB-45C drone controller on 15 June 1952. The aircraft was transferred and flew from Patrick AFB, FL as a drone controller from 17 November 1952 until 7 April 1953 when it returned to Hawthorne, CA. On 12 August 1953 8005 crashed at Los Angeles IAP.

B-45C 8006....

B-45C **8006** was completed by North American on 7 December 1949 and delivered to and accepted by the USAF at WPAFB on 10 January 1950. It was assigned to the AMC's All Weather Flying Center at WPAFB. On 24 February 1950, the aircraft was practicing touch-and-go landings at Dayton AFB (today known as Cox-Dayton International Airport in Vandalia, OH). Just after becoming airborne, the starboard nacelle exploded and separated from the aircraft. Then the ensuing fire burnt the front spar, surrounding skin and structure and the starboard wing failed entirely. The investigation discovered that the delivery pilot had issues with the number 4 engine fuel regulator and fuel bypass valve which had not been repaired before this flight took place.

Photo 25: The wreckage of 8006 lies just north of the town of Vandalia, OH. (Author's Collection)

B-45C 8007....

Photo 26: 8007 is shown on the ramp at EAFB after North American had installed the fuselage mounted air brakes. (Steve Ginter Collection)

B-45C 8007 was completed by North American Aviation at their Long Beach Plant on 15 November 1949. It was also delivered to and accepted by the USAF at MAFB on 15 November 1949. It was bailed to NAA and delivered to Los Angeles IAP a few days later in December 1949 and remained there until 14 April 1950. On 19 April 1950 it was re-assigned to the APGC and assigned to the 3200 PTG at Eglin AFB, FL. During this period at NAA and at Eglin, 8007 became the first of three B-45s ever to be fitted with dive brakes. North American performed the modifications to install two fuselage mounted air brakes. Their purpose was to slow the aircraft when pulling out of a dive for delivery of close air support munitions. The Air Staff was still seeking a long-term mission for the B-45 and the idea of using a turbojet-powered aircraft for the role of close air support was very attractive. Response time to the combat area would be good, as would the payload and time on target. Flight test were made at Muroc/Edwards. Then the aircraft was flown and assigned to 3200nd PTG at Eglin. Records are few about what was fired, but it is known that 2.75-inch folding fin rockets were launched from retractable racks in the forward bomb bay. Unfortunately, while accuracy was good and the other attributes were positive, the airframe could not withstand the higher G loadings and significant damage was forecast if the trials continued.

Shortly thereafter the aircraft was transferred to Northrop to support another important project, the SM-62 Snark International Cruise Missile. On 20 April 1952, 8007 was bailed to NAC and delivered to Hawthorne Municipal Field, Hawthorne, CA by AMC for conversion into a second drone controller for the Northrop Snark Program. It was re-designated as an EDB-45C on 15 June 52. On 8 April 1953 the aircraft was relocated to Los Angeles IAP and re-bailed to Northrop as the sole EDB-45C drone controller for the SM-62 Snark Program replacing EDB-45C **8005** through 28 August 1958. In accordance with DoD instructions, it was re-designated as a JDB-45C on 30 November 1956. This was the new DoD designation for aircraft that had no-permanent conversions to a role other than that for which they had been produced. Between 9 October 58 and 30 April 1959, SBAMA at NAFB prepared **8007** for display at the USAF Museum at Wright-Patterson, but **8001** was selected instead. **8007** was sent to Davis-Monthan AFB and the 2704th Air Force Aircraft Storage and Disposition Group for storage, reclamation, and disposal.

Photo 27: 8007 is shown in its markings as a JDB-45C late in its career. (David Menard Collection)

B-45C 8008....

B-45C **8008** was completed by North American at Long Beach on 31 January 1950 and accepted by and delivered to the USAF on 9 February 1950 at WPAFB. The aircraft was assigned to the 2750th Air Base Wing at Wright-Patterson on 10 February 1950 and designated as an EB-45C by the 2750th between 5 October and 8 November 1950. It was bailed to NAA at Long Beach for modifications to carry Special Weapons as a B-45C between 23 November 1950 and 27 February 1951. It was transferred to the AFSWC, KAFB, and assigned to the 4925th Special Weapons Group on 5 March 1951. **8008** was re-designated as an EB-45C by the 4925th Special Weapons Group on 14 August 1951. It completed a Depot-Level maintenance program at SBAMA, NAFB and was then transferred to the ARDC, Rome Air Development Center at Griffiss AFB, NY between 1 October 1954 and 7 November 1956. On 8 December 1956 **8008** was re-designated as a JB-45C by the Rome Air Development Center. On 31 December 1956, the aircraft was Struck-off-Charge from USAF records and transferred to the US Army at Picatinny Arsenal, Morris County, New Jersey for munitions tests. (The author was unable to locate any images of 8008 in service.)

B-45C 8009....

B-45C **8009** was completed by North American at Long Beach on 26 January 1950 and accepted by the USAF on 22 March 1950 at Long Beach. It was bailed by AMC to the GE Co. at Long Beach on 20 April 1950. The aircraft was modified at Long Beach as an airborne turbojet test platform and flew from there and EAFB until July 1951. It is believed that it commonly flew flight tests with various GE J47 engines including the J47-17, their first after-burning model. On 8 July 1951 8009 was transferred to Schenectady Airfield, NY and bailed by AMC, to the GE CO. Electric Co, GE Engines at Schenectady Airport for turbine engine development work.

Photo 28: This image shows 8009 having retracted the J79 fully into its belly and closed its bomb bay doors as fully as possible. This was essential if the test aircraft was to land safely and no damage made to the J79. (The Gerald Balzer Collection via Tony Landis and HQ AFMC/HO)

Photo 29: This image shows 8009 carrying a prototype J79 engine in the extended position where the engine could be started and run during powered tests. (The Gerald Balzer Collection via Tony Landis and HQ AFMC/HO)

At this point, 8008's primary test engine was the GE J79 which would ultimately power the McDonnell F-4 Phantom II family, the B-58 Hustler, the F-104 Starfighter, and the NAA A-5 and RA-5C Vigilante. The aircraft was sent to SBAMA for Depot-Level Maintenance between 1 May and late April 1953 and then again bailed to GE Engines, Schenectady Airport, NY for turbine engine development work from 29 June 1953. Per DoD directives, it was re-designated as a JB-45C on 4 December 1956 by GE Engines, Schenectady, NY. It was re-designated as an EB-45C to 13 February 1957. On 30 April 1957 8009 was Struck-off-Charge by the USAF and transferred to the Naval Air Development Unit at NAS South Weymouth, MA. It was then assigned to Project LINCOLN to develop high speed radar tracking systems. Its disposition from the US Navy is unknown.

Photos 30 and 31: These images show 8009 testing the extension and retraction system in its bomb bay carrying a J47 in 1951. The first image shows the bomb bay doors fully open so there would be no interference with the engine extending or retracting. The second image shows the bomb bay doors retracted around the extended engine as they would have been in flight during tests. (Craig Kaston via the Steve Ginter Collection)

Photo 32: 8009 was photographed on the ramp at NAS South Weymouth south of Boston, MA shortly after the aircraft arrived from GE Engines, Schenectady, NY in May. 1957. (Jim Burridge via the Steve Ginter Collection)

B-45C 8010....

B-45C **8010** was completed by North American at Long Beach on 31 January 1950. The USAF accepted the aircraft on 14 April 1950, delivered it to the AMC and assigned it to the 2750th Air Base Wing on the 21 April 1950. It was designated as an EB-45C by the 2750th on 1 May 1950 and served at Wright-Patterson until 11 June 1952. On that date the aircraft was bailed by the AMC to the Pratt & Whitney Aircraft Company at Hartford, CN for turbojet engine development and test work until 31 Aug 58. On 30 November 1956 **8010** was redesignated a JB-45C by Pratt & Whitney, and on 10 April 1957 **8010** was re-designated as an NB-45C, having been converted to a configuration that could not be returned to its production configuration. On 22 June 1962, the aircraft was Struck-off-Charge by the USAF and assigned to the US Air Force Museum for static display as the last airworthy B-45 known to exist.

Photo 33: 8010 is shown in this image testing a Pratt & Whitney J57, one of the most successful and dependable turbojets developed and produced by Pratt & Whitney. The engine powered some of the most important aircraft in the USAF inventory including the Boeing B-52 and KC-135, the McDonnell F-101, and the North American F-100. The US Navy used it to power their outstanding F-8 Crusader. (The Steve Ginter Collection)

Photo 34: 8010 was imaged here straight and level testing a Pratt and Whitney J52 for the US Navy's A-4 Skyhawk, A-6 Intruder, and EA-6B Prowler. It also powered the AGM-28 Hound Dog cruise missile carried by the B-52C through G models. (The Steve Ginter Collection)

Photo 35: 8010 was photographed in 1971 at the USAF Museum wearing the markings it carried late into its flying career with Pratt & Whitney and upon its arrival at the Museum. The various aircraft seen throughout the background were part of the USAF Museum Collection. At the time if the photo, they were being repositioned in Hangar 1 and 9, the Museum's Annex, and the Museum's Air Park in Area B of Wright-Patterson AFB to gain more inside storage area and display space outdoors. (The Gerald Balzer Collection via Tony Landis via HQ AFMC/HO)

Conclusion....

Between May 1949 and May 1962, the B-45C-1 made history in the USAF atomic weapons program; served as an excellent research and flight test aircraft introducing a huge series of turbojets that revolutionized the jet engine business and service; and made commercial passenger and freight service a world-wide operation as well. The Tornado also had the structural flexibility to allow multiple projects to be conducted simultaneously. As with all the other members of the family, it suffered losses. One was due to a design issue shared with all the other models; and two were due human errors. But nonetheless C-model persevered longer than all other models of the B-45 family.

Photo 36: 8009 cruised at high altitude in this test flight of the GE J79, soon to power the McDonnell Douglas F-4 Family. (Craig Kason via the Steve Ginter Collection)

Photo 37: The shinny B-45C-1 8005 was photographed shortly after its delivery from North American to Edwards AFB in 1950. (AFFTC History Office via Steve Ginter)

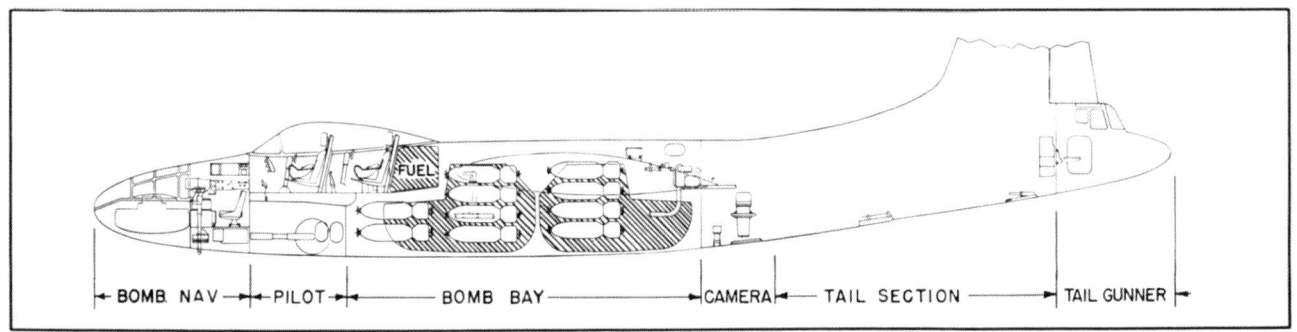

Drawing 2: This Standard Aircraft Characterisics Summay inboard profile shows the similarities between the B-45A-1 and the B-45C-1. (NMUSAF)

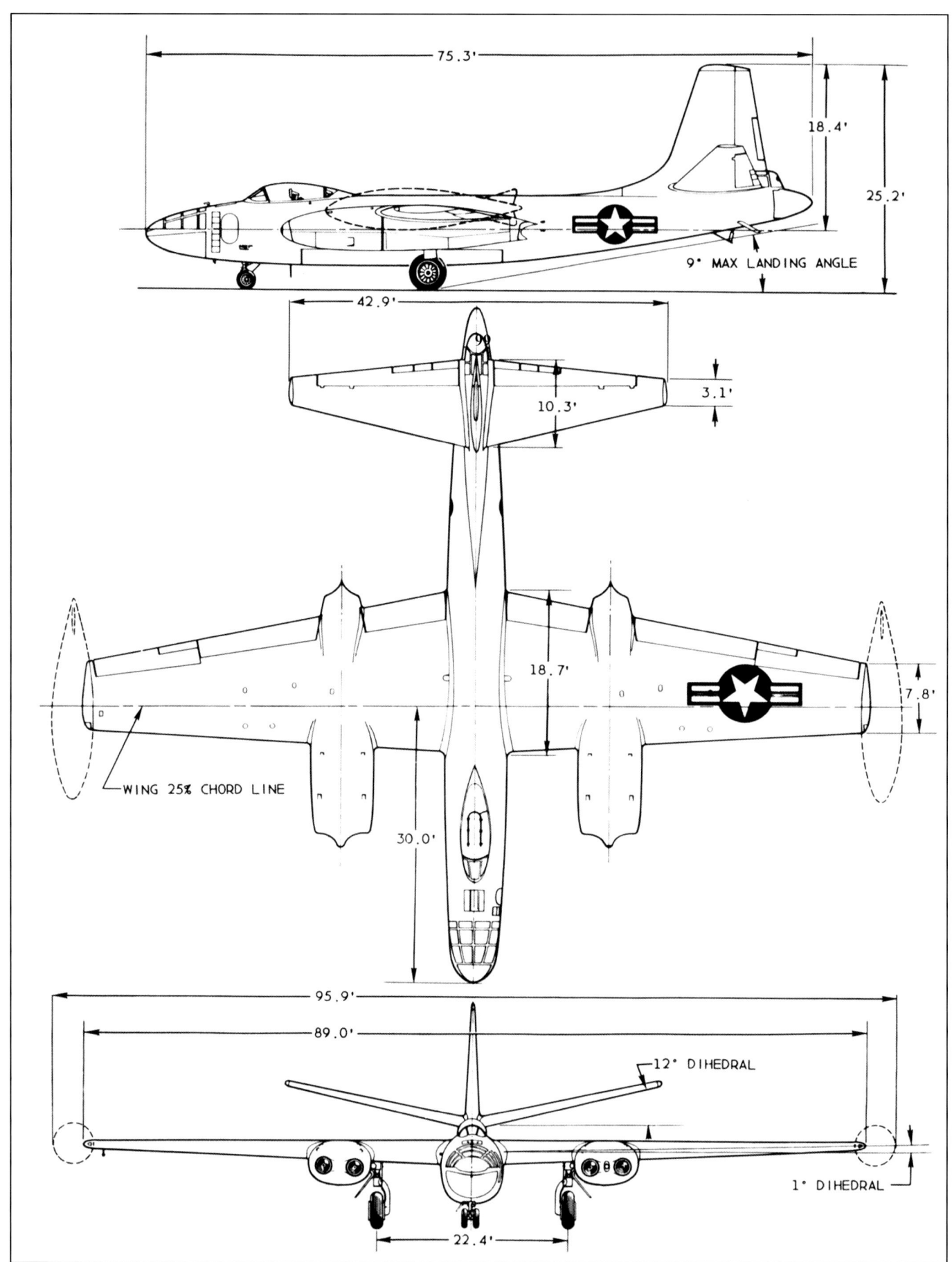

Official USAF Three View Drawing of the B-45C

CHAPTER 5 – THE RB-45C DESCRIPTION

Production....

The RB-45C was the last production version of the B-45 family. The same contract that had authorized the ten B-45Cs, AC-18000, also called for the conversion of 33 B-45Cs on the production line into RB-45Cs. Its first flight was in April 1950. They were designed to be high altitude, high speed photo-reconnaissance aircraft to be assigned to the Strategic Air Command. With the advent of the Soviet MiG-15 FAGOT fighter, the vulnerability of the RB-29 and RB-50 fleet was now serious, and SAC needed higher performance aircraft to survive missions deeper into "denied" territory within the Soviet Union, the People's Republic of China, and their Allies. Few however realized just how soon that mission and those capabilities were going to be needed in a real war.

Photo 1: RB-45C 8024 was photographed on a post-completion acceptance flight over southern California in July 1950 before its assignment to Barksdale AFB, LA. (The Gerald Balzer Collection via Tony Landis from HQ AFMC/HO)

Structural Differences....

The RB-45C and the B-45C shared the same basic airframe, tip tanks, landing gear, and similar layouts. The differences lie in the reconnaissance fit and significant changes to mission capabilities.

Photo 2: This image shows a head-on view of the first RB-45C off the production line, taken at Edwards AFB, CA. (The Gerald Balzer Collection via Tony Landis from HQ AFMC/HO)

Photo 3: This is a view of the forward port side of 8011 at Edwards. A small window was installed in RB-45C forward crew entry door to check for fire in the nacelle area before bailing out. (The Gerald Balzer Collection via Tony Landis from HQ AFMC/HO)

Photo 4: This image shows the port side fuselage of 8011 with the port side tri-metrogon camera door open. (The Gerald Balzer Collection via Tony Landis from HQ AFMC/HO)

Photo 5: This view shows the aft port side of 8011. Note the original turret fairing designed for the XB-45s is still in use. (The Gerald Balzer Collection via Tony Landis from HQ AFMC/HO)

Photo 6: This image shows the aft portion of RB-45C 8011. (The Gerald Balzer Collection via Tony Landis from HQ AFMC/HO)

Photo 7: This image was taken from the aft starboard side of 8011 and shows some of the differently shaded camera doors in the aft fuselage. (The Gerald Balzer Collection via Tony Landis from HQ AFMC/HO)

Photo 8: This image shows the complete starboard side of 8011. (The Gerald Balzer Collection via Tony Landis from HQ AFMC/HO)

Photo 9: This imaged shows a ¾ view of the forward starboard side of 8011. (The Gerald Balzer Collection via Tony Landis from HQ AFMC/HO)

Photo 10: RB-45C 8012 was photographed on the ramp at Boeing Field, Seattle, WA in 1950 before the start of in-flight air refueling certification tests. (The Gerald Balzer Collection via Tony Landis from HQ AFMC/HO)

Photo 11: RB-45C 8012 was imaged on the Boeing Field ramp with a set of two 1,125 gallon JP-4 fuel tanks which when filled, weighed approximately 14,850 pounds. It appears the bird has been fueled as the dihedral doesn't show in the wings. (The Gerald Balzer Collection via Tony Landis from HQ AFMC/HO)

Photo 12: This image shows the approximate size of the ground crew needed to pre-flight an RB-45C prior to engine start. The two carts labeled "C-21-B" were auxiliary power units to provide electrical power to the aircraft prior to engine start. The aircraft had no on-board auxiliary power units and had to have these units to turn the engine starters. Each nacelle had two auxiliary units fitted, one for each engine. Once the engines were started, engine driven generators provided aircraft power. Then the two aux power units would be moved to the opposite nacelle and engines Numbers 3 and 4 started. Generators/alternators were switched on after each engine start. (David Menard via Steve Ginter)

Photo 13: A graceful image of an RB-45C over the Pacific Coast of California highlights the plan view of the aircraft. (Gerald Balzer Collection via Tony Landis from HQ AFMC/HO)

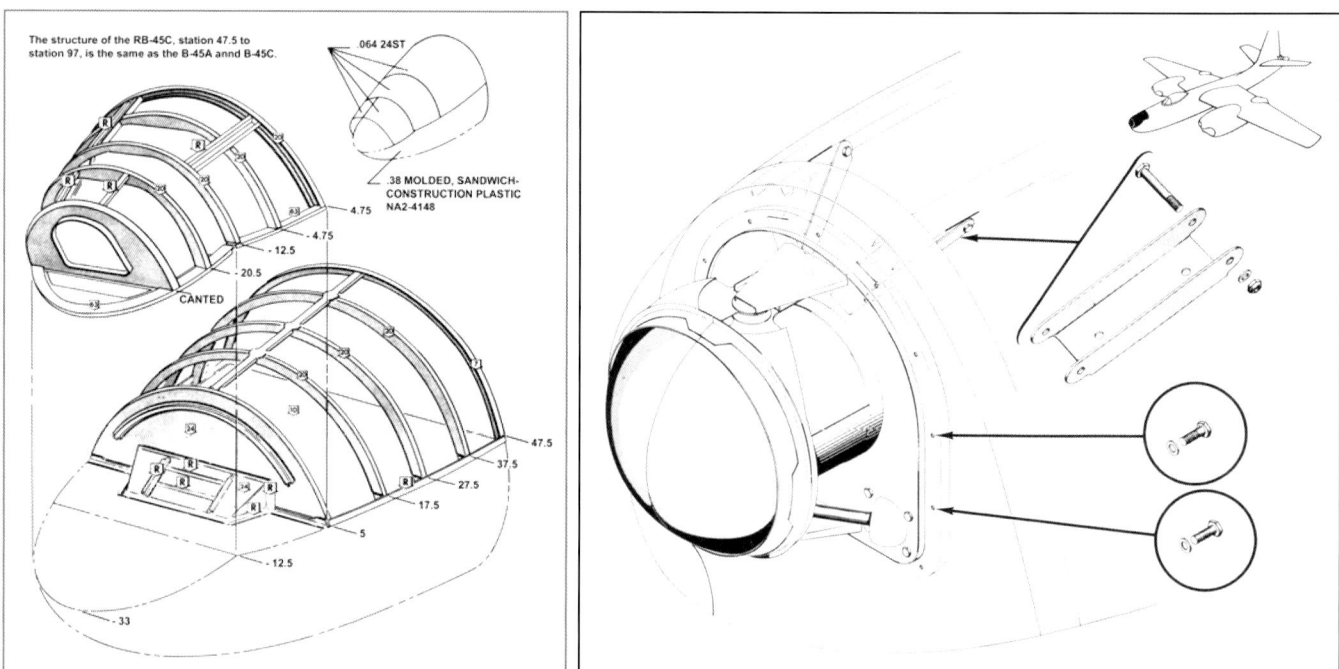

Photo 14: RB-45C 8011 cruises at medium attitude over the southern California desert with its camera doors open during a test flight prior to hand-over to the USAF in 1950. (Gerald Balzer Collection via Tony Landis from HO AFLC/HO)

Structure Changes....

The basic structure of the RB-45C was based on the B-45C and had more than enough weight carrying capacity to handle the required load. The more important and complicated issue was weight and balance. The camera suite and systems necessary to ensure the cameras would work at their required altitude, temperatures, and multiple exposures were extremely complicated and heavy. Access for maintenance and adjustments for operations was difficult and complex and ranged from one end of the aircraft to the other.

Drawing 1 and 2: The first set of structural changes dealt with mounting the forward oblique cameras in the nose of the aircraft. (NMUSAF)

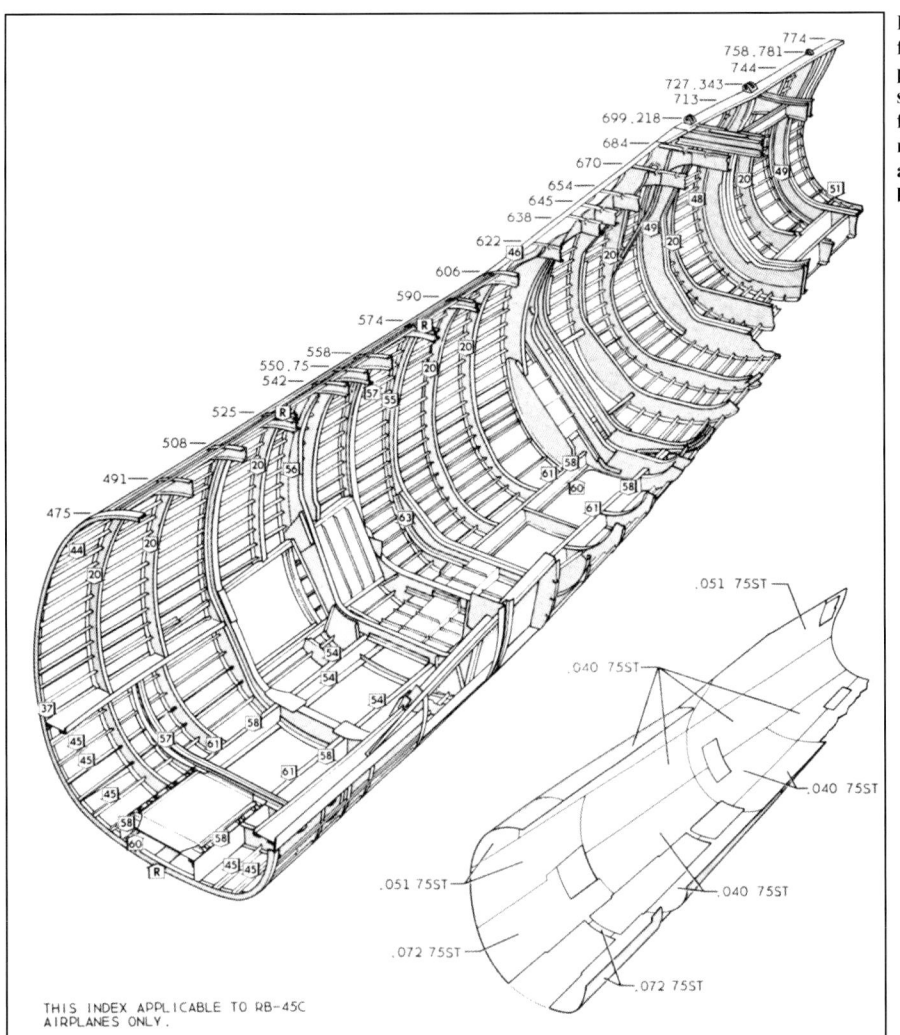

Drawing 3: The second change dealt with the aft fuselage from the aft bomb bay bulkhead to the pressure bulkhead just in front of the tail gunner's station. This section mounted all five of the aft fuselage-housed camera stations and their magazines. The area however was not pressurized and could not be entered until the aircraft was below 10,000-feet mean-sea-level. (NMUSAF)

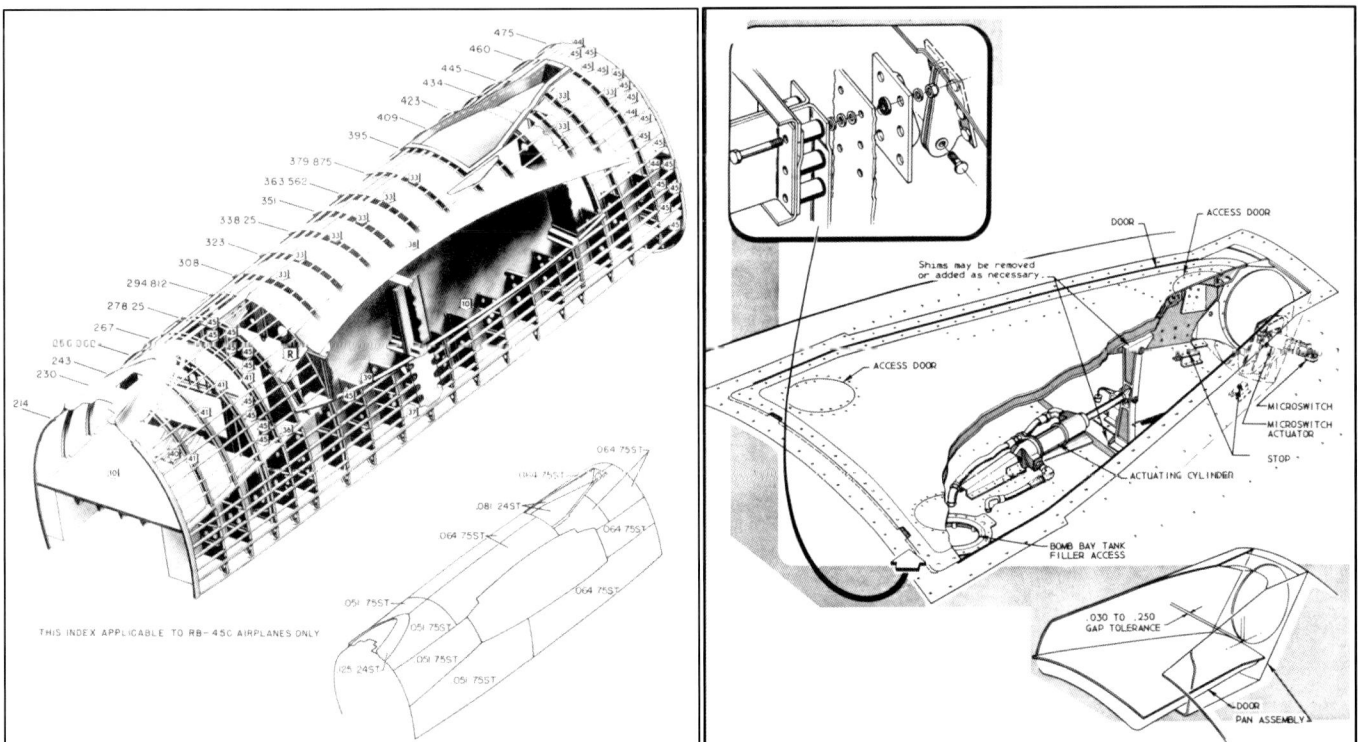

Drawing 4: The last airframe structural change was made to the RB-45C just at the trailing edge of the wing. This was the location of the new In-Flight-Refueling (IFR) receptacle. It took advantage of the B-45C's single point ground refueling system to fill every tank. (NMUSAF) Drawing 5: This image shows the location and details of the RB-45C IFR receptacle. (NMUSAF)

Fuel System....

The RB-45C and the B-45C could both carry two 1,125-gallon jettisonable wing fuel tip-tanks, a 355-gallon self-sealing fuselage tank just behind the cockpit, and one self-sealing 318-gallon non-jettisonable tank, and one non-self-sealing 872-gallon jettisonable fuel tank in the fore and aft bomb bays. There were self-sealing tanks in each wing. If the mission required the aircraft carrying photo-flash bombs, some or all four bomb bay fuel tanks would be left behind for room to carry up to 25 226-lb. T-68 photo-flash bombs for night photography. As built, the RB-45C was not equipped to carry 500-gallon Fletcher tanks.

Range was the key consideration in the development of the RB-45C and to extend the maximum range as far as possible, it was equipped with the Boeing-developed Flying Boom In-Flight-Refueling System. The single-point-refueling-system, also installed in the B-45C, made it much easier to install and use the in-flight-refueling (IFR) system which could transfer fuel to any tank in the aircraft. This became the ultimate answer to increasing the mission radius in any aircraft. The RB-45C could receive fuel from the Boeing KB-29P and the KC-97A Flying Boom equipped tankers at a rate of approximately 300 gallons per minute. However, the aircraft was not equipped to receive fuel from any drogue-equipped tanker such as the Boeing KB-50J or K.

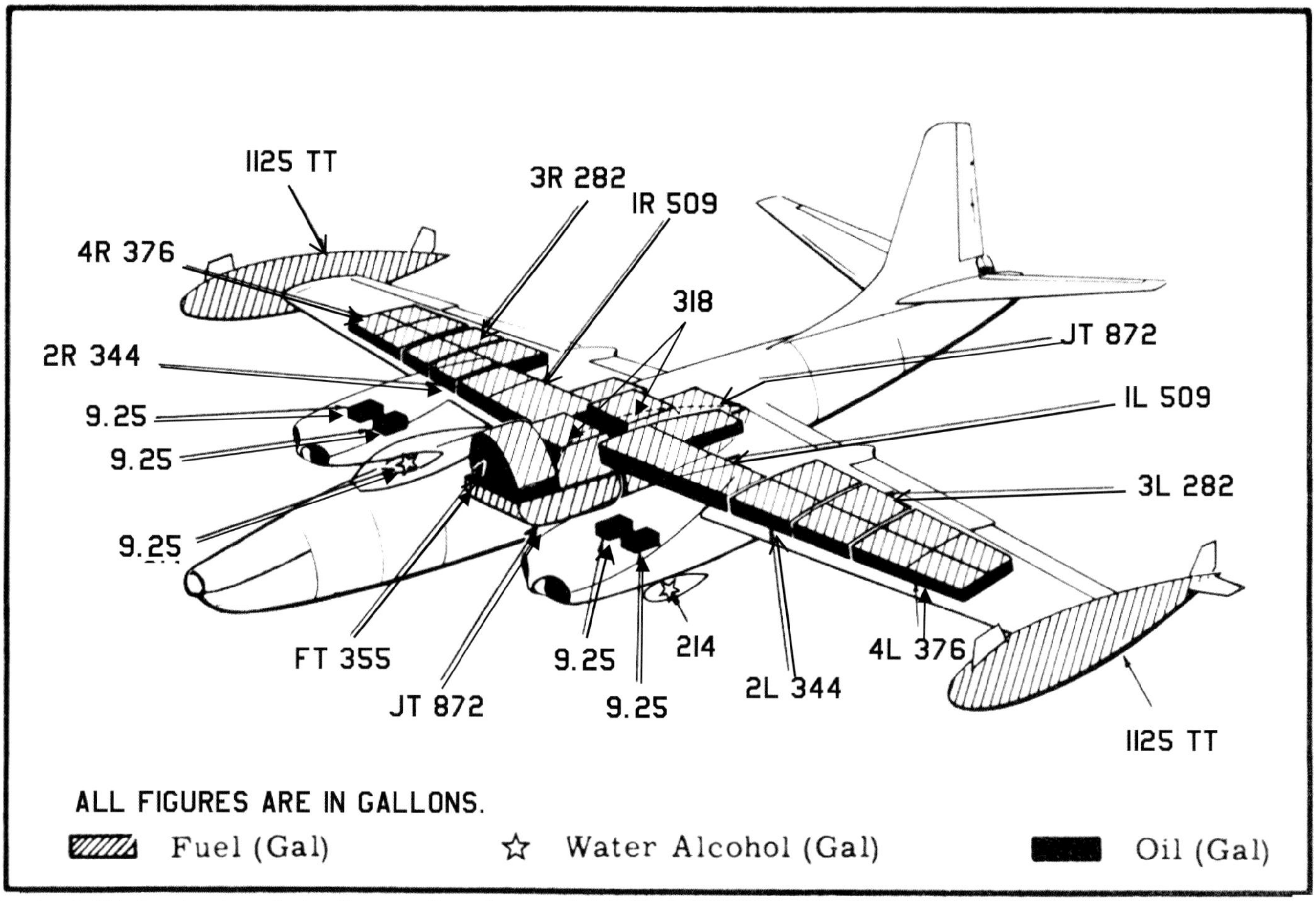

Drawing 6: This drawing shows the location, capacity and names of all fuel tanks and the number of cells in each tank in the RB-45C-1. Capacities are in gallons and are listed in bold three-digit numbers. FT was the 355-gallon single-cell self-sealing fuselage tank; the two JT tanks were each 872-gallon jettisonable, but non-self-sealing, single-cell lower bomb bay tanks, one carried in each bomb bay if the space was not needed to carry photo flash bombs. Each bomb bay could also carry one 318 gallon self-sealing tank in the upper bomb bays, again if the space was not needed to carry photo-flash bombs. Tanks 1 through 4L were the left-wing self-sealing tanks and 1 through 4R were the right-wing self-sealing tanks. The number of cells in each tank indicated by the number of rectangles in each tank. As built, the RB-45C could not carry the 500-gallon Fletcher drop tanks under each nacelle. 214-gallon water/alcohol drop tanks could be carried, one under each nacelle. The 1125-gallon tip-tanks were jettisonable. A gallon of JP-4 weighed approximately 6.84 pounds, therefore the RB-45C-1 carried a maximum fuel load of 7,633 gallons, or 52,209.7 pounds of fuel. The four numbers leading to black boxes on the nacelles are the 9.25-gallon capacities of the oil tanks in each nacelle, one dedicated to each engine. (NMUSAF)

Photo 15: RB-45C 8024 cruises at medium attitude over the Los Angeles area during a test flight prior to hand-over to the USAF in 1950. (Gerald Balzer Collection via Tony Landis from HO AFLC/HO)

Photo 16: This photo centers on the RB-45C's In-Flight-Refueling (IFR) receptacle. It is the dark panel outlined in black lines just forward of the wing's trailing edge. The dark panel was the receptacle's watertight slipway door, which hinged downward at the front allowing the tanker's boom to enter the receptacle and pass fuel to the receiver. The receptacle itself was beneath and just aft of this panel just in front of the short black line across the fuselage. (NMUSAF)

Photo 17: 8012 has just made contact with KB-29P 44-86363 on an early RB-45C In-Flight-Refueling Certification mission over central Washington State, with flaps partially extended to slow while in formation with its tanker KB-29P in June 1950. (Courtesy Craig Kaston via the Steve Ginter Collection)

A USAF contract with Boeing and North American covered the work to insure the RB-45C could be converted into an IFR receiver and to conduct all necessary modifications to all 33 RB-45C airframes. RB-45C **8012** conducted all the compatibility and certification tests on missions flown from Boeing Field, Seattle, WA between June and December 1950 with KB-29P 44-86363.

Photo 18: After the receptacle was fitted to 8012, a series of fitting tests were conducted with the two aircraft (RB-45C and KB-29P) ground-towed together as closely as possible. The goal was to insure the KB-29P boom could safety reach and fit into the receptacle with enough vertical space above the canopy. (Alan Griffith)

Photo 19: Moving into the eastern agricultural regions of Washington State, 8012 approaches the tanker from below to enter what was called the observation position. This location, below and with the nose just slightly aft of the KB-29P's boom allowed the boom operator to give the entire receiver a good safety check. Checks included fuel leaks, insuring the slipway door was fully open and no ice or liquids inside, and that the receptacle was completely clear of the same hazards. Once the checks were complete, the boom operator would direct the receiver forward and upward towards the boom to be in position for its insertion into the receptacle. (Craig Kaston via the Steve Ginter Collection)

Photo 20: Taken over the desert in central Washington State, 44-86363 and 8012 made numerous missions to get the RB-45C certified, often multiple sorties in one day. The flaps are dropped on 8012 to reduce the RB-45's airspeed to where the KB-29 could maintain a safe formation with the RB-45C. In this image, it is easy to see that the B-29, the USAAF's largest operational bomber at the end of WWII, was not that much larger than the new reconnaissance jet derived from the USAF's first jet powered light bomber. (Steve Ginter)

Although the Initial Operational Capability or IOC of the RB-45C was scheduled for the summer of 1951, the world situation was dramatically accelerating the need for it. Within about two weeks of **8012**'s arrival at Seattle War broke out on the Korean Peninsula. by July and August the situation was going badly for the unprepared United Nations Coalition. After the UN gained the upper hand and drove the North Koreans out of South Korea - Chinese forces entered the war and the Soviets introduced the swept wing MiG-15 into combat - in late 1950. This meant that SAC and the Far East Air Force's (FEAF) all-important reconnaissance capability using existing propeller aircraft was nullified. Getting the RB-45C into the theater along with its accompanying in-flight-refueling tankers had become a high priority.

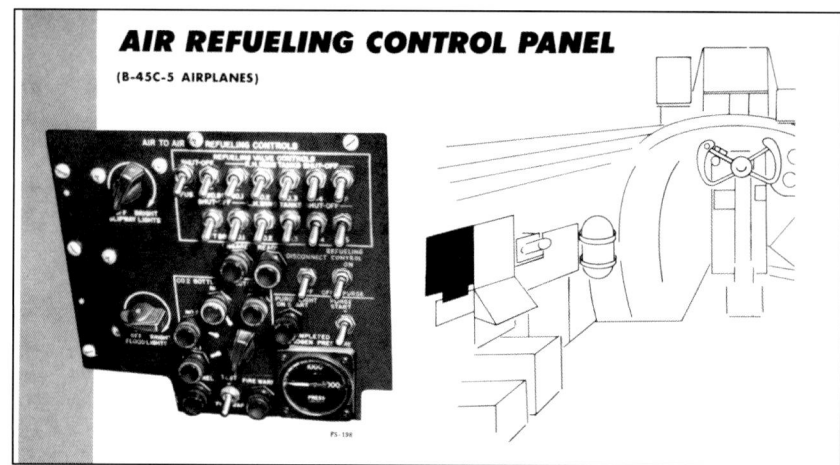

Photo 21: This drawing shows the RB-45C receiver's IFR control panel. It was located on the co-pilot's radio panel above the crew passageway to the bomb bay. The panel controlled which tanks would receive fuel from the tanker by the two rows of switches on the upper right of the panel. The switch in the down position was on to receive fuel, while the up position was to close that tank off.

The middle switches on the right were the emergency disconnect switches and the system master switch. The lights and lower switches controlled the CO_2 system which exhausted all JP-4 residue and fuels from the receiver's fuel lines after a refueling and could be used under pressure to purge the fuel lines in an emergency.

The two rheostat switches controlled the floodlights on the slip-way door and receptacle during night air-refueling operations. (NMUSAF)

The Air Refueling Control Panel in **Photo 21** allowed the receiver aircraft to configure the fuel lines to direct fuel to the proper tanks in a manner that would maintain the center of gravity and wing balance as the receiver became heavier. The boom operator had no control over where the fuel went within the receiver, only to fly the boom in response to turbulence and the receiver's maneuvers. The boomer's other essential job was to direct the receiver via radio or lights on the tanker's belly to make maneuvers to help keep the boom engaged in the receptacle and to keep the boom within certain safety bounds and the two aircraft physically separated. Sets of colored lights were arranged on the bomb bay doors of the tanker that gave the receiver guidance on maneuvering up or down, left, or right, and forward or aft during radio silence air refueling.

Photo 22: 8012 is now in contact with the tanker and likely taking on fuel over the mountainous territory of western and northwestern Washington. This gave both crews the opportunity to check out the RB-45C's maneuverability and the sensitivity of its controls. The goals were to trim the aircraft and adjust the control surface boost systems to make sure it was not too easy to over-control the aircraft. To do this the aircraft had to seek out terrain and weather that produced turbulence so they could get first-hand experience of how the RB-45C would respond. Just in front of the windshield above the photo-navigator was an adjustable sunscreen for that station. (Craig Kaston via Steve Ginter)

Photo 23: This image shows the terrain and the altitude that 44-86363 and 8012 were flying at when Photo 20 was taken. (Courtesy The Boeing Aircraft Company)

Interior Stations....

As with all the other B-45 models, the RB-45C had four crew stations, the photo-navigator in the nose, the pilot and co-pilot under the jettisonable canopy in ejection seats, and a gunner in the very aft end of the fuselage. The canvas safety harness mostly for maintenance personnel was installed in the aft portion of the pilot's passageway just in front of the pressure bulkhead leading to the front bomb bay. While the crew could enter the bomb bay in flight if the rest of the aircraft were depressurized, the crew could not enter the camera compartment in flight for emergency maintenance above 10,000-ft. above ground level.

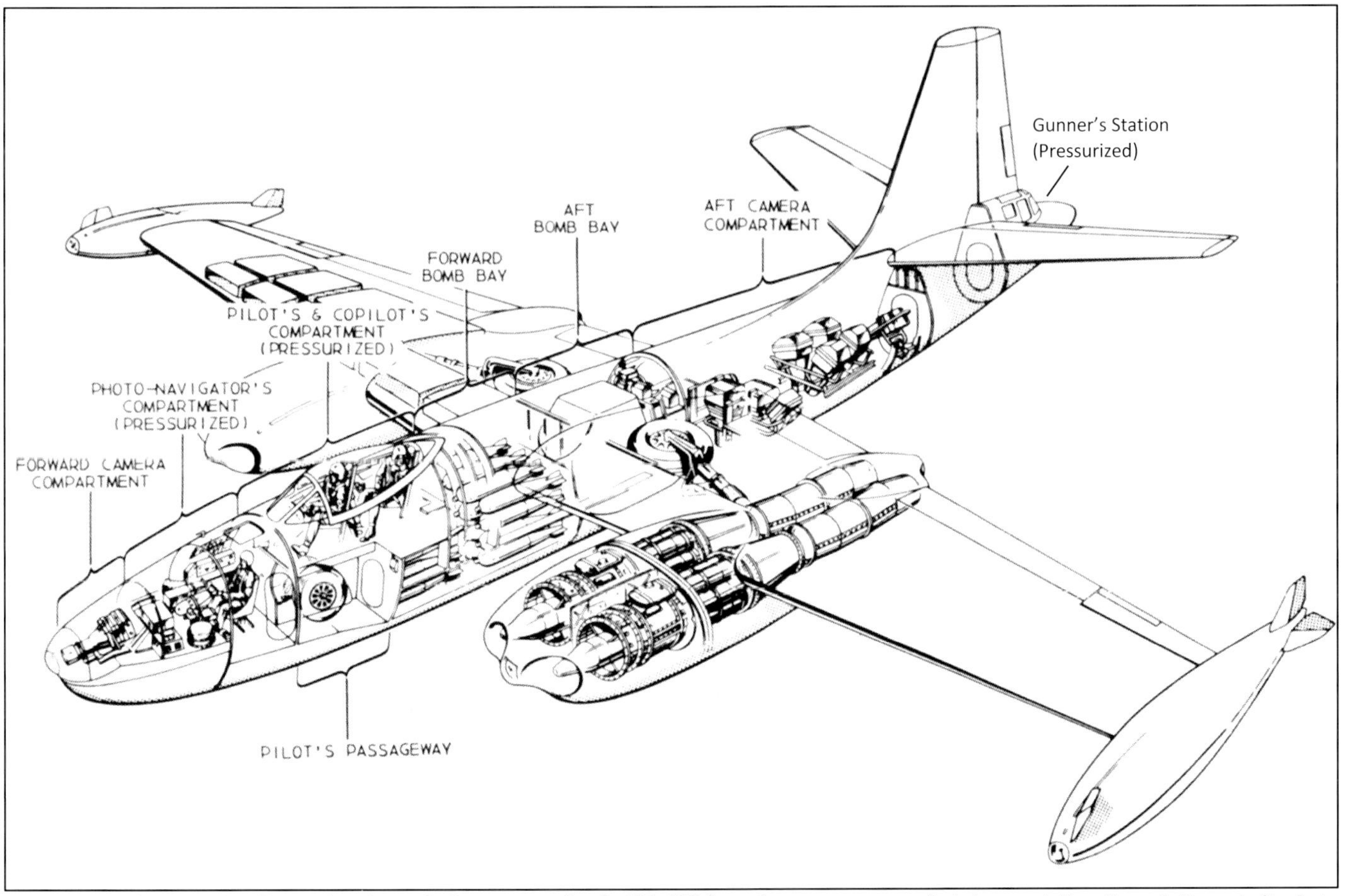

Drawing 7: This orientation illustration shows the key systems for the RB-45C from the nose camera and photo-nav station, through the cockpit, photo-flash weapons in the forward bomb-bay, camera stations in the aft bomb-bay, and gunner's station. The drawing also shows a good layout of the engines and their multi-angled exhaust ducts. (NMUSAF)

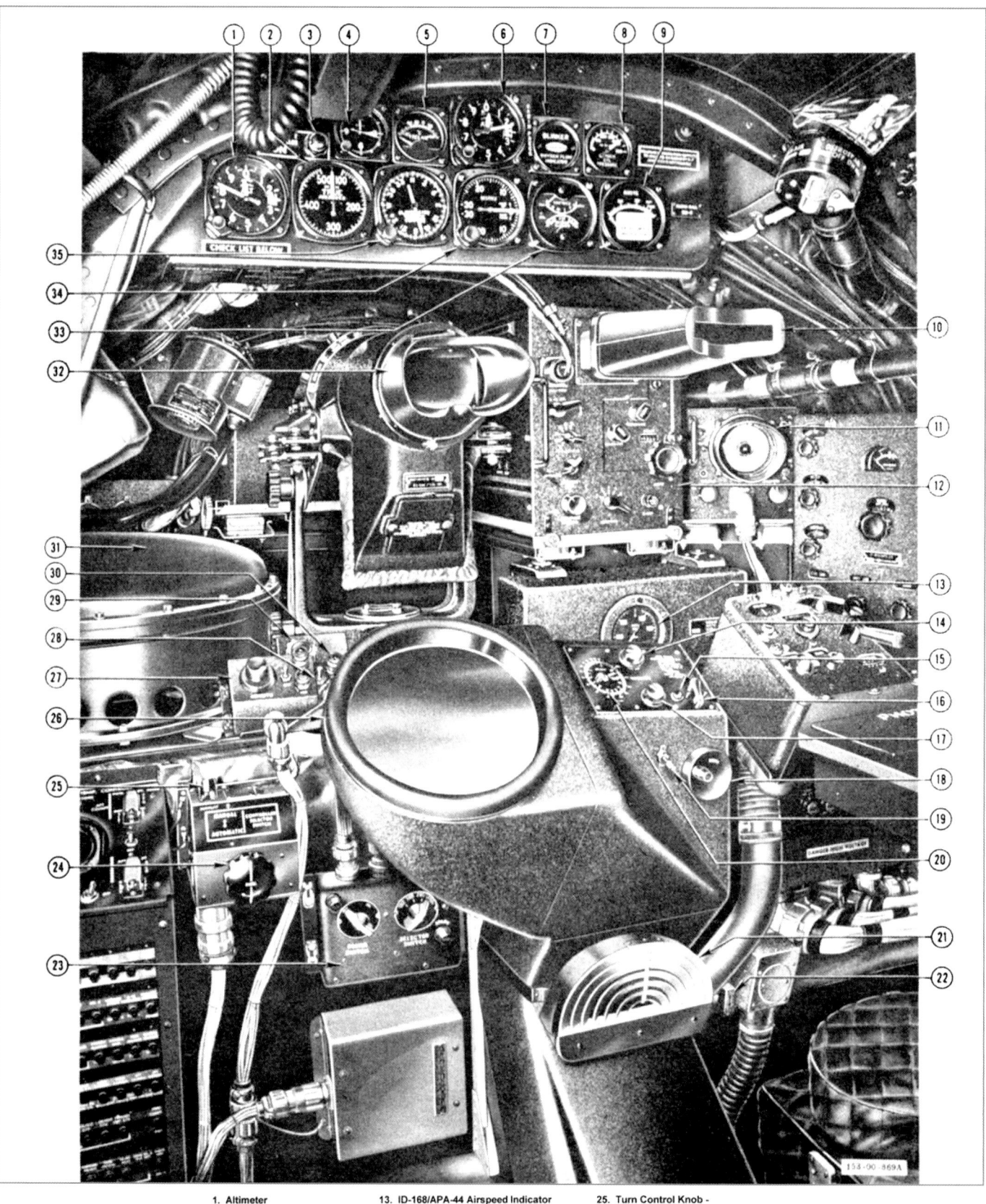

1. Altimeter
2. True Airspeed Indicator
3. Auto-pilot Controller Indicator
4. Clock
5. Free Air Temperature Gage
6. Cabin Pressure Altimeter
7. Oxygen Flow Indicator
8. Oxygen Pressure Indicator
9. ID-166/APA-44 Monitor Voltmeter
10. Visor for R-249/APN-9A (Loran) Receiver-Indicator
11. SCR-718C Radio Altimeter
12. R-249/APN-9A (Loran) Receiver-Indicator
13. ID-168/APA-44 Airspeed Indicator
14. Photo-viewfinder Wide-angle Rheostat
15. Photo-viewfinder Grid Indicator Light
16. Photo-viewfinder Grid Change-over Lever
17. Photo-viewfinder Grid Rheostat
18. Photo-viewfinder Drift Control Crank
19. Photo-viewfinder Power Switch
20. Photo-viewfinder Ground Speed Dial
21. Heat and Vent Floor Outlet
22. Shutoff Valve for Photo-viewfinder Outlet
23. Azimuth Controller for A-28 Stabilized Mount
24. P-3B Camera Control Box
25. Turn Control Knob - Automatic Pilot
26. Controller Selector Switch - Automatic Pilot
27. Photo-viewfinder
28. C-416/APS-23 Control Unit
29. Camera Initiation Switch
30. Photo Operation Indicator Light
31. RT-124/APS-23 Receiver-Transmitter
32. ID-218/APS-23 Indicator
33. Radar Steering Meter
34. Inclinometer
35. Remote-indicating Compass

Photo 24: This image shows the RB-45C Photo-Nav panel as installed with the photo-viewfinder in the center and the AN/APN-23. (NMUSAF)

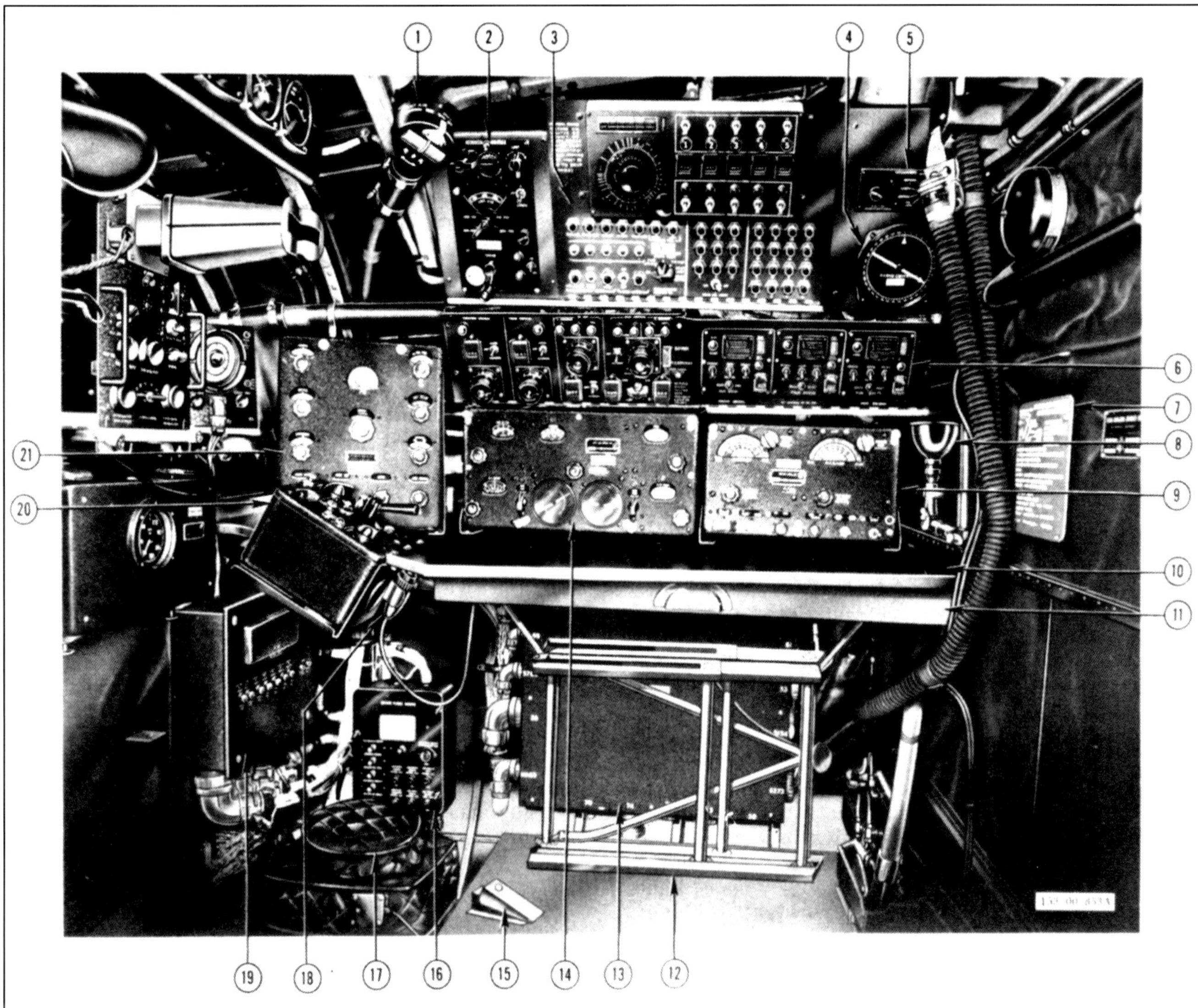

1. Heat and Vent Outlet and Astrodome Shutoff Valve
2. AN/ARN-6 Radio Compass Control Panel
3. Secondary Camera Control Panel
4. Radio Compass Indicator
5. Interphone Panel
6. Primary Camera Control Panel
7. Landing Gear Emergency Operation Instruction Placard
8. Chart Table Light
9. CP-22/APA-44 Mileage Computer
10. Chart Table
11. Chart Table Drawer
12. Main Entrance Ladder
13. J-166/APA-44 Junction Box
14. CP-21/APA-44 Ballistics Computer
15. Microphone Switch
16. Radar Junction Box
17. CN-66/APS-23 Gyroscope
18. Manual Bomb Release Switch
19. J-218/APS-23 Junction Box
20. C-293/APA-44 Tracking Control
21. C-413/APS-23 Control Unit

Photo 25: This image shows the starboard side of the photo-nav station with many of the camera secondary controls installed in this location. (NMUSAF)
Photo 26: This image shows the camera system defrost and temperature controls. NMUSAF Photo 27: Shows the camera secondary control panel. (NMUSAF)

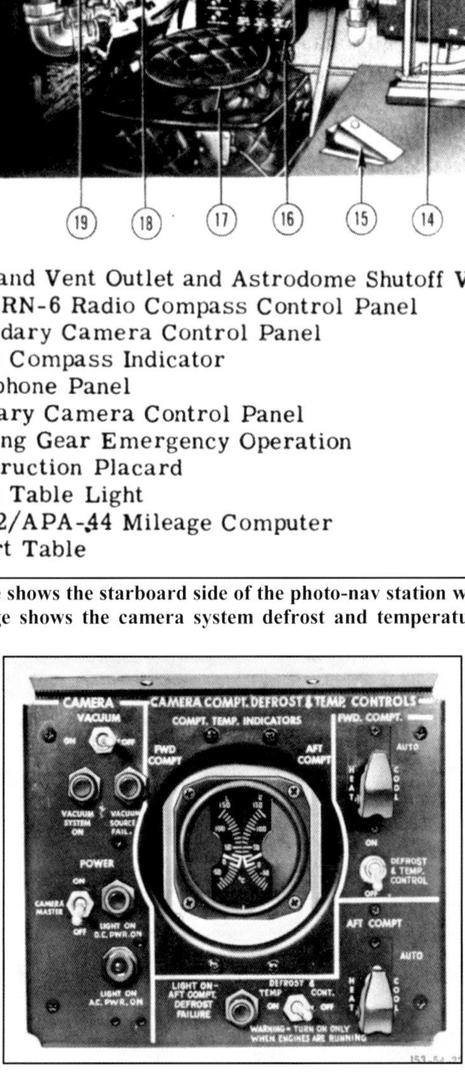

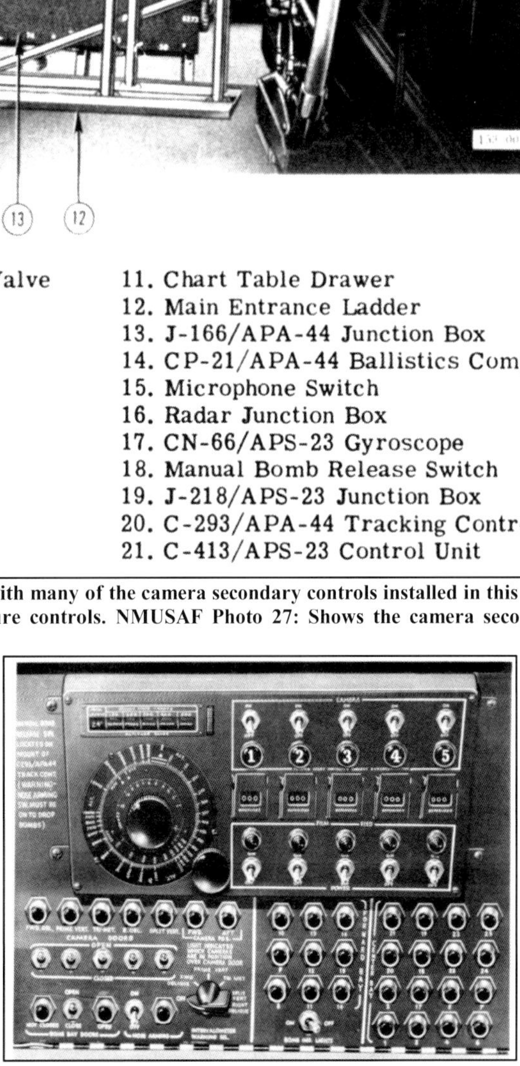

The camera systems installed in the RB-45C required significant oversight and attention. Systems which regulated temperature, pressurization, and vacuum controlling the film moving over the camera lens and platens, were monitored constantly. Film in all the cameras was constantly moving to provide multiple exposures as the aircraft moved. Vacuum holding the film in place would release freeing film to move, then increase to hold the film in place for the camera to expose it. The length of time was driven by factors such as the exposure of the image, itself driven by the airspeed and altitude of the aircraft. Once the exposure was complete, the vacuum would release - and the film would move hiding the exposed image and placing an unexposed portion of film in place for the next exposure. Exposure time could vary greatly depending on the type of film, altitude, time of day and other factors such as visibility. What is not generally realized is that the camera suite on the RB-45C was very nearly as capable as that carried by the SAC RB-47. The main difference was range and a slight variation in maximum altitude and airspeed.

Many of the structural changes mentioned earlier were made to make room or access available to fit these types of systems and their controls. Visually, though these changes were small, their complexity was significant and one of the reasons that the IOC for the RB-45C was not due until mid-1951.

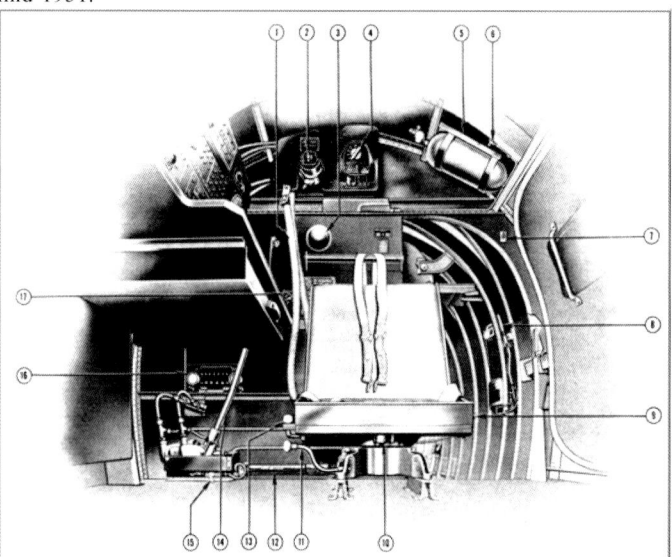

1. Oxygen Hose
2. Oxygen Regulator
3. Alarm Bell
4. Heated Clothing Control Panel
5. Portable Oxygen Bottle
6. Refiller Valve – Portable Oxygen Bottle
7. Passageway Light Switch
8. Camera Emergency Hydraulic Controls
9. Shoulder Harness Lock Control
10. Seat Swivel Control
11. Seat Vertical Control
12. Lock Release Cylinder – Emergency Landing Gear Lowering System
13. Emergency Escape Control
14. Hand-pump – Emergency Landing Gear Lowering System
15. Manual Check Valve – Emergency Landing Gear Lowering System
16. SN-47/APS-23 Synchronizer
17. Chart Table Light Rheostat

1. Main Entrance Hatch Control
2. Main Entrance Hatch
3. Main Entrance Hatch Emergency Release
4. Fire Extinguisher
5. Parachute Static Line
6. Radar Pressurization Control Panel
7. First-aid Kit
8. AN/APN-68 Radar Control Panel
9. Camera Electrical Control Panel
10. Disposal Canisters
11. Camera Circuit Breaker Panel
12. Hand Axe

Photo 28: This image shows the aft bulkhead of the photo-nav station and the chair which would rotate counter-clockwise to give the photo-navigator good access to the forward crew access door and emergency escape door. (NMUSAF) **Photo 29:** This image shows the forward crew access doors and emergency escape hatch. (NMUSAF)

Photo 30: RB-45C 8024 climbs over the southern California coastal area south of Los Angeles on an acceptance test flight in 1950. These flights were designed to test all operational systems in the aircraft before the mission systems, such as the reconnaissance systems were given a thorough work-out and the aircraft was readied for turn-over to the USAF. (Gerald Balzer Collection via Tony Landis from HQ AFMC/HO)

1. Fluorescent Light
2. Auto-pilot Engaged - Power-off Indicator
3. Auto-pilot Trim Indicator
4. Radar Steering Meter
5. Radio Compass
6. Flap Position Indicator
7. Landing Gear Door Indicator Light
8. Landing Gear Position Indicators
9. Landing Gear Position Indicator Light
10. Forward Engine Compartment Fire Warning Indicators
11. Fire Extinguisher Switch
12. Aft Engine Compartment Overheat Indicators
13. Fire Detector System Test Switches
14. Clock
15. Accelerometer
16. ATO Indicators
17. Oxygen Pressure Gage
18. Oxygen Flow Indicator
19. Water Injection System Indicators
20. Water Injection System Switch
21. Stand-by Compass
22. Remote-indicating Compass
23. Attitude Gyro
24. Bomb Door Indicators
25. Rate-of-Climb Indicator
26. Tachometers
27. Exhaust Temperature Indicators
28. Hydraulic Pressure Gage
29. Hydraulic Fluid Level Indicator
30. Fuel Pressure Gages
31. Aileron Trim Indicator
32. Rudder Trim Indicator
33. Elevator Trim Indicator
34. Instruments Inoperative
35. Cabin Pressure Altimeter
36. Control Column Release Lever
37. Free Air Temperature Gage
38. Turn-and-Bank Indicator
39. Altimeter
40. Fuel Flowmeter and Totalizer Indicator
41. Elevator Trim Auxiliary Switch
42. Refueling Boom Disconnect Switch
43. Camera Initiation Switch
44. Auto-pilot Release Button
45. Radio Microphone and Interphone Button
46. Nose Wheel Steering Trigger Switch
47. Fuel Gages - Right Wing Tanks
48. Fuel Gages - Left Wing Tanks
49. Fuel Gages - Large Bomb Bay Tanks
50. Wing Tip Tanks "Empty" Indicators
51. Fuel Filter De-icer Switch
52. Fuel Filter Ice Warning Indicator
53. Fuel Gage - Fuselage Tank
54. Oil Temperature Indicators
55. Air-to-Air Refueling Indicators
56. Camera Pre-exposure Warning Light
57. Photo Operation Indicator Light
58. Oil Pressure Gages
59. Pilot's Check List
60. Glide Path and Runway Localizer Indicator
61. Marker Beacon Indicator
62. Airspeed Indicator

Photo 31: This image shows the layout of the pilot's forward instrument panel in the RB-45C. (NMUSAF)

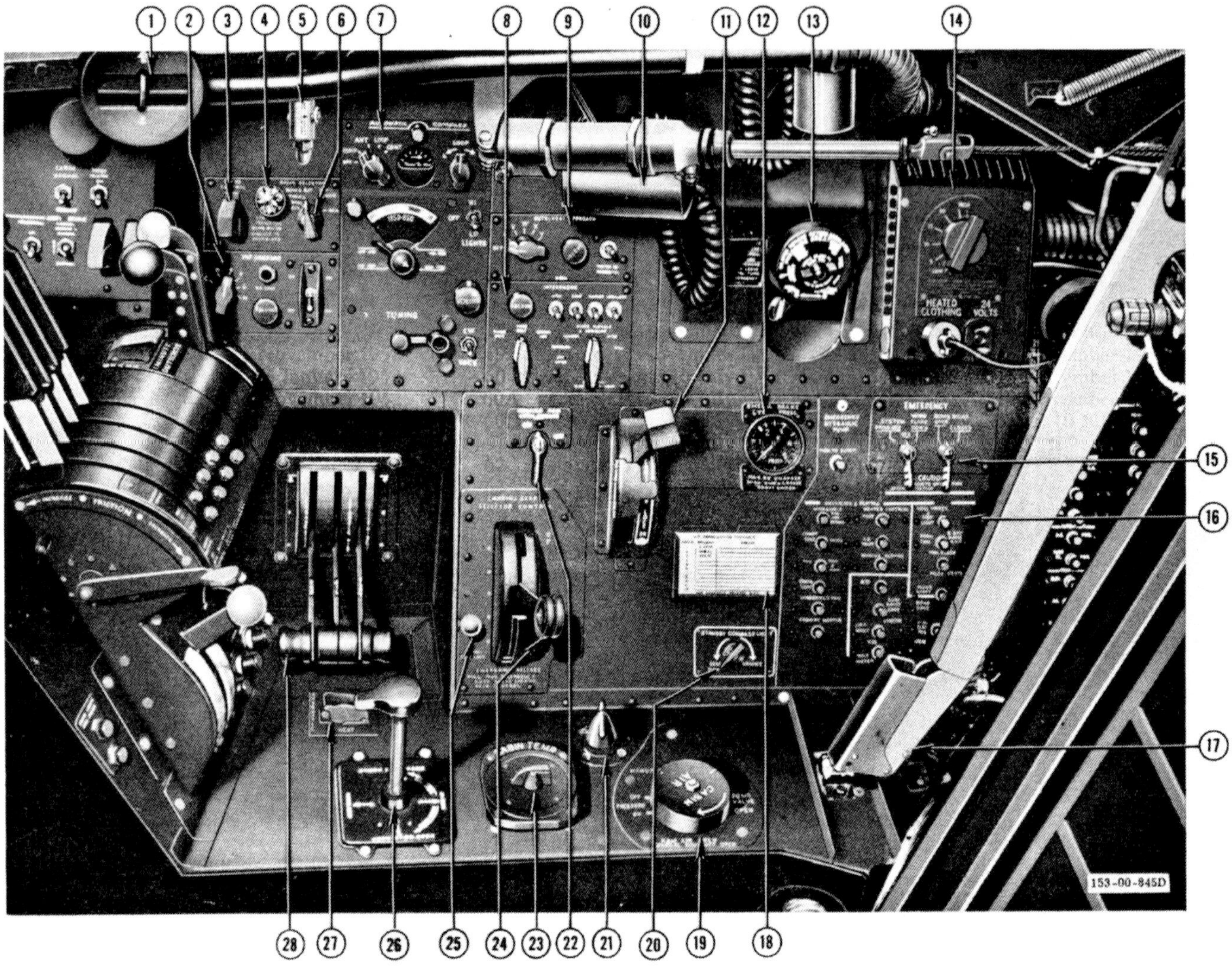

1. Heat and Vent Outlet
2. Command Radio Control Panel
3. Bomb and Fuel Tank Salvo Control Switch
4. Salvo Circuit Indicator Light
5. Pneumatic System Bleeder Valve
6. Bomb and Fuel Tank Salvo Selector Switch
7. Radio Compass Control Panel
8. Interphone and Mixer Switch Panel
9. Instrument Approach Control Panel
10. Cockpit Light
11. Controls Lock
12. Emergency Brake Hydraulic System Pressure Gage
13. Oxygen Regulator
14. Heated Clothing Control Panel
15. Emergency Hydraulic Selector Valves
16. Circuit Breaker Panels
17. Canopy Manual Emergency Release
18. Radio Frequency Channel Card
19. Cabin Air Valve
20. Compass Light Rheostat*
21. Wing Tip Tank Mechanical Release
22. Hydraulic Main System Pressure Control Valve
23. Cabin Temperature Rheostat
24. Landing Gear Selector Control
25. Landing Gear Lock Emergency Release
26. Trim Tab Control Stick
27. Cabin Temperature Manual Override Switch
28. Auto-pilot Engaging Handles

Photo 32: This image shows the starboard side panels of the pilot's cockpit of the RB-45C highlighted by the engine throttles and auto-pilot controls. (NMUSAF)

1. Auto-pilot Release Button
2. Radio and Interphone Microphone Button
3. Elevator Trim Auxiliary Switch
4. Rudder Bars
5. Copilot's Check List
6. Auto-pilot Engaged - Power-off Indicator
7. Oxygen Flow Indicator
8. Airspeed Indicator
9. Turn-and-Bank Indicator
10. Attitude Gyro
11. Remote-indicating Compass
12. Rate-of-Climb Indicator
13. Fluorescent Light
14. Altimeter
15. Oxygen Pressure Indicator
16. Heat and Vent Floor Outlet

Photo 33: This image shows the layout of the co-pilot's forward and starboard panels. Note that the co-pilot has a small set of instruments with a very limited capability especially in instrument flight conditions. (NMUSAF)

1. Heated Clothing Control Panel
2. Cockpit Light
3. Oxygen Regulator
4. Generator Switches
5. Ammeters
6. Heat and Vent Outlet
7. Automatic Pilot Controller
8. A-C Voltmeter Selector
9. A-C Voltmeter
10. Alternator Selector Switch
11. Alternator Warning Lights
12. Alternator Master Switches
13. Canopy Manual Emergency Release
14. Canopy Defroster Outlets
15. Alternator Selector Control Switch
16. D-C Voltmeter
17. Disposal Container
18. Main Hydraulic System Filler Hose
19. Portable Oxygen Bottle
20. Landing Gear and Door Emergency Manual Controls
21. Pneumatic System Pressure Gage
22. Pneumatic System Filler Valve
23. Main Hydraulic Reservoir Filling Hand-pump
24. Bomb Door Manual Control
24A. Surface Control Boost Test Shutoff Switch
25. Surface Control Boost Test Switches
26. D-C Voltmeter Selector
27. Generator Overvoltage Lights
28. A-C Frequency Meter
29. Inverter Switch
30. Inverter Warning Lights
31. Circuit Breaker and Fuse Panel
32. AN/APX-6 IFF Transpondor
33. AN/APX-6 IFF Control Panel*
34. Interphone Amplifier Auxiliary Gain Control
35. Cockpit Light:
    Fluorescent Light - AF48-11 Through AF48-34
    Red-White Floodlight - AF48-35 and Subsequent
35A. SA-3/A Inertia Switch
35B. IFF Destructor Test Indicator Lights
36. Alarm Bell

Photo 34: This is the co-pilots port-side panel located just above the pilot's passageway leading aft to the bomb-bay. The co-pilot's primary mission, besides relieving the pilot on long range missions, was to handle communication on a variety of equipment. The starboard side of the co-pilot's station was largely equipped with communications gear. (NMUSAF)

J47 Changes, and Assisted Take-Off or ATO....

Although the RB-45C used the same GE J47 engines as the B-45C, its J47-15s were capable of using water-injection to increase thrust during take-off and at low speeds. A water/alcohol mix was normally injected either at the compressor inlet or in the diffuser just before the combustion chambers. Adding water increased the mass and density of the air, and therefore increasing the amount of oxygen* and allowing more fuel to be burned in the combustion chambers. This also increased the mass of air being accelerated out of the engine by the turbine, increasing thrust. The injection also served to cool the turbines. Since temperature was normally the limiting factor in turbine engine performance at low altitudes, the cooling effect allowed the engine run at higher RPM with more fuel injected and more thrust created without overheating. *Oxygen does not burn, it only supports combustion.

The drawback of the system was that injecting water quenched the flame in the combustion chambers somewhat, as there was no way to cool the engine parts without also cooling the flame. This led to unburned fuel out the exhaust and a characteristic trail of dense black smoke. Jettisonable 214-gallon distilled water and alcohol drop tanks could be attached underneath each nacelle. A small amount of alcohol was added to the mix to insure the water did not freeze on cold nights prior to take-off. Distilled water was used to reduce the chances of damaging contamination within the engine.

Photo 35: This image shows the general size of the 214-gallon water/alcohol tanks in relations to the starboard nacelle. (NMUSAF) **Photo 36:** This image shows a head on view of the tank, showing how rugged the tank as, and how little streamlining was used in its construction. (NMUSAF)

Photo 37: This image shows the port side of the port tank and gives detail of the two mounting struts and four stabilization braces. (NMUSAF via the Author) **Photo 38:** This image shows the starboard side of the port tank mounted under the port nacelle. (NMUSAF)

Auxiliary Take-Off....

The RB-45C was the second member of the family to prepare for the use of liquid-fueled propellant Auxiliary Take-Off pods. The RB-45C was expected to operate from less than adequate runways overseas and ATO was considered important. The RB-45C specifically used two Aerojet LR13-AJ-3 rocket pods, two generations beyond what was first flown on the XB-45 some two years earlier. One was carried under each nacelle

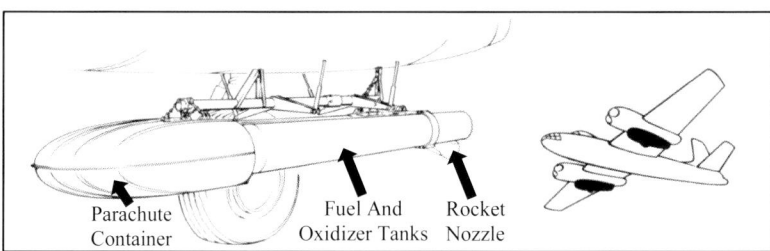

Drawing 8: This is a Tech Data drawing of the Aerojet XLR13-AJ-3, two generations beyond the models tested on the first XB-45. The area beneath the B-45 drawing shows the mounting location of the pods. (NMUSAF)

and each pod produced 4,000-lbs. of thrust for 40 seconds. The propellant was concentrated nitric acid and a 65% xylidine/35% gasoline mix. The propellant was hypergolic, meaning the components spontaneously ignited when they came into contact with each other. Both pods were jettisonable, these using parachutes stored in fairings at the forward end of the pods structure.

Photo 39: A pair of Aerojet LR13-AJ-3 are being fueled on their transport trailer on the ramp. Extreme caution had to be used during this action due to their explosiveness and the caution needed during the movement of their handling trailer. (Alan Griffith) Photo 40: The handling trailer was moved by hand toward the port nacelle of RB-45C 8033 for a test flight. (Alan Griffith)

Photo 41: The handling trailer is moved directly under the nacelle and adjusted to raise into the attachment lugs for the pods under the nacelle. (The Alan Griffith Collection) Photo 42: This image shows a direct head-on view of the 4,000-lb. ATO pod loaded onto the starboard nacelle of RB-45C 8033 at Edwards AFB in November of 1950. (NMUSAF)

Photo 43: 8033 has ignited the two LR13-AJ-3 pods and just broke ground on the main runway at Edwards Main Base on November 3, 1950. (SDAM via Craig Kaston)

The test series on the LR-13-3 mirrored that of the XLR-13-1 in 1947. The most important tests were the simultaneous ignition of the pods at equal power, and that they were located as near to center of gravity of the aircraft as possible. Then came the jettison tests to be sure both pods separated cleanly and simultaneously and did not impact the wing, aft fuselage, or the empennage. As the XLR-13-1 pods did on the XB-45, there were no serious issues encountered with any of test criteria. That said, the toxicity of the propellants was the main short fall. As with the Red Fuming Nitric Acid (RFNA) oxidizer and dinitrogen tetroxide N_2O_4 in the -1 pod, the concentrated nitric acid and a 65% xylidine/35% gasoline mix in the -3 pod was just as hazardous to handle, fill, and transport. The danger near populated areas, or base facilities would be the deciding issue in not using the pods operationally. Therefore, the water-injection system of the J47s became the primary ATO provision for the RB-45C fleet.

Photo 44: 8033 climbs to clear simulated obstacles after take-off. (The Steve Ginter Collection) **Photo 45:** 8033 begins to retract its landing gear after a positive rate of climb was established. (Steve Ginter)

Photo 46: The image shows 8033 at high altitude ready for simultaneous release of the pods in one of the most important tests that was performed to insure the LR-13-3s could be used safely on the RB-45C. (Courtesy Craig Kaston via the Steve Ginter Collection) **Photo 47:** At this instant the cameras caught the pods separated and clearing the empennage as they passed under the horizontal stabilizers. The test was a success. (Steve Ginter)

Aerial Photography Equipment....

Photo 48: RB-45C 8011 was imaged on a North American acceptance check flight which included an operational check of the reconnaissance cameras on board. Five of the seven camera positions are visible. (Steve Ginter)

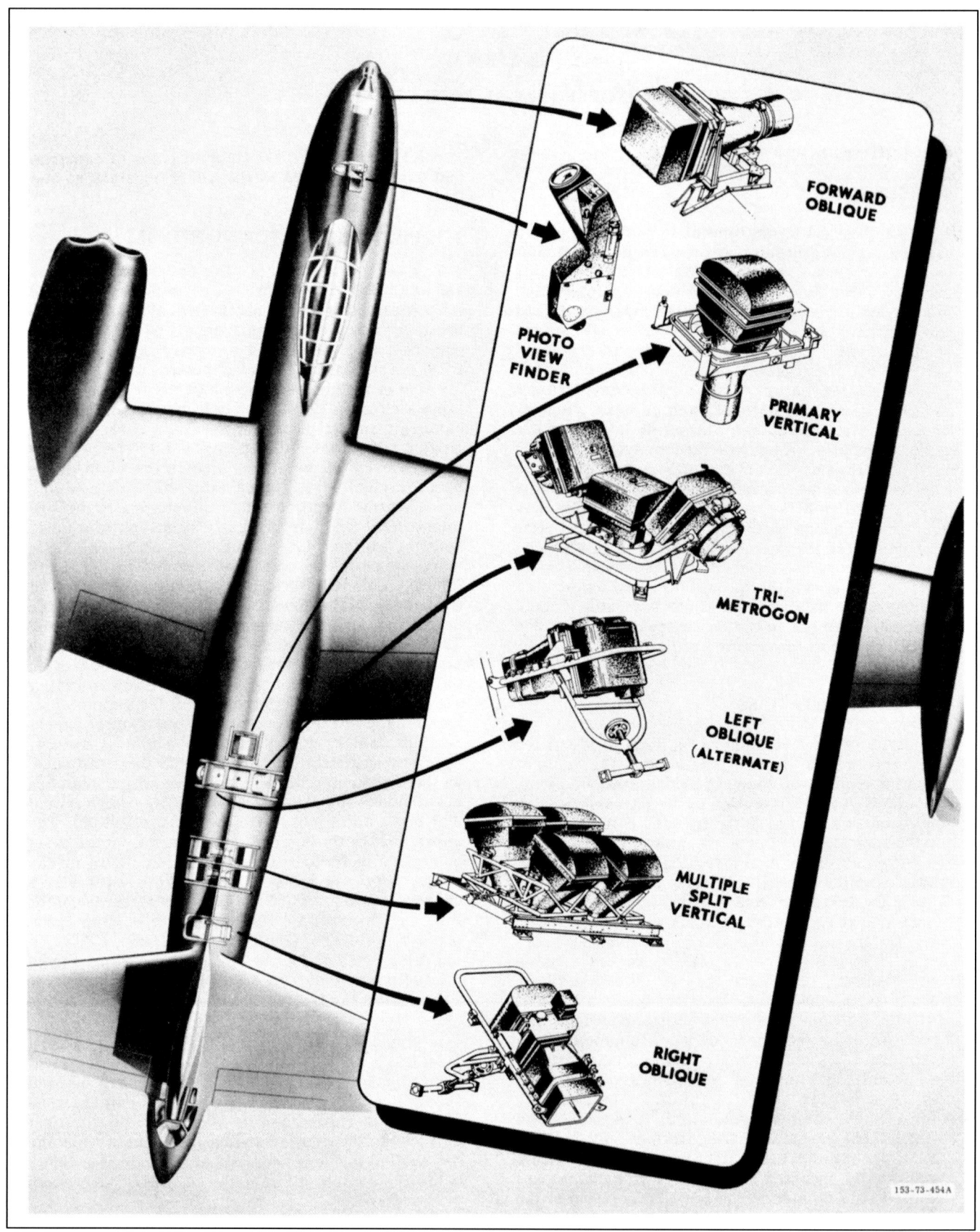

Drawing 9: This drawing depicts the locations of all the major reconnaissance camera locations on the RB-45C from the nose to tail. The first station was the forward oblique which gave the aircraft a still or motion picture image capability against targets directly in front of the aircraft. The next station aft, in the forward section of the aft fuselage just behind the aft bomb bay, was the primary vertical camera station which could use close-up lenses for detailed imagery of targets as the aircraft passed over them at altitude. The next station was the tri-metrogon station which used three cameras to create imagery for detailed and accurate images from horizon to horizon. The next station was an oblique camera location that gave a detailed side views of specific targets as the aircraft flew past. The next station was the multiple split-vert or vertical which could use two to four cameras to create images which over-lapped, and when viewed with special viewing lenses created very detailed stereo images that presented in detail the vertical heights within those images. Highly accurate charts with very precise height data were made for ground forces to plan maneuvers and for aerial forces to accurately conduct air strikes using good height information. (NMUSAF)

STATION		CAMERA TYPE	USE	LENS CONE	MAGAZINE	MOUNT
FORWARD OBLIQUE	ONE	K-22A	Day Reconnaissance	12" or 24"	A-9A ▲	*
	ONE	A-6 35-MM	Aerial Motion Picture	25 mm 50 mm 250 mm	400 Ft Capacity	*
PRIMARY VERTICAL	ONE	T-11	Mapping	6"	Self-contained	A-28 (Stabilized)
	ONE	K-22A	Day Reconnaissance	6", 12" or 24"	A-14 A-9A ▲	A-28 (Stabilized) or A-27A
	ONE	K-17C	Day Reconnaissance	6", 12" or 24"	A-14 A-9A ▲	A-28 (Stabilized) or A-27A
	ONE	K-38	High Altitude Day Reconnaissance	24" or 36"	A-8B	A-28 (Stabilized)
	ONE	K-37	Night Reconnaissance	12"	A-14 A-9A ▲	A-28 (Stabilized) or A-27A
	ONE	S-7A	Low Altitude High Speed Strip	88 mm stereo 7 " stereo 20 " stereo 6 " single	Self-contained	A-28 (Stabilized)
TRI-METROGON	THREE	K-17C	Charting	6"		
LEFT OBLIQUE (ALTERNATE)	ONE	K-22A	Day Reconnaissance	12" or 24"	A-9A ▲	*
	ONE	S-7A	Low Altitude High Speed Strip	88 mm stereo 7 " stereo 20 " stereo 6 " single	A-9A ▲ Self-contained	*
MULTIPLE SPLIT VERTICAL	TWO (forward)	K-38	High Altitude Day Reconnaissance	24"	A-8B	*
	TWO	K-37	Night Reconnaissance	12"	A-14	*
	TWO (aft)	K-38	High Altitude Day Reconnaissance	24"	A-8B	*
RIGHT OBLIQUE	ONE	K-22A	Day Reconnaissance	12" or 24"	A-9A ▲	*
	ONE	S-7A	Low Altitude High Speed Strip	88 mm stereo 7 " stereo 20 " stereo 6 " single	Self-contained	*

*CONTRACTOR FURNISHED

▲ A-5A magazine may be used as a substitute.

Drawing 10: This tech data image illustrates the selection of cameras, lenses, and film magazines available to carry out missions based on detailed mission planning requirements. This diagram makes clear how complicated the planning for a RB-45C mission was, not only for the mission planners but also the ground crews. Correct cameras, mounts, lenses and exposure and timing settings had to be made on the aircraft precisely at each station to insure a successful mission. Lens cone selections made sure the target images would be clear and included as much detail as possible. Magazine and mount selection made sure the sortie would have enough film to cover the required targets. Simple, reconnaissance sorties were not. (NMUSAF)

Photo 49: One of the best-known and informative images of the RB-45C was taken at Barksdale AFB, LA in 1951 of a 91st Strategic Reconnaissance Wing aircraft. It showed the almost universal Cyclops unitary opening in the nose showing the effect of a way too enjoyable party the night before. It was a public affairs photo to show the wide variety of cameras, lenses, and magazines the photo-Tornado could carry. This image will give some indication of the complexity that went into planning, preparing, and carrying out every operational RB-45C mission. (David Menard)

Photo 50: RB-45C 8037 was imaged on the ramp at Long Beach Airport just prior to its turnover to the USAF on 11 Dec 1950. (NMUSAF)

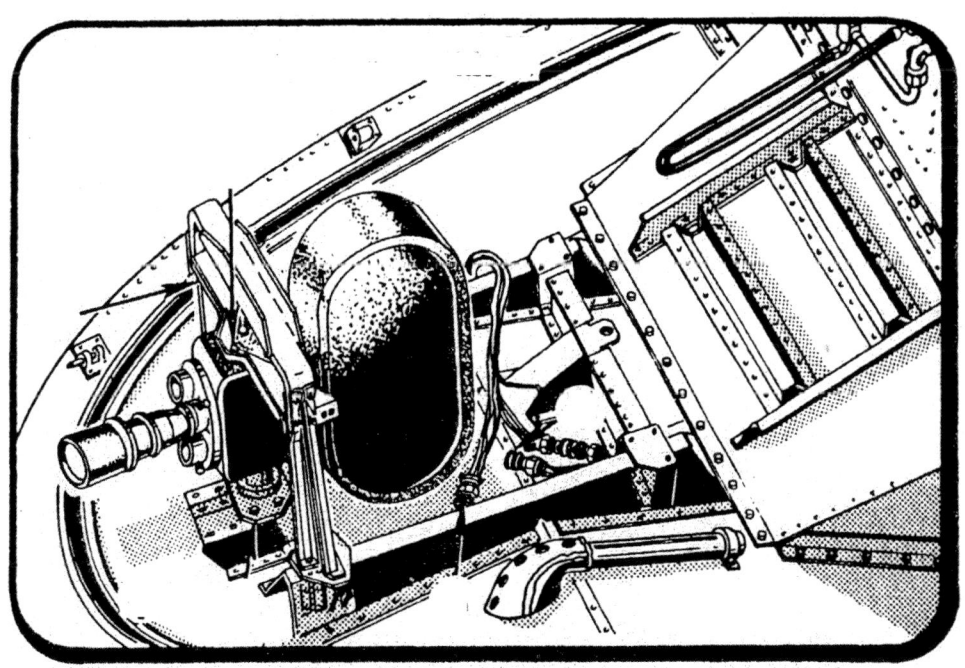

Photo 51: The single most representative feature of the RB-45C was the forward oblique camera position by which this model was universally known. This RB-45 has yet to have its Cyclopes eyeball painted with party eye paint, but it would not be long in coming when the aircraft reached Barksdale AFB, LA. (NMUSAF)

A-6 CAMERA WITH 250 MM LENS CONE
(200-FOOT MAGAZINE)

Drawing 11: This drawing shows the installation of an A-6 motion-picture camera in the forward oblique camera station of an RB-45C. (NMUSAF)

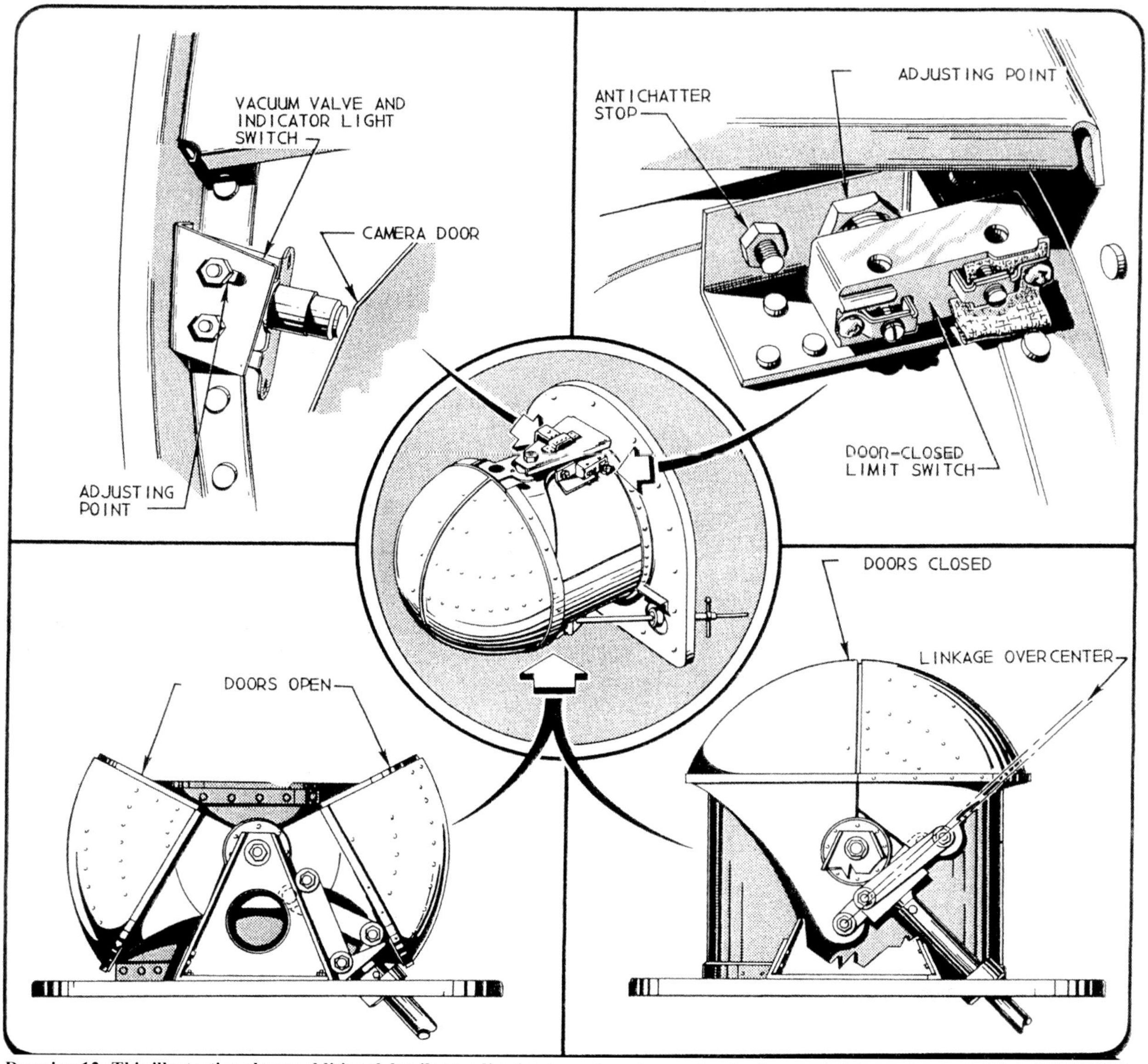

Drawing 12: This illustration shows additional details on adjusting the external cover for the forward oblique camera position. (NMUSAF)

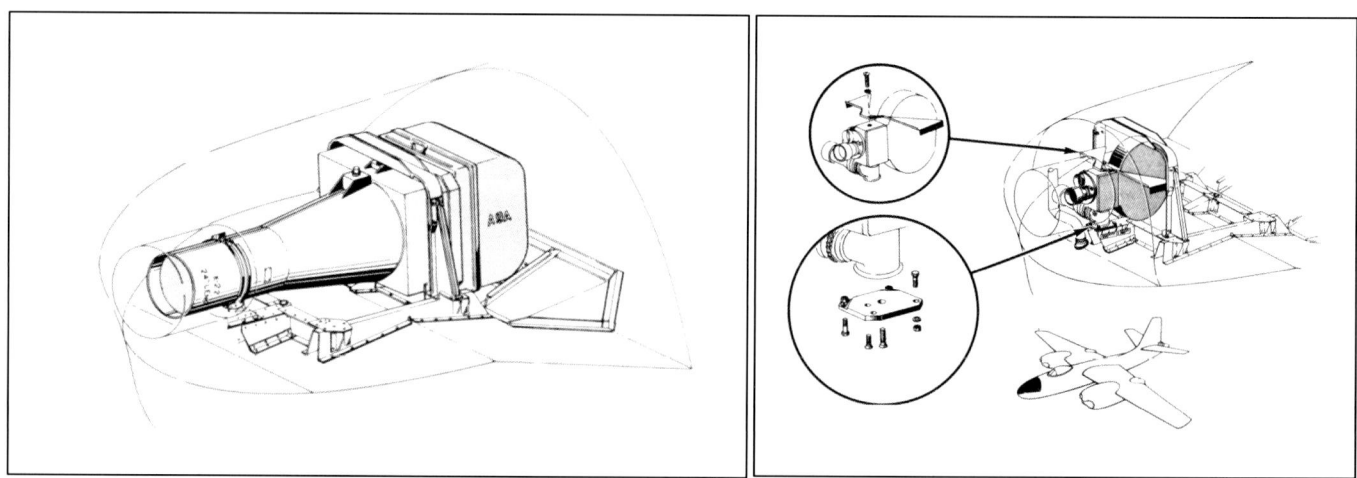

Drawing 13: This drawing shows details of the K-22A day-reconnaissance camera with a 24" lens and cover. (NMUSAF) **Drawing 14:** This drawing shows the details of the A-6 35mm aerial motion picture camera which could also be mounted at the forward oblique station. (NMUSAF)

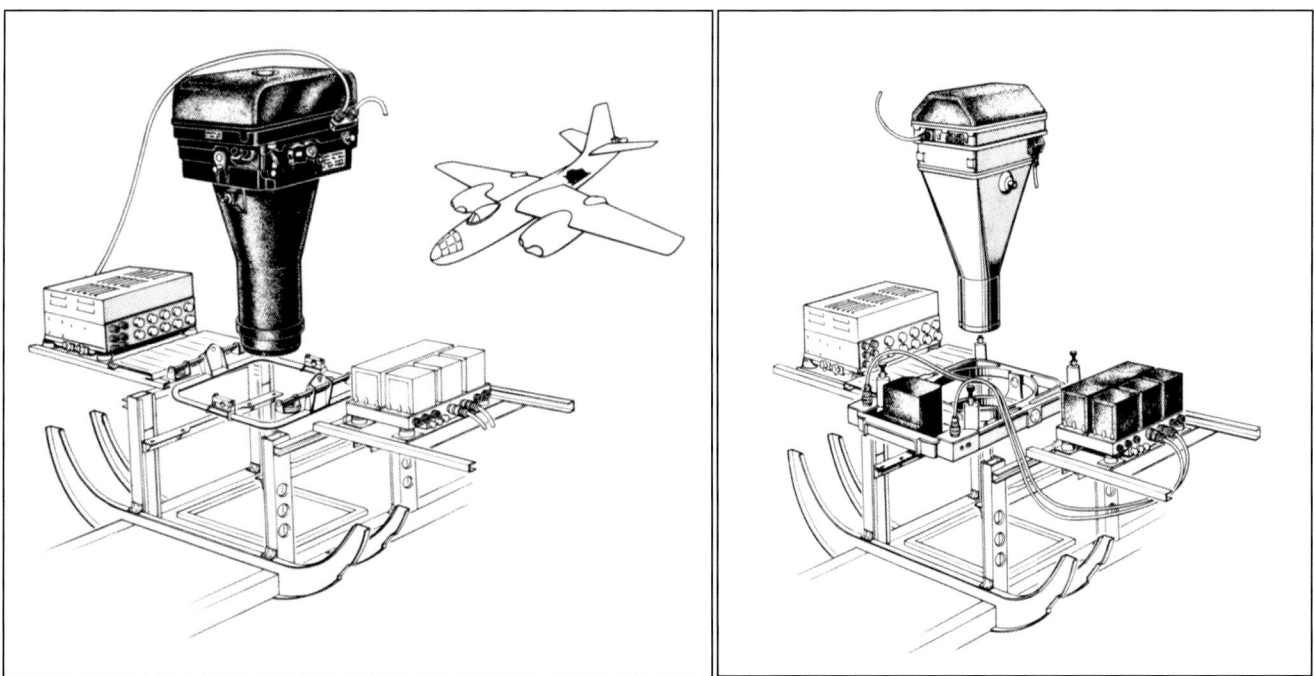

Drawing 15: This is the Vertical Camera position just aft of the wing in the lower aft fuselage with a K-38 high altitude day reconnaissance camera and 24" lens and cover installed. (NMUSAF) **Drawing 16:** This is also the vertical camera station with a smaller K-17 day light reconnaissance in a stabilized mount which minimized blurring of the images from blurring due to turbulence. (NMUSAF)

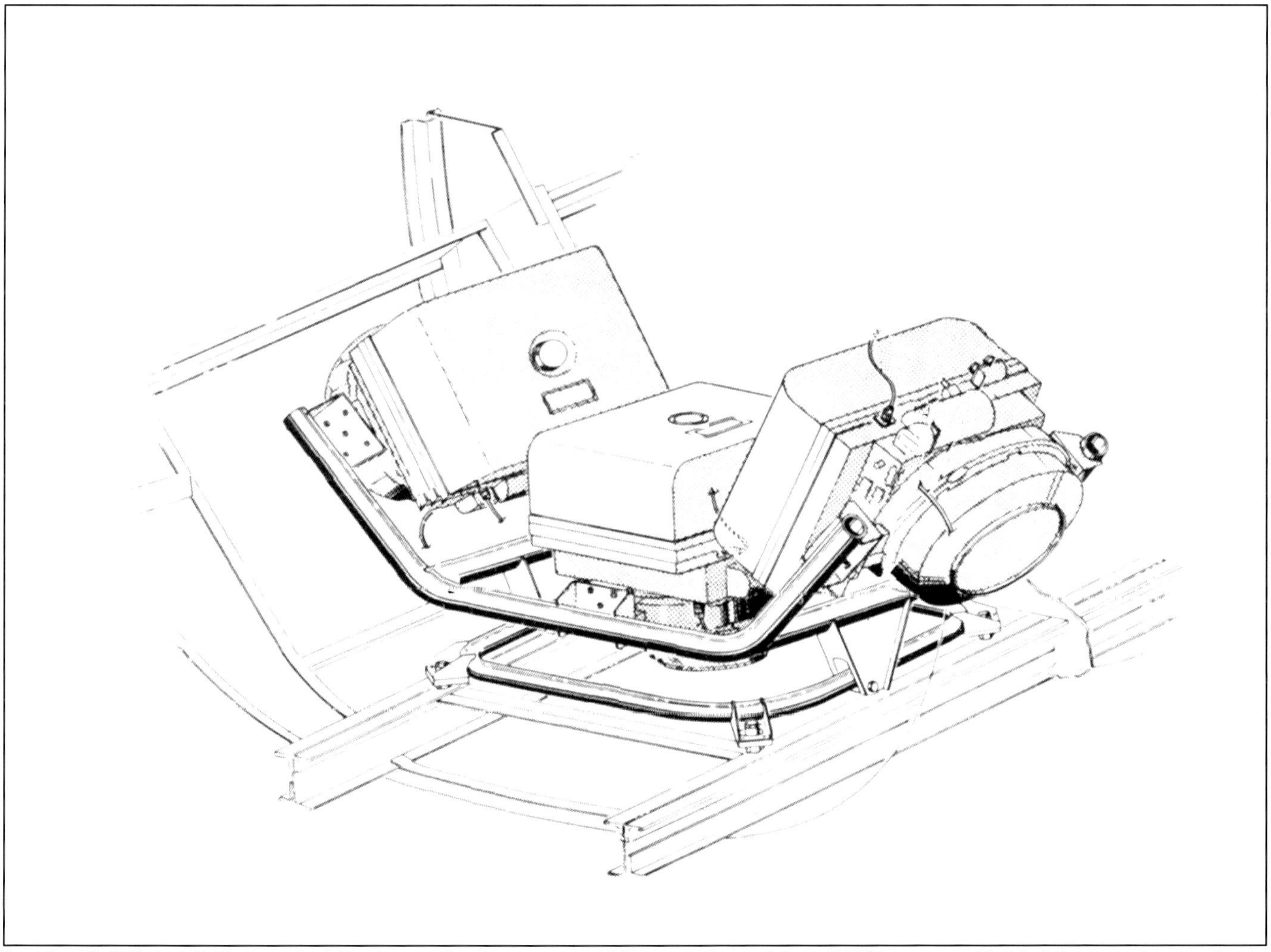

Drawing 17: This installation was the tri-metrogon charting camera which was made up of three K-17C cameras with 6 inch lenses and covers. It gave horizon to horizon imagery coverage from its station just aft of the primary vertical station. Its purpose was to produce high quality, detailed imagery for the creation of very meticulous navigation charts with precise vertical data. (NMUSAF)

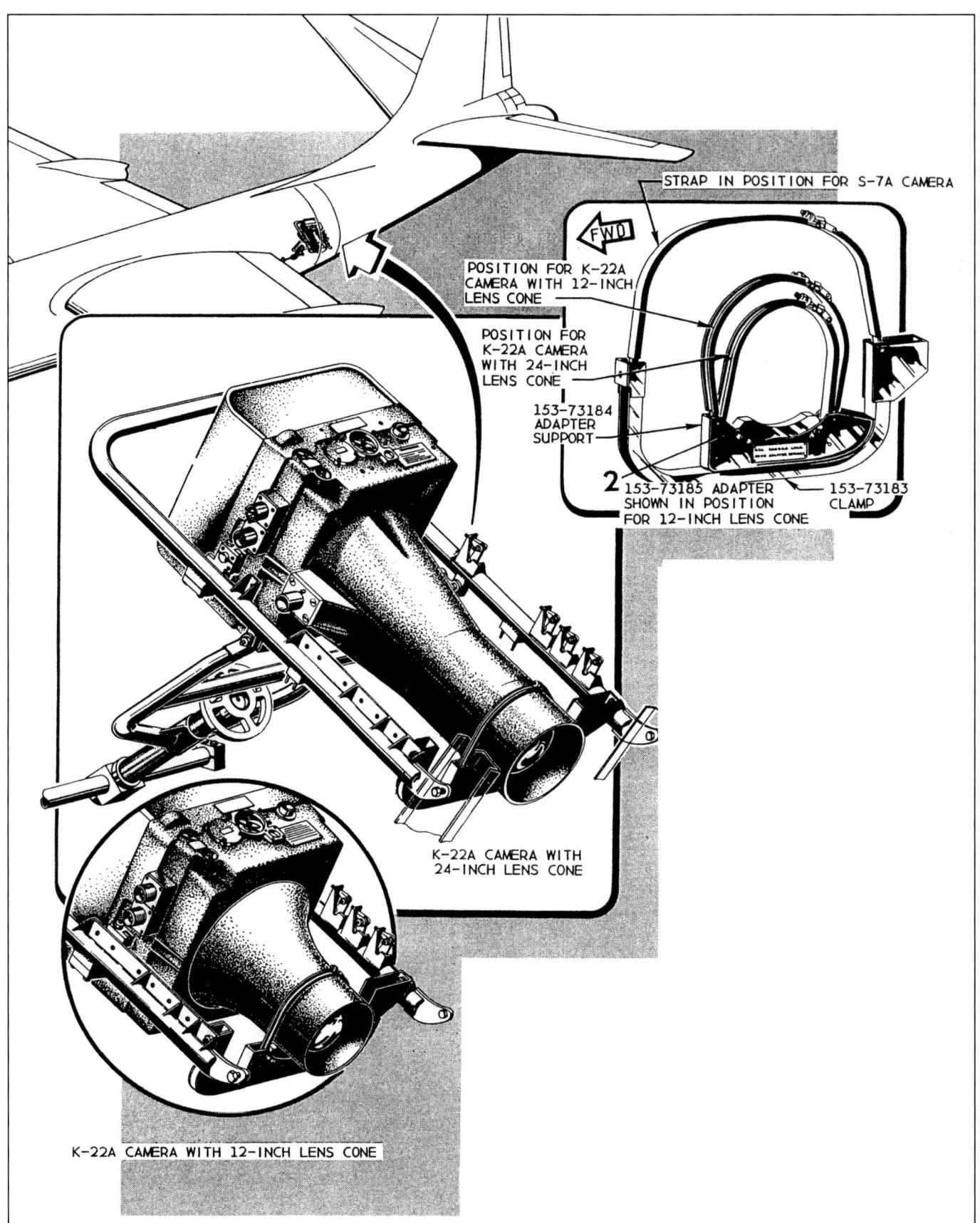

Drawing 18: This drawing shows two of the variety of cameras that could be carried in the left oblique position just above the tri-metrogon station on the RB-45C. This drawing shows a K-22 camera with a 12-inch lens and cone in the closest location and a K-22A camera with a 24-inch lens and cover in the center drawing. The metal straps used to fix the cameras and lenses are shown in the right-hand drawing. The different lenses allowed more or less magnification for each camera on the photo run. No access to the aft fuselage below 10,000-ft Above-Ground-Level (AGL) did limit the amount of film that could be carried to whatever each magazine loaded could hold, establishing the maximum area that could be to be covered on each sortie. (NMUSAF)

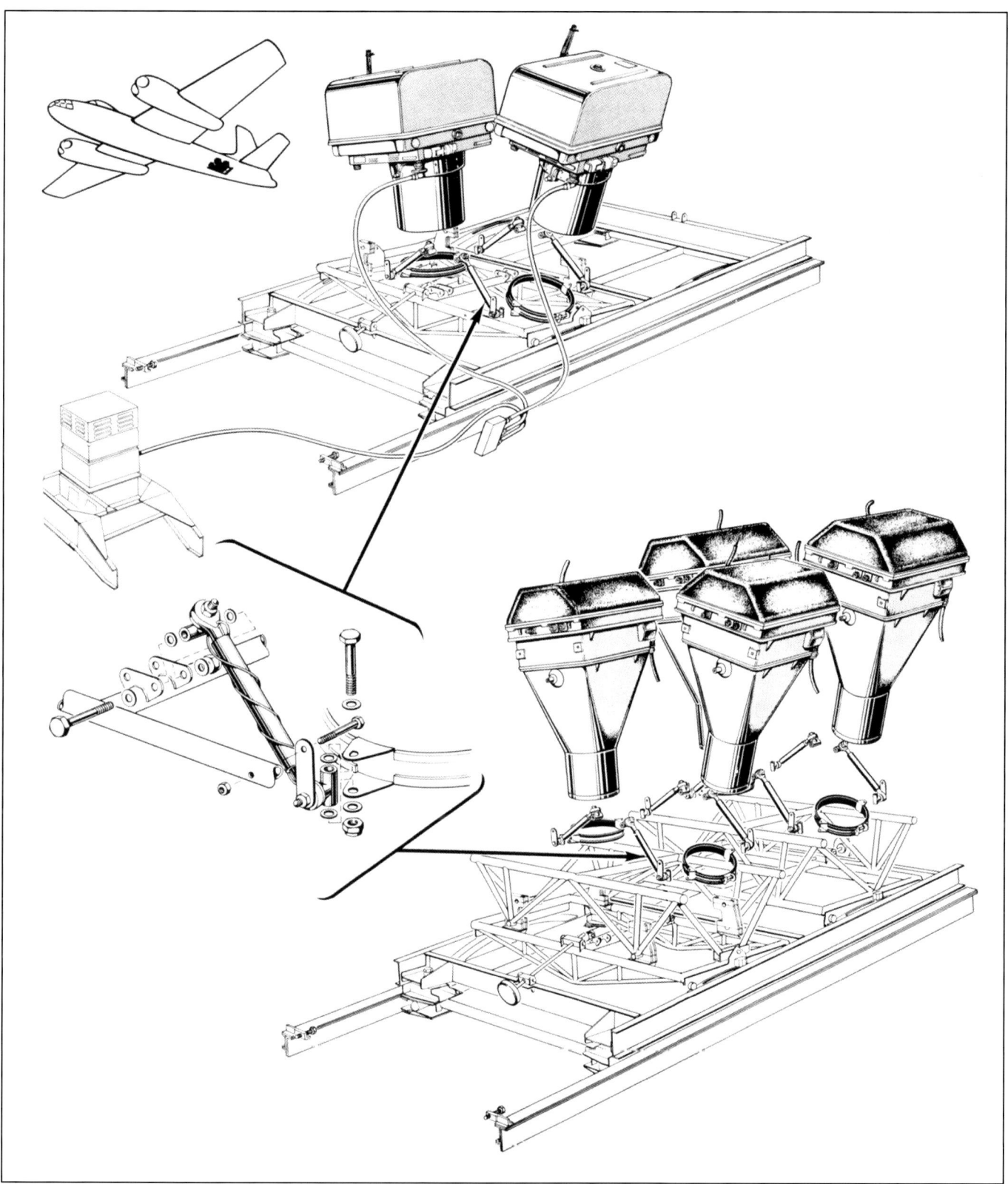

Drawing 19: The RB-45C was well equipped to accomplish both high and low altitude mapping missions. These drawings cover the multiple split-vertical mounts, the upper one for night imagery using two K-37 night reconnaissance cameras fitted with 12-inch lenses and covers. The lower drawing for daylight reconnaissance mounted four K-38 high altitude cameras with 24-inch lenses and covers. Imagery from each pair of cameras partially overlapped, and when viewed with special optical equipment allowed the over-lapping area to produce an image of visual presentation of vertical relief. The different vertical heights in the area could be traced and assigned a specific matching elevation. These specific elevations could then be tracked across and throughout the image. This effort produced an area with a set of specific elevations from lowest to highest across a given area and a chart with accurate elevation information over that area was produced. Note the amount of structure required to support the cameras and lenses along with the power and vacuum supplies.

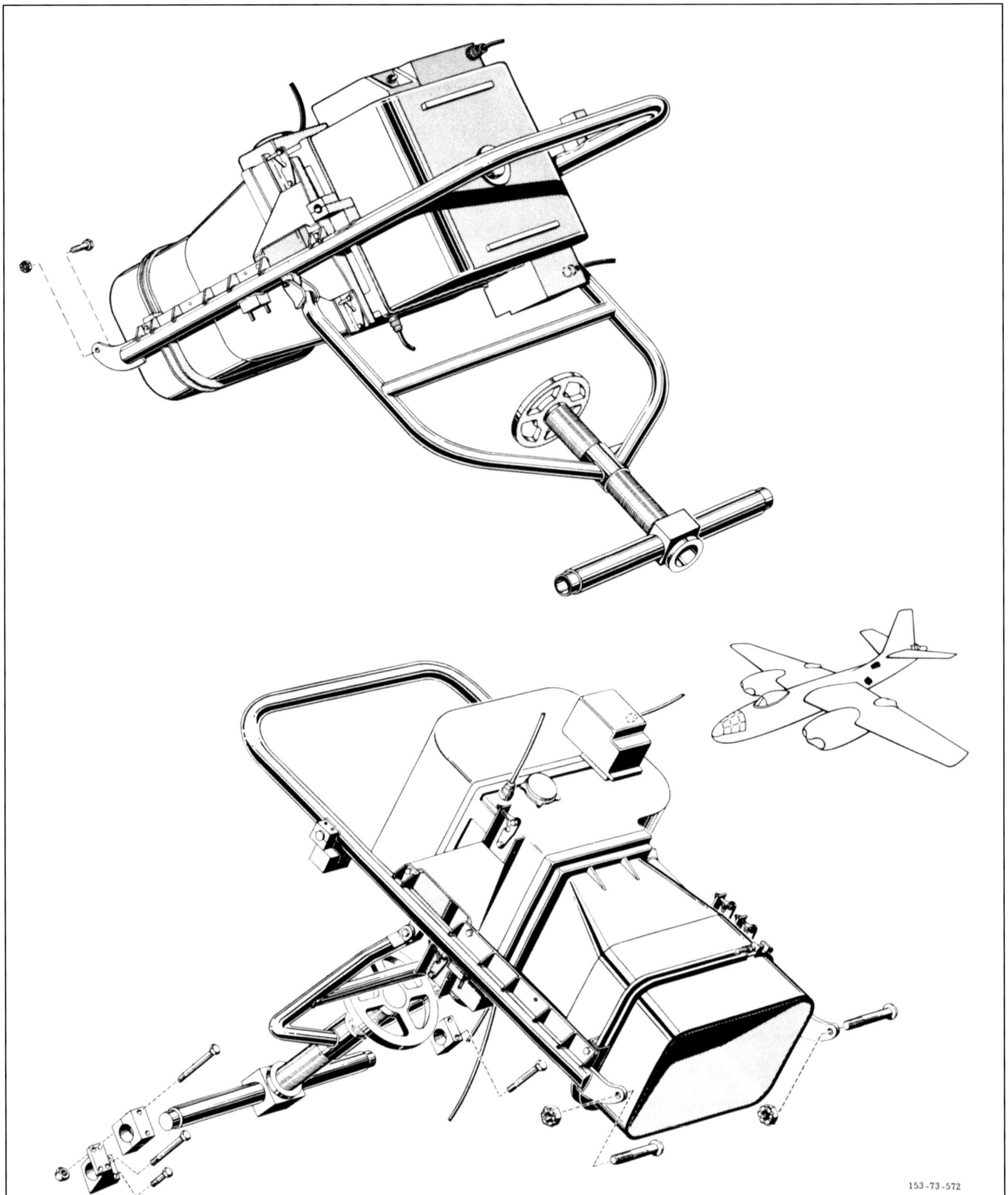

Drawing 20: This drawing covers both the left and right oblique cameras, in this illustration both fitted with S-7A low altitude high speed strip cameras capable of exposing film at low altitude over 500 miles per hour. Their purpose was to capture enemy combat forces in fast, low altitude sorties that reduced the effectiveness of anti-aircraft artillery defenses.

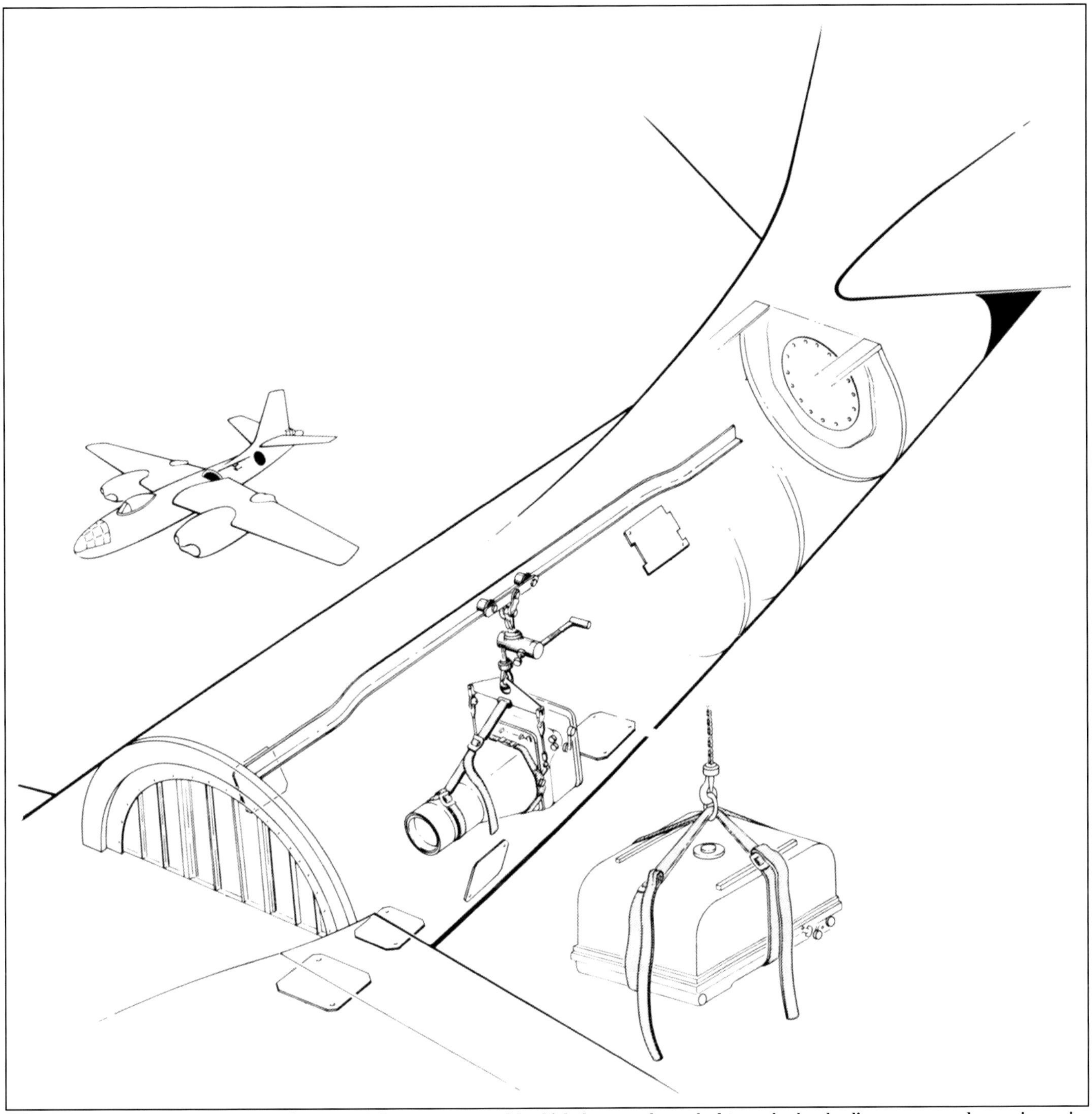

Drawing 21: Given the weight of the equipment and the narrow area with which the ground crew had to work when loading cameras and magazines prior to mission launch, an overhead rail system was installed to move cameras, lenses, and magazines fore and aft on the rear fuselage. The drawing shows all six of the camera hatches which were covered by doors for protection of the lenses and cameras. If required, pressure bulkheads at the aft end of the bomb bay and just in from of the gunner's station area could be opened and resealed if required. (NMUSAF)

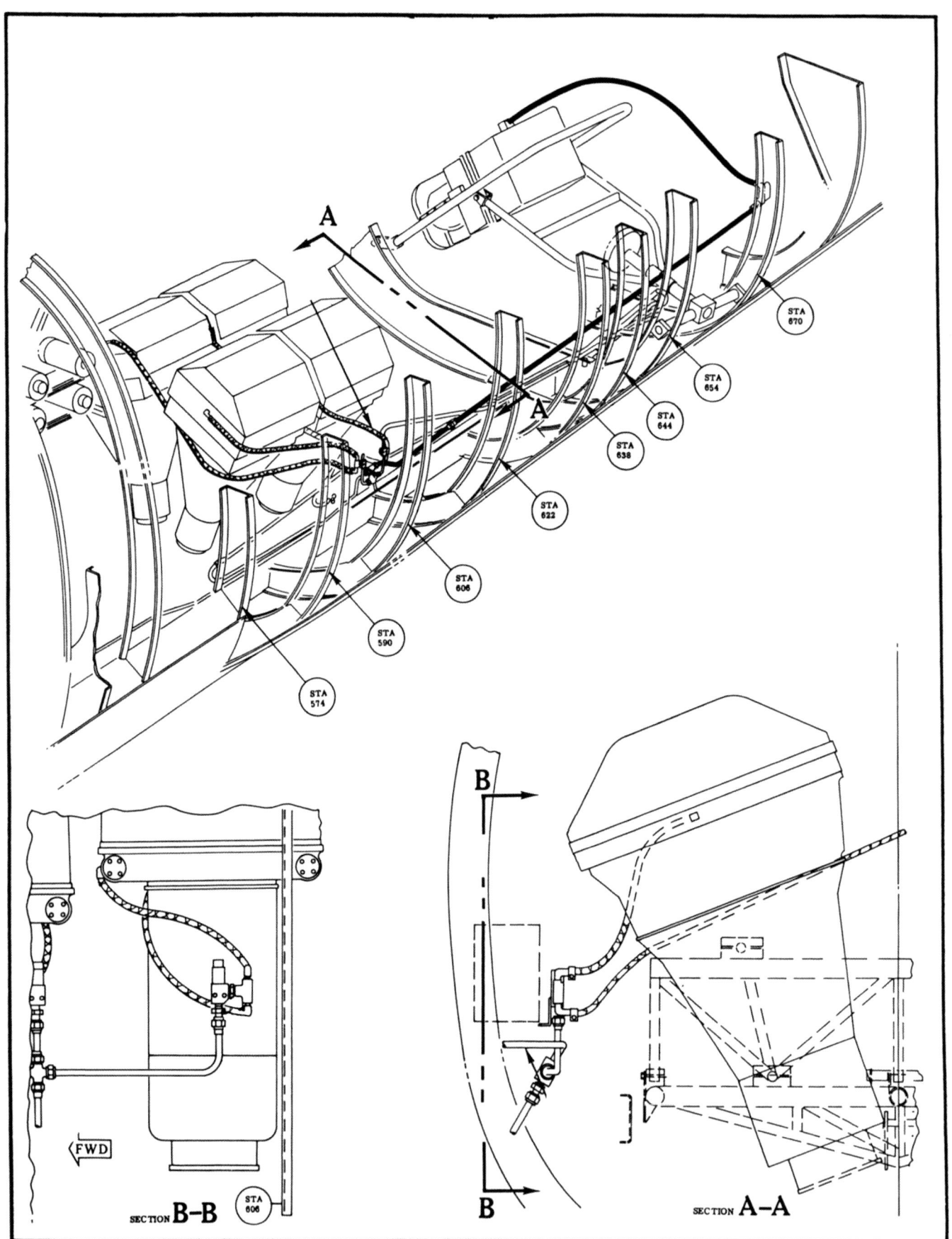

Drawing 22: This drawing gives some idea of the amount of structure in the aft fuselage which supported the camera stations and how complicated the routing of the vacuum tubing system was in the area. The right oblique camera was at the upper center of the drawing while the four-camera split vertical mount was at the lower left-hand center. The vacuum lines were shown in lined tubing. (NMUSAF)

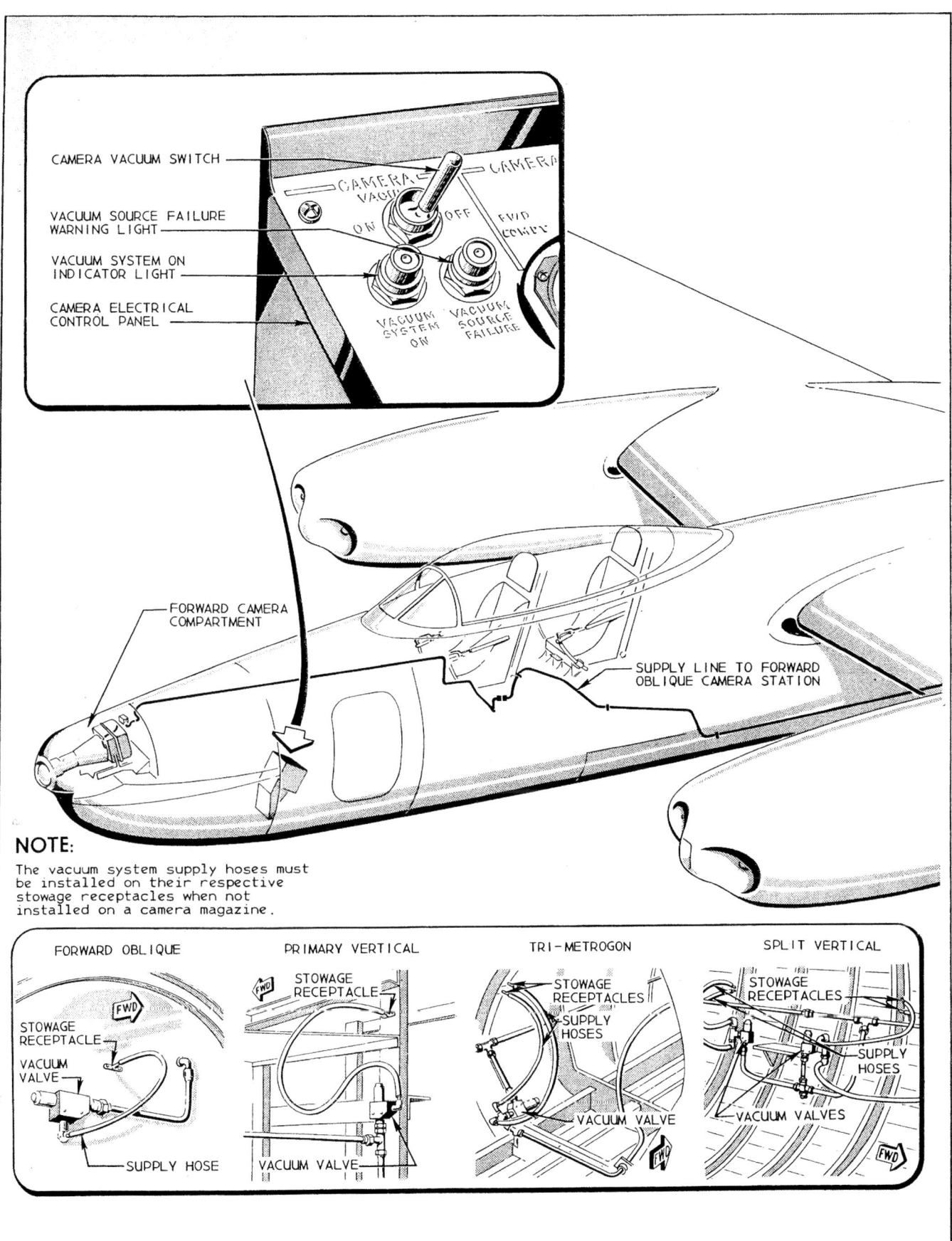

Drawing 23: This drawing is sheet one of the layout and controls for the aircraft vacuum system which insured the film was well held to the lenses during exposure. The controls for the system were operated by the photo-navigator and located next to the photo viewfinder. The details involved in installing these details were one of the reasons that the RB-45C Initial Operational Capability was so far beyond its final assembly dates. (NMUSAF)

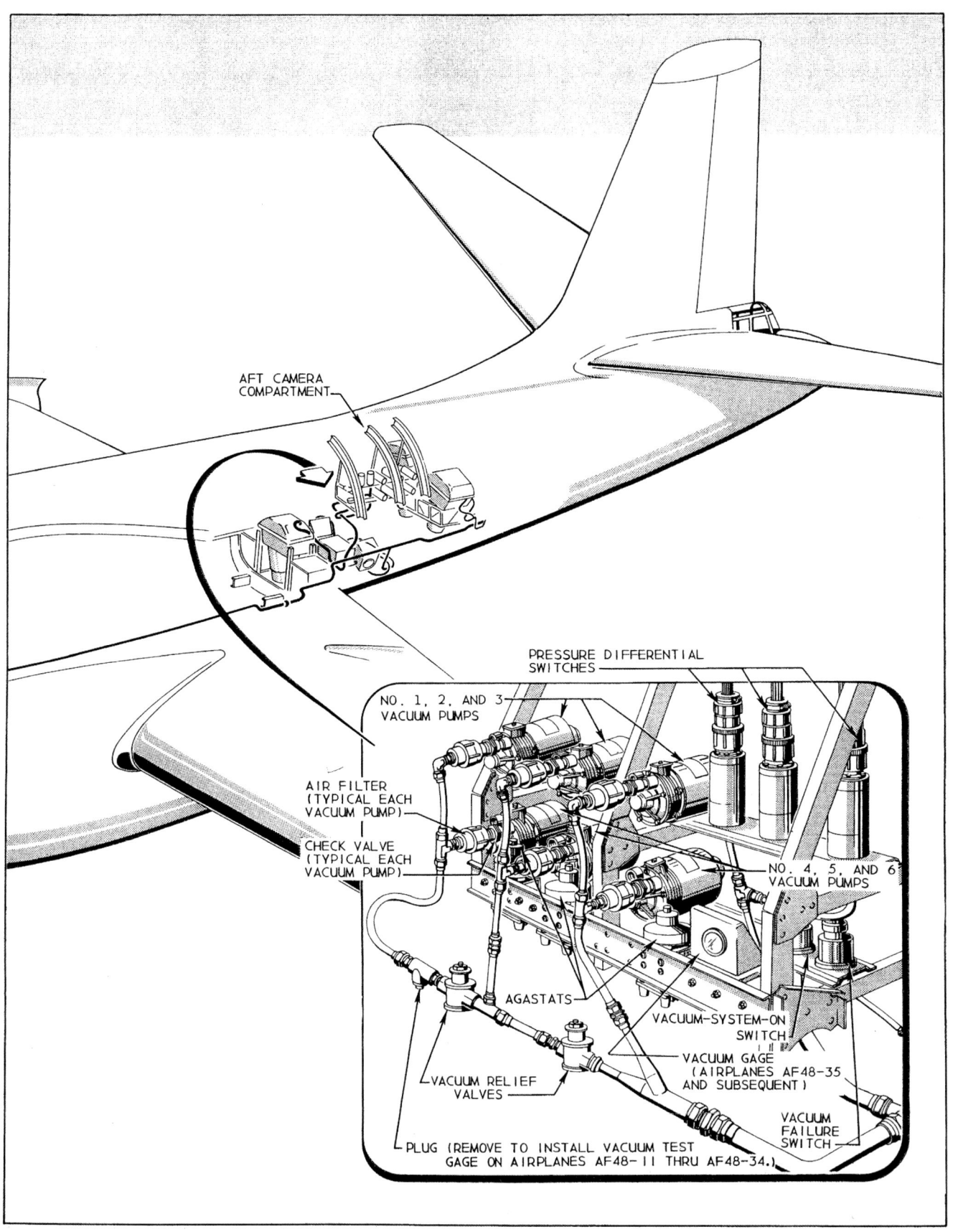

Drawing 24: Sheet 2 of the camera vacuum system shows the extent of the vacuum pumps in the aft fuselage which maintained the vacuum suction on all eight camera stations in the aircraft. (NMUSAF)

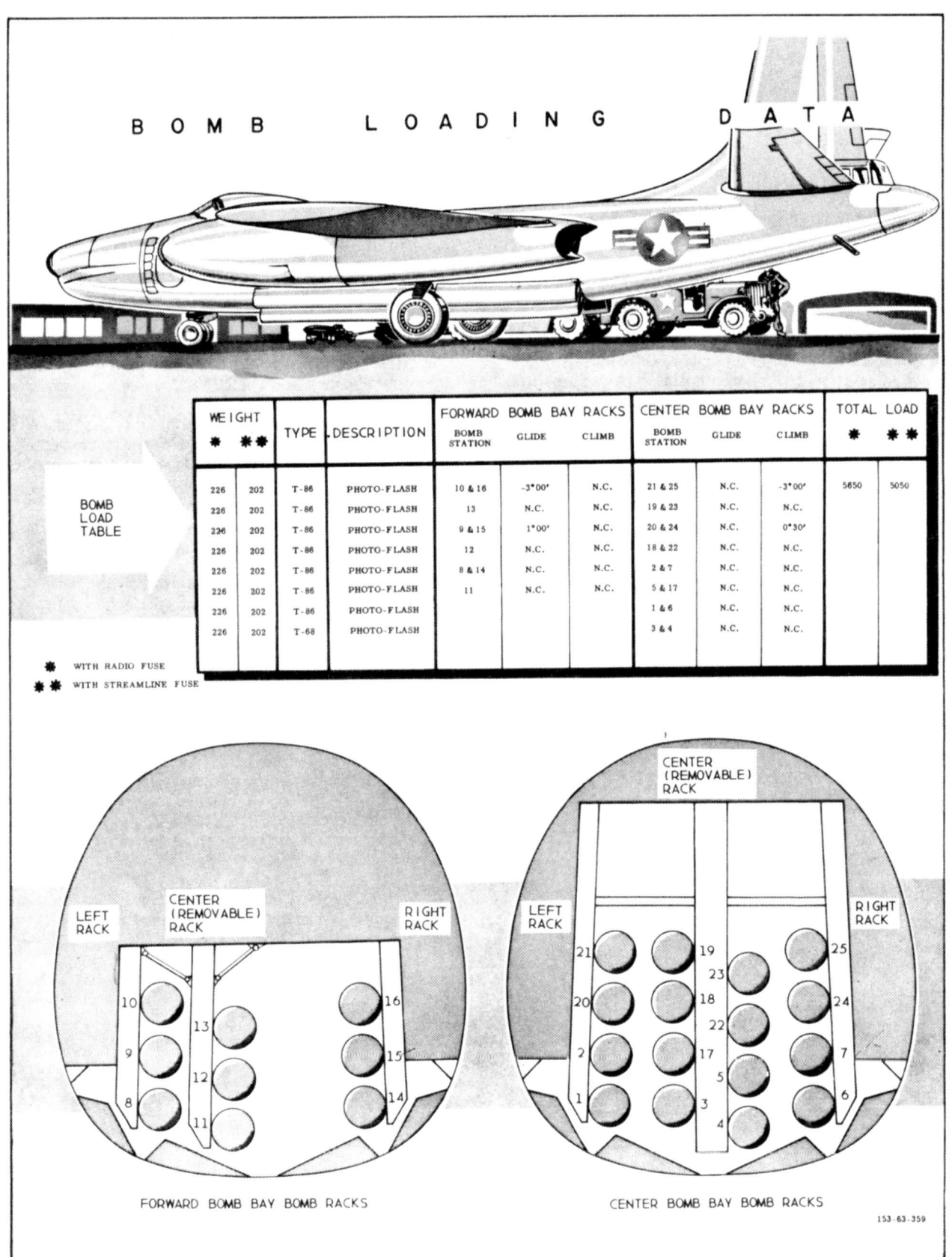

Drawing 25: The RB-45C tech data carried specific guidance like its bomb carrying cousins on how to carry the T-86 and T-68 photo-flash weapons. 25 weapons could be carried in the forward and center bomb bay racks. Like the explosives carried by the B-45As, they were released in a special order to maintain weight and balance limits and to ensure they went off to provide illumination for the exposures. (NMUSAF)

Photo 52: The T-86 and T-68 photo flash bombs presented some of the very few issues with the RB-45C imagery system. Soon after they were in regular use, significant vibration was identified when the bomb doors were opened at high altitude and high speed. The vibration was so bad that imagery was distorted and in some cases, unusable. (NMUSAF)

Photo 53: Operational testing of the RB-45C was conducted at Edwards AFB beginning in May of 1950. The USAF crew in this picture was Major T.B. McGuire, Major C.H. Barnett, and Capt. Charles (Chuck) Yeager. The NRB-45C which flew the missions was named *Evil Eye Fleagle*, an extraordinary character with amazing optical skills from Al Capp's Comic Strip, Lil' Abner. (NMUSAF)

Photo 54: This image was looking southwest over Long Beach harbor area and San Pedro. It was taken from 30,000 ft. and 500 miles per hour in June 1950 by *Evil Eye Fleagle*. (NMUSAF) **Photo 55:** This image looked due south from just north of what was building to become Los Angeles International Airport and the former Hollywood Park Racetrack directly towards Torrance and the Peninsula of Rancho Palos Verdes. (NUSAF)

Photo 56: This image, taken at the North American Long Beach plant, shows an RB-45C preparing for pre-acceptance engine run ups prior to the delivery to and acceptance by the USAF in 1951. (Steve Ginter)

Photo 57: RB-45C 8024 was photographed during a June 1950 pre-acceptance flight from the North American Long Beach plant. (Steve Ginter)

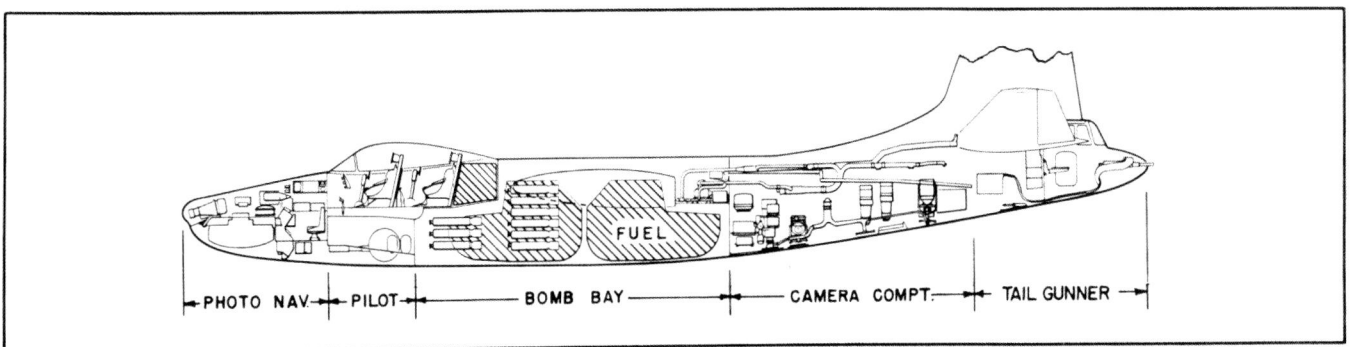

Drawing 26: The inboard profile of the RB-45C aircraft.

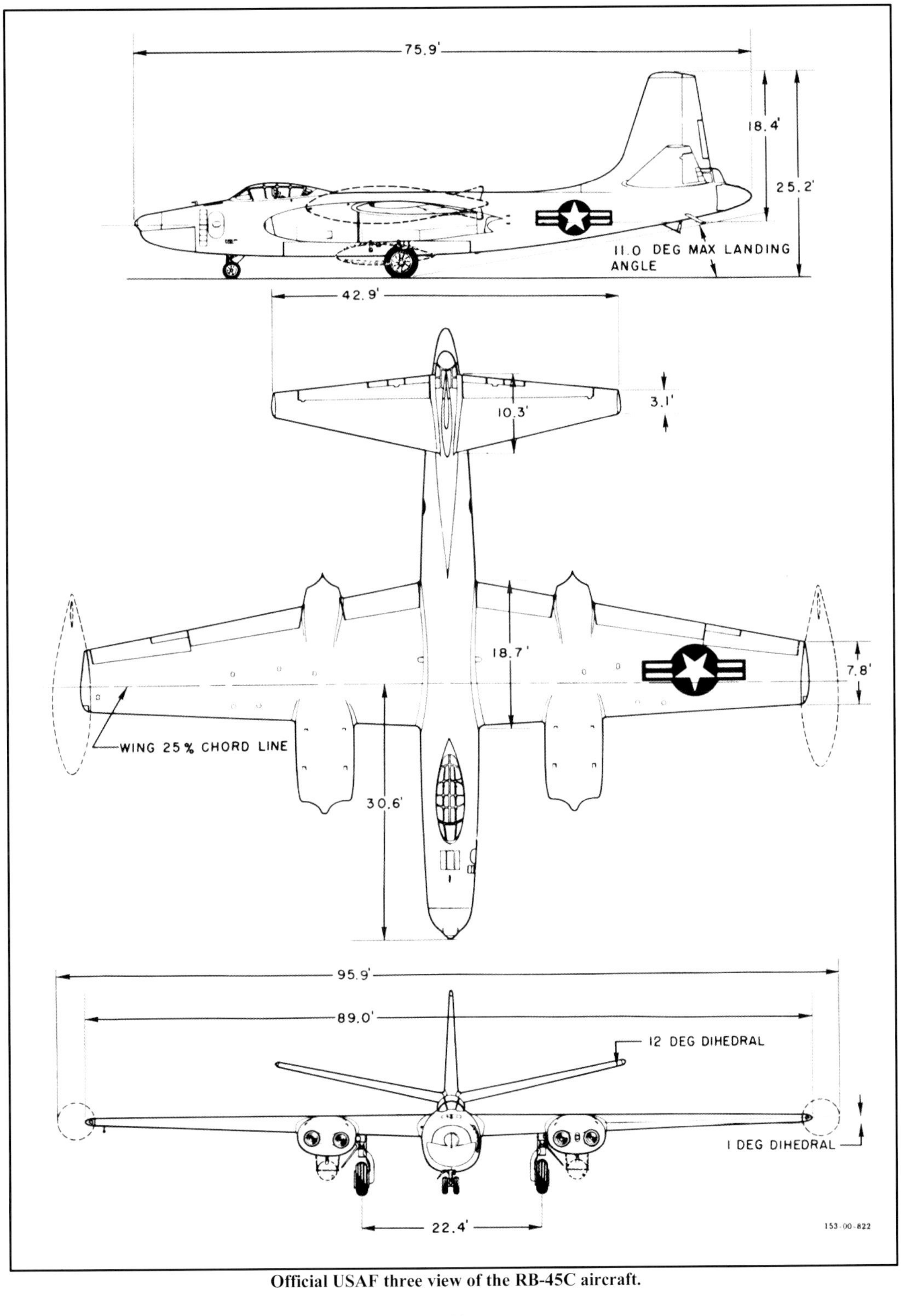

Official USAF three view of the RB-45C aircraft.

CHAPTER 6 – THE B-45A IN SERVICE

Part 1: The B-45A-1 enters service....

The first 22 B-45A-1s powered by GE J35 turbojets jet found scattered homes as they began to arrive in service beginning in April 1948. Many were delivered directly to San Bernardino AFB, CA, and the San Bernardino Air Materiel Area (SBAMA) to start setting up and organizing the depot that would service the fleet for the next ten years or more. Much of this work was orientation of personnel on the very new and unique operation of turbojet powered aircraft. The next step for the depot was to begin installation of communications and navigation electronics supplied by the military as Government Furnished Equipment (GFE) that had not been installed during final assembly of the B-45A-1s.

The next group of dedicated aircraft was intended to be delivered to the operational aircrew training contingent at Muroc AFB, CA. This did not occur and most of the B-45As at Muroc were sent to Wright-Patterson AFB, OH to be modified as additional operational test aircraft. The next training location was to be Biggs AFB near El Paso, TX, but the high field elevation, high ambient temperatures, and relatively short runway made Biggs unsuitable. Instead, the Air Staff decided to select Barksdale AFB, LA as the training base, as well as the home for the only USAF B-45A operational bombardment group to be formed. Seven B-45A-1s assigned to the 47th Bombardment Group (L) for Light based on range, of the Continental Air Command (CAC) and their subordinate combat units, the 84th and 85th Bombardment Squadrons (L), were now part of CAC's Tactical Air Command. They began arriving at Barksdale from SBAMA in April 1949. While Barksdale was not a perfect location, with its high ambient temperature in the summer, it had little or no high terrain in the area and a runway almost 10,000-feet long, every inch of which would often be used on take-off.

Photo 1: B-45A-1 7022, the last B-45A-1 was photographed at Norton AFB, San Bernardino Air Material Area, just prior to its delivery to the 47th Bombardment Group (L) at Barksdale AFB, LA on 16 August 1949. (The Gerald Balzer Collection via Tony Landis from HQ AFMC/HO)

The time at Barksdale from March 1949 through October 1949 was difficult at best. Lack of sufficient supplies, ground support equipment such as auxiliary power units for engine starting, cranes for engine changes, and trained personnel, aggravated a situation that all but precluded any form of operational capability. But somehow flying was accomplished, and the B-45A-1 was slowly becoming operational. This portion of the story would continue with new B-45A-5s that began arriving just before a planned move of the unit to Langley AFB, VA in November of 1949. Perhaps, in an unintentional way, HQ USAF made a decision that made the confusion about the B-45's introduction absolutely unavoidable. The new unit at Langley would be created from the 363rd Tactical Reconnaissance Group, with one Tac Recon squadron flying piston-powered Douglas RB-26s. Then two turbojet Bombardment units, the 84th and 85th Bombardment Squadrons were assigned. Finally, the 47th Bombardment Group would be disestablished.

In October 1949, the first former Continental Air Command B-45A-1 was transferred to the Strategic Air Command's 91st Strategic Reconnaissance Wing at Barksdale AFB, LA for air crew training in advance of the expected arrival of RB-45Cs by late 1950. Flight training was also to be accomplished at other locations to meet the needs of other Major Commands than SAC. One such were three B-45A-1s assigned to the 363rd Tactical Air Reconnaissance Group at Langley, apparently in anticipation of RB-45Cs being assigned there in the future as well. The Air Force Special Weapons Center also received its first turbojet trainer in April 1952.

The last B-45A-1s assigned in mass were nineteen aircraft assigned to the 2600th Target Tow Squadron (TTSq) at Biggs AFB, TX; the 2750th TTSq at Yuma AFB, AZ; the 1st TTSq at Biggs AFB, TX; the 2nd TTSq at Castle AFB, CA; the 4th TTSq at George AFB, CA; and the 17th TTSq at Yuma AFB, AZ. Most of these aircraft were not equipped with external or retractable tow targets, but served instead as high altitude targets for US Army Automatic Anti-Aircraft Radar and Long-Range Early Warning Radar targets throughout the United States as well as the Royal Canadian Air Force (RCAF) in Canada.

Photo 2: EB-45A-1, 7007, assigned to the 1st TTSq was imaged during a temporary duty assignment to General Mitchel Field in New York City in December 1949 for high altitude target work against U.S. Army radar units along the U.S. east coast. (Gerald Balogh Collection via David Menard)

Testing....

A significant number of other B-45A-1 transfers were made to WPAFB for modifications to carry test and evaluation equipment along with recording gear for the hundreds of flights that awaited the Operational Test and Evaluation (OT & E) fleet at bases such as Wright-Patterson, Muroc, and Eglin, and at NAA at Inglewood. Ten B-45A-1s were converted into EB-45A-1s by Wright-Patterson and NAA for testing work and used at Wright-Patterson, North American at Inglewood, Muroc, Griffiss, and Eglin.

Four more aircraft were converted in July 1950 into EDB-45A-1 drone controllers at Wright-Patterson and used at Holloman AFB, NM by the 2754th Experimental Wing from July 1950, and by the ARDC at Patrick AFB, FL from August 1952 through February 1953. One EDB-45A-1 was bailed to Northrop at Hawthorne, CA were it was lost in May 1952. NACA at Langley AFB lost B-45A 7021 near Langley AFB in August 1952. The Rome Air Development Center at Griffiss AFB, NY received its first EB-45A-1 in September 1952 and one was ultimately transferred to the U.S. Navy and flown as a drone controller at China Lake Naval Air Station and North Base at EAFB.

Photo 3: B-45A-1 7015, assigned to the 3200nd PTG at Eglin AFB, FL conducted B-45A environmental tests in extreme climates from Florida in the summer to Canada and Ladd AFB, AK just south of Fairbanks between December 1948 and March 1949. (The Gerald Balzer Collection via Tony Landis from HQ AFMC/HO)

Specific testing projects continued to qualify the B-45A to carry and use the Aerojet **XLR13-AJ-1** liquid fuel rocket powered ATO system. The effort confirmed that the B-45A-1 and -5 design and construction made no changes that disqualified the use of the units on the A- model fleet. Nonetheless, transfer of these liquid-fueled rocket-powered boosters to operational units and locations, never occurred due to overwhelming hazards in their use and handling.

The new B-45A's performance, even quipped with the GE J35, was a level of improvement never seen in the industry. For the first-time, sustained speeds close to 75% of the speed of sound at altitudes nearly doubling those obtainable during the war were possible. Now technicians, engineers, developers, and production outlets had the actual means to research, test and produce flyable examples of concepts, systems and aircraft considered near fantasies just a few months in the past. Among the first elements of the aeronautics family to begin to push the envelope hard were those working on new electronically guided air-to-air weapons. Hughes Aircraft Company based at Culver City, CA won a six-contender contract to develop "some type of air-to-air guided missile". After a long, troubled and at times far less than successful period, the effort finally resulted in the development of the Convair F-102A, Hughes GAR-1/2/3 and 4 interceptor missiles, MA-1 and later MC-3 fire control system and antenna. The "Deuce" as it became known, led to the best interceptor fielded by the USAF during the Cold War, the Convair F-106 Delta Dart.

Photo 4: B-45A-1 7014, bailed to the Hughes Aircraft Company based at Culver City, CA, was the airborne platform which flight-tested a number of interceptor-carrying radar units. These antennas were housed in the circular, bulbous compartment in the nose of the aircraft with the controls and scopes housed in the bomb/nav compartment, and electronics in the aft fuselage. These efforts led to the target-acquisition and missile and rocket fire-control radars used on the F-89 Scorpion, the F-94C Starfire, and F-102 Delta Dart among many others. (Craig Kaston via Steve Ginter)

The next area of development and testing significantly supported by the B-45A was the creation of an American family of operational long-range, heavily armed, and atomic capable drones or self-controlled and navigated unmanned aircraft. This project, the MX-775 began in March of 1946 and had the overall goal of developing an unmanned, long range, air breathing guided missile for the U.S. Air Force. Three broad development phases comprised the engineering effort of the weapon system: (1) a guidance system capable of programming automatic missile flight along a prescribed trajectory; (2) an atomic armed missile airframe suitable for long-range intercontinental strategic air operations; and (3) related ground support equipment.

Northrop soon realized that the airframe and ground support equipment were within the range of reasonable possibility. But the development of an un-manned automatic star-tracking navigation system capable of guiding a strategic missile with a 5,000-pound atomic warhead to a target over 5,000 miles away seemed like science fiction. Over the next two years a concept known as the "time-specified trajectory" guidance method was developed and proven by Northrop. With the flight plan completely specified, the bulk of the computation would be done in advance by ground-based computers. Onboard equipment was thus kept to a minimum primarily that which was needed to transmit the computer-based guidance commands to the missile guidance system and carry them out. By 1953 the program had been named *SNARK* by Jack Northrop from Lewis Carroll's fictional tale of a snake/shark creature.

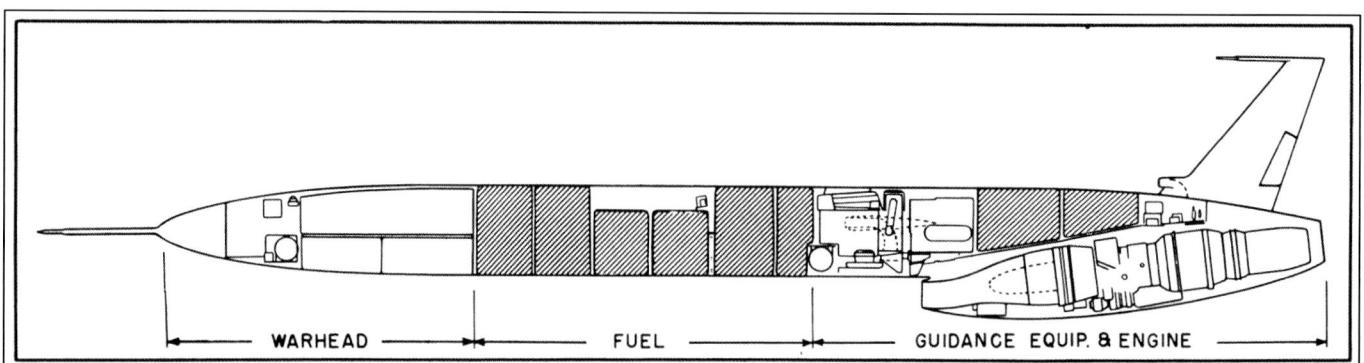

Drawing 1: An inboard profile of the Northrop SM-62 shows the Pratt & Whiney J57 main powerpalnt in the aft fuselage with fuel tanks occupying the aft two-thirds of the airframe. Take-off and initial flight was boosted by two 130,000 pound short duration solid fuel rocket boosters. The missile was armed with a single W39 3.8 megaton thermo-nuclear "hydrogen" weapon. (NMUSAF)

The first aerial prototypes were known as N-25s, a smaller, shorter range version of the airframe which later became the *SNARK*. They flew from 1949 through the summer of 1952 at Holloman AFB, NM. The next airworthy airframe was the Northrop N-69 with nearly the same dimensions and performance as the *SNARK*. The first flight of the N-69 SNARK was on 6 August 1953, followed by the first warhead release and drop on 26 September 1955. The first missile flight entirely under automatic guidance and control was on 13 September 1956; and the first long duration missile flight using daytime star tracking was in 14 Nunemebr 1956.

The B-45A joined the program in 1950 facing two major test objectives. First, two B-45A-1s (after one air accident loss) **7002** and **7008** were modified to hold navigation autopilots that could read the paper tape printouts from the ground-based *SNARK* navigation computers that created the flight plans and remotely fly the EB-45As. These flight plans were designed to be read and followed by the Northrop N-25 test versions of the

SNARK. The first objective was to ensure that the ground computers could produce readable flight plan tapes, and that the autopilots on the N-25 could read and accurately follow them.

Photo 5: An operational prototype of the SM-62 *SNARK* was imaged in 1958 at Patrick AFB, FL. The *SNARK* had an operational range of 5,500 miles and a circular error probable of 8,000 feet for its warhead. The solid rocket boosters were mounted one on each side of the fuselage beneath the wings, and a single JP-4 fuel tank for the J57 was mounted relatively close to the fuselage under each wing. The *SNARK*'s maximum cruise speed was 650 mph, and its maximum service ceiling was over 50,000 feet. (DoD) Photo 6: EDB-45A-1 7008 entered the *SNARK* test program on 10 June 1951 after its conversion at Wright-Patterson AFB. It served the program the longest until 10 February 1959. It was photographed here at the Castle Air Museum after its recovery from the China Lake Naval Air Station and its beautiful and accurate restoration by the Museum. Note the late presentation of the aircraft's tail number, 0-70008. The leading 0 before the dash stood for the fact that the aircraft was older than ten years. (Steve Ginter Collection)

Using guidance control hardware from the N-25 prototypes, the EB-45As loaded the flight plans and ensured its autopilots could read the data from the N-25 equipment onboard and then fly the routes as planned. The flight plans covered routes between Los Angeles, CA and Shreveport, LA and were flown at all times of the day and night. The routes covered approximately 1,400 miles one way and the aircraft would refuel at Barksdale AFB, and return to Los Angeles immediately.

After these initial tests were completed, the EB-45As were converted into drone control aircraft by the 2750th Air Base Wing at Wright-Patterson and Northrop Aviation at Hawthorne, Ca in mid-1951. Now the aircraft could take over direct control of the N-25, N-69 or SM-62 aircraft if needed in an emergency. At this point both EDB-45As were transferred to Holloman AFB, NM to load and fly plans involving more maneuverability, airspeed, and altitude changes, and monitor the performance of the missile autopilots. The second objective was now primary, to evaluate the ability of the N-69 and SM-62 navigation systems to operate accurately over longer distances, over all types of terrain and ensure the automatic guidance and control system would function at any time.

As the ranges got longer, the program was transferred to Patrick AFB, FL where the SM-62s would be launched from Cape Canaveral and escorted by the EDB-45As as far as the Tornado's fuel capacity would allow its range to permit the ability to assume control of the SM-62's flight if required. The last test vehicle was the operationally configured SM-62 *SNARK* missile. The first operational prototype missile (the N-69E configuration) was flown on 20 June 1957; and the first full test range flight with a warhead delivery was on 31 October 1957. The first launch of an operational prototype SM-62 (from Patrick AFB, FL) by USAF personnel was conducted on 27 June 1958. The *SNARK* went on Strategic Air Command Alert at Presque Isle AFB, MN between 27 May 1959 and 25 June 1961. The EB-45As and EDB-45A-1s supported the *SNARK* program from 19 October 1950 to 30 November 1958. They flew from Northrop's facilities at Hawthorne, CA and Los Angeles International Airport, as well as Holloman AFB, NM, Barksdale AFB, LA, and Patrick AFB, FL. While the *SNARK* program was outdated by the time of its introduction into service, the program directly led to the guidance system used on the SR-71 reconnaissance aircraft and to the navigation system used on the C-5A Galaxy.

Training....

One of the major goals the USAF hoped to accomplish with the B-45A-1s was training crews to fly and conduct short range, high altitude, and high-speed conventional bombing missions with the Tornado. By the end of September 1949 eleven J35-powered B-45A-1s were at Barksdale AFB along with the initial group of B-45A-5s. All the aircraft were very basic airframes with no radar, fire control systems or self-defense armament. While they were airworthy, the engines were in extremely poor maintenance condition due to the logistic systems inability to provide critically needed parts, tools, and support equipment like cranes and tractors. These shortfalls were made more significant because J35 time between overhauls ranged from 10 to 25 hours for each engine. The missions consisted of 1.5-to-2.5 hour dead reckoning day and night navigation missions across Texas, Louisiana, Arkansas and Mississippi. The one benefit the sorties provided was accustoming the crews to the speed at which mission checkpoints features happened. 450 to 500 mph was a lot faster than 200 to 225 mph. The transfer to TAC and relocation to Langley AFB did not improve the maintenance situation. Langley had little in the way of permanent facilities that could handle the B-45As and unique challenges were placed upon the maintenance crews. Hangar design did not allow them to simply tow the B-45s into the facility. The nose of the B-45A had to be raised with jacks, then the aircraft teased forward into the hangar for bad weather maintenance.

Five other B-45A-1s were reassigned to Target Tow Squadrons. One B-45A-1 remained at Barksdale, assigned to the 91st Strategic Reconnaissance Wing in October of 1949 to prepare crews to fly the RB-45C. Four B-45A-1s were reassigned to the 363rd Tactical Reconnaissance Group at Langley. To date, no records have been located to explain this reassignment.

The B-45A-1 in summary....

The J35 powered B-45A-1 was less than successful. The engine did at last become a production success for Allison, but it was underpowered when applied to the B-45A. Parts, tools, and specialized equipment to service the jet failed to build significant confidence in the airframe or its engines. And while the aircraft was good to fly and responsive, problems concerning airspeed control on approach was a major issue. The aircraft did provide good support in testing and research areas, especially in support of the *SNARK* program for Northrop, but no overall missions assigned to B-45 quelled the overall lack of support for the aircraft at the Air Staff level. That lack of support and the issue of its further operational service were still open matters as the second variant of the aircraft, the B-45A-5, entered service with the USAF in March of 1949. No significant operational mission had emerged for the B-45A-5, even though a few were assigned to an operational bombardment unit. With no operational radar navigational unit or few if any alternate means of all-weather bombardment, the aircraft was not a significant combat planning factor for the service. Finally, the amalgamation of two units with drastically different operational missions and mission aircraft at Langley AFB, VA, reflected the Air Staff's almost total lack of concern for getting the B-45A up to operational par in a reasonable time.

Part 2, the B-45A-5 enters service....

GE's "bet the company's existence" decision turned out to be a significant success. The J47 was one of the most successful turbojets produced and one of the most successful power plants in aviation history. Of course, it would take several years to resolve many of the basic issues with the engine and its production, and it was the afterburning versions that were the most successful. These versions unfortunately were never adapted to the B-45 family. As with the B-45A-1, the B-45A-5 entered service in a test mode, establishing the first fully functioning turbojet powered light Bombardment Group in the USAF.

The first of the bombardment B-45A-5s began arriving at Barksdale in March of 1949 and by 1 September 1949, 29 B-45A-5s had been assigned to the 47th Bombardment Group (L). On 17 October 1949 the 47th Bombardment Group was deactivated and their 34 B-45A-5s were reassigned or delivered directly to the 363rd Tactical Reconnaissance Group at Langley AFB, VA. Unfortunately, the Group had lost four aircraft, all B-45A-5s and three crews in their six months at Barksdale in 1949. All accidents were due to parts or systems that did not function according to their design or standards. Once at Langley, the 84th and 85th continued their routine of day and night dead-reckoning training sorties.

Photo 7: B-45A-5 7084 was one of the last bombardment B-45A-5s assigned to Barksdale AFB, LA.

Photo 8: 7084 was assigned to the 47th Bombardment Group when photographed at the annual Detroit Air Show. At this time, it was also assigned to 84th Bombardment Squadron at Langley AFB, VA. (Mr. Balogh via David Menard)

Part of the decision to relocate the B-45As to Langley was to make room for the 91st Strategic Reconnaissance Wing to arrive at Barksdale within a year or so and to make room for a new wing of SAC B-47B bombers. And while the 47th Bombardment Group had been gaining assigned airframes, the 91st Strategic Reconnaissance Wing had gained 8 more B-45A-5s at Barksdale for a total of nine aircraft trainers between December 1950 and December 1951, awaiting their first RB-45Cs to arrive in late 1950. Again, these B-45A-5s were not fully equipped with navigation and communications electronics that would match the RB-45C, but their performance did equal their forthcoming mounts. And although the Wing's Initial Operational Capability (IOC) would not be until mid-1951, SAC was very anxious to get the unit operational as soon as possible to improve their recce capability over the RB-29s and RB-50s. By mid-October 1950, the Air Staff had ceased their plans to fully equip the 363rd Group with any more operational aircraft and began to store new B-45A-5s as they came off the assembly line at Long Beach. The first of 32 B-45A-5s were sent to storage beginning on 28 February 1950 on the ramp at the Long Beach to be cocooned and stored until 23 April 1951. And until that date the operational development history of the B-45A had come to a temporary halt once again.

Photo 9: This B-45A-5 was one of eight aircraft assigned to the 91st Strategic Reconnaissance Wing at Barksdale in early 1950. (NMUSAF)

Testing....

Immediately upon receipt from the plant at Long Beach, the Air Materiel Command put the new higher performance B-45A-5s to work. Air Proving Ground Command and the 3200 Proving Ground Group at Eglin AFB, FL was a big recipient. Eglin had already received 3 B-45A-1s and in March 1949 it began to receive three additional B-45A-5s, one B-45C-1, and one RB-45C-1. It retained B-45s until September 1954. Air Proving Ground Command was responsible for testing various operational ideas to verify their feasibility before they were introduced into the operational units. For the B-45 program one of the major questions from the Air Staff was whether the B-45 could be converted into a successful and effective close air support platform supporting the U.S. Army.

At least one airframe, B-45C-1 **8007**, was slightly modified at Edwards AFB, CA then flown to Eglin AFB, FL for air to ground weapons tests. At least one pair of 20mm cannons were mounted in the nose or forward fuselage, and some variety, likely 2.75 inch folding fin rockets or 3-inch high velocity aircraft rockets (HVAR), were mounted within a bomb bay rack. An unknown number of weapons release tests were flown at the ranges on Eglin. While the weapons did release safely, the stress of the higher G maneuvers and the weapons releases on the B-45C's airframe caused measurable damage and the Proving Ground engineers determined that the B-45 could not withstand the operational strain of routine close air support missions. The program was rapidly ended, and no other forward-firing weapons tests were ever conducted on the Tornado.

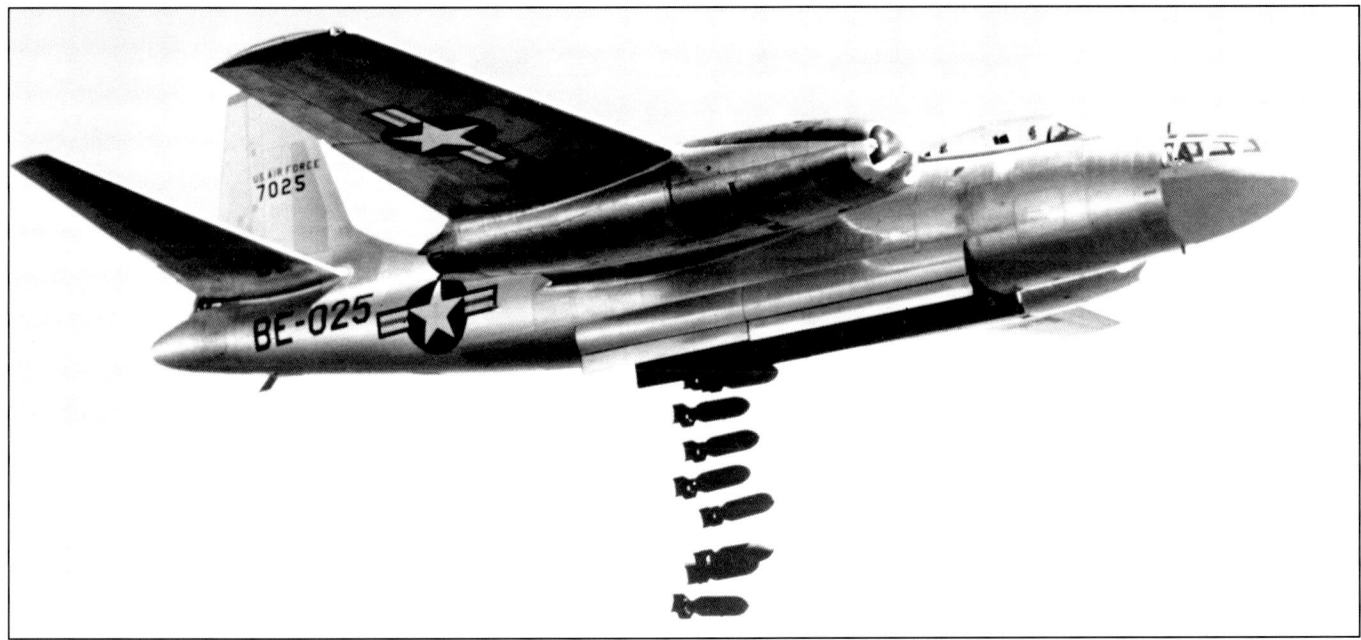

Photo 10: One of the most important Air Proving Ground Group B-45A-5s was 7025 which performed many if not all the weapon's separation and clearance tests for the program. The aircraft's record was dropping an entire string of 500-lb. bombs, 29 in total, at over 500 mph on a totally internal drop. (Gerald Balzer Collection via Tony Landis from HQ AFMC/HO)

Early Joint Service Efforts....

Less than a month after arrival at Langley AFB, VA two B-45A-5s were sent TDY, one to Eglin AFB, VA and one to Brookley AFB, AL. The aircraft were to provide ground exhibits and tours, as well as flying displays and attack simulations, for the Air Indoctrination Course II. This was a Tactical Air Command initiated and funded program of presentations at the Eglin AFB Ranges on 10/11 October, 17/18 October, 24/25 October, and 27/28 October 1949. The guests were U.S. Army, Navy and Air Force students and instructors from 21 Service Schools numbering some 3,000 personnel who attended the complete course. An additional 2,500 military and civilian personnel could attend only the air demonstrations. The program, which was one of the first of what would be called today "joint service" programs, was informative and well received.

he B-45A-5s were grounded just before and during the presentations, and additional B-45A-1 and -5s were sent to Eglin and Brookley from Langley for cannibalization purposes to keep enough aircraft airworthy to meet the scheduled flying programs. At the end of each second-day's

Course segment, the B-45s and other USAF and USN fighters conducted a major low-altitude simulated attack against industrial and naval targets in the Mobile Bay area for the attendees. These exercises demonstrated the potential impact of a major, coordinated, large scale aerial attack. Many of the B-45A attacks were conducted using ground-based Radar Bombing Direction using AN/MSQ-1 radars, which became COMBAT SKY SPOT to direct all-weather conventional missions throughout Vietnam, Laos, and Cambodia in the Southeast Asia War.

Photo 11: B-45A-5 7054 is towed past a large group of attendees during a display session at the Air Indoctrination Course II at Eglin AFB, FL Auxiliary Field Number 2 in October 1949. (NMUSAF)

Photo 12: Two fire guards stand watch as B-45A-5 7055 starts number one and two engines at Eglin AFB, FL Auxiliary Field 2 during Air Indoctrination Course II in October 1949. (NMUSAF)

TAC and the 84th Bombardment Squadron participation in the Course were expensive and labor intensive with multiple engine and parts exchanges over the month. The B-45A-5 had at least two, fleet-wide, groundings during the Course. Adding the groundings to the maintenance man-hour load, normal supply issues, and ill-equipped and multi-separated operating locations with no in-place support available for either aircraft or turbojet engines, made keeping aircraft available for the missions, difficult at best. What was originally planned to involve only two aircraft ultimately involved five or more for over 20 days.

Photo 13: B-45A-5 7060 sits on the ramp at Brookley AFB, Mobile, AL between flying demonstrations over Eglin and Mobile during the Air Indoctrination Course II in October 1949. (Author's Collection)

To the Races....

On 18 August 1951 eight USAF jet aircraft, three light bombers and five jet fighters were on the Edwards AFB, CA ramp very early in the morning having final fueling and other maintenance checks completed. These aircraft would soon take part in the first all jet-powered Bendix Transcontinental Trophy Race between Edwards and Cleveland, OH. At 0630, Pacific time, the crews arrived and made their final checks with maintenance personnel. By 0656 all aircraft had started engines and taxied into their takeoff order on the Edwards AFB runway. At that moment Miss Robin E. Grove, "Miss Edwards of 1951" waved the starting flag, and the jets began their takeoff rolls. The first of three B-45A-5s, **7084** commanded by Lt. Col George Thabault was off the ground at 0657 for the 1,900 nautical mile route.

Photo 14: Taken around 0600 hours this image shows the five of the eight contestants and all three B-45As in the race. The aircraft have been placed in their start-up positions and flight and maintenance crews were completing the last of their pre-flight and take-off inspections. (Air Force Flight Test Center History Office [AFFTC] /HO via Steve Ginter)

The second B-45A-5, **7026**, assigned to the 3200nd Proof Test Wing on temporary duty from Eglin AFB, CA was commanded by Major Leo Dykes and was the third aircraft to launch. The final B-45A-5 to launch was **7058** assigned to the 47th Bombardment Wing at Langley AFB, VA commanded by Capt. Bernard Watts, and was the 4th race entry to launch. Based on time Lt. Col. Thabault finished third in **7084**; Major Dykes finished fourth in **7026**, and Capt. Watts finished sixth in **7058**. Three different types of aircraft participated in this race: the F-86E, the straight-wing F-84E and the B-45A-5 light bomber.

It was an interesting exercise that showed that the performance of the light bomber was not significantly less than the top fighters in the USAF at the time. From an Air Force official operational view, the race was a performance and chemical review of JP-3 as the possible standard USAF Jet fuel. JP-3 contained a large amount of allowable aromatics in the fuel to widen the range of temperatures at which it could be refined. With these wide ranges of allowable variences during ptoduction, JP-3 was somewhat unstable for use in a number of different turbojet engines. Therefore JP-4, a 50/50 mix of kerosene-gasoline became the primary military and civilian jet fuel for many decades.

Photo 15: 7084 was the first aircraft in the air for the 1951 Bendix Transcontinental Trophy Race. At this the time of the race it was assigned to the 47th Bombardment Wing at Langley AFB, VA, in these markings shown. (D. Balogh via Dave Menard)

Photo 16: B-45A-5 7026 was the third aircraft launch that morning. The pilot of the heavily-loaded aircraft with a full load of fuel was just beginning to retract the landing gear as aircraft approached the departure end of the runway. (NARA II via Tommy Thomason)

Photo 17: B-45A-5 7058, the fourth Bendix Transcontinental Trophy Race contestant to launch, was commanded by Captain Bernard Watts. (NARA II via Tommy Thomason)

Other events that also had a wide public affairs importance for the Air Force were air shows at major airports in major cities across the nation. One such show, at Chicago, IL, involved two B-45A-5s from Langley AFB, one on static display and one flying aerial demonstrations daily, attracted a crowd of nearly 300,000 visitors on one weekend.

Photo 18: B-45A-5 7036 participated in the July 1949 airshow in Chicago at Orchard Field Airport, currently O'Hare International Airport, which hosted over 300,000 visitors in one weekend. 7036 shares the ramp with a famous and historical aircraft, the Boeing B-50A 46-10, *Lucky Lady II* which completed the first non-stop round the world flight on 2 March 1949 taking off and landing at Carswell AFB, TX. The building next to 7036 is the former Douglas Aircraft WW II C-54 plant which at this time stored most captured WW II enemy aircraft held by the Smithsonian Institution. (B. Kemp via David Menard)

Target Towing, Part 2....

Once more the mission of target tug was high on the list for the B-45A-5 to carry out. Five B-45A-1s were modified with a single reel target tow system: TB-45A-1 **7006**, **7009**, **7015**, **7018**, and **7019**. These five aircraft were still in service when the first TB-45A-5s began to be delivered in April 1950. At least one additional B-45A-1, 7020, was modified with a dual reel system at Yuma AFB, AZ. Power to the reels was provided by two engine driven hydraulic pumps, one each on engines 1 and 3 controlled by the pilot. The single reel system reel held up to 12,000 feet (or 2 nautical miles) of armored cable. The aircraft could also carry a single 1,174 gallon jettisonable fuel tank in the forward section of the bomb bay. Twenty B-45A-5s were modified into TB-45A-5s by the SBAMA. The Target Tow Squadrons that were to receive them included the 1st, 2nd, 4th, 17th, and the 4750th Target Tow Squadrons.

Photo 19: TB-45A-5 7063 was converted to a TB-45A early in its career and initially equipped with a single reel tow unit when first assigned to tow duties at the Air Force Flight Test Center at Edwards AFB, CA Their mission was to simulate Soviet Atomic Bombers which posed a threat to the mainland U.S. (Peter Bowers via Steve Ginter)

Photo 20: This image shows the single reel cable exiting the fuselage just to the starboard side of the tail skid on TB-45A-5 7063 at Edwards. The cable reel was in the aft bomb bay compartment and was rigged through the aft portion of the unpressurized fuselage where it exited the aircraft. It was then guided to a circular guard that kept the banner aft of the target tow operator's station. This insured the cable did not interfere with the aft fuselage or empennage. The box mounted to the heavy structure just in front of the banner guide was the cable cutter. (Craig Kaston)

Most, if not all, of the TB-45A-5s converted to target tows, as well as some of the surviving TB-45A-1s were eventually equipped with a dual reel target tow system. These aircraft were fitted with two Mk 8 Mod 0 reel units in the aft bomb bay, again powered by two engine driven hydraulic pumps on engines 1 and 3. Each reel could carry 6,000 feet of 3/16 inch armored cable and two 6 x 30-foot or 9 x 45-foot banners in externally mounted protective metal sleeves, which were retracted during cruise to and from the target area. As with the single reel systems, the duel unit aircraft could also carry a 1,200-gallon fuel tank for additional range. Aircraft flown against targets towed by the TB-45A-1 and A-5s included the Lockheed F-94A, B and C Starfire; the North American F-86D, and L Sabre; the Northrop F-89C, D, and H Scorpion; and the early model Convair F-102A Delta Dart and Lockheed F-104A Starfighter.

Photo 21: This image shows 7063 carrying a target banner and a three-dimensional corner reflector at Edwards AFB, CA. The radar reflector gave the intercepting aircraft a relatively realistic radar return from the reflector during the intercept without using the tug as a radar target. The single reel banners were trailed behind the tug aircraft for both take-off and landing and produced considerable drag throughout the tug's flight. (Craig Kaston)

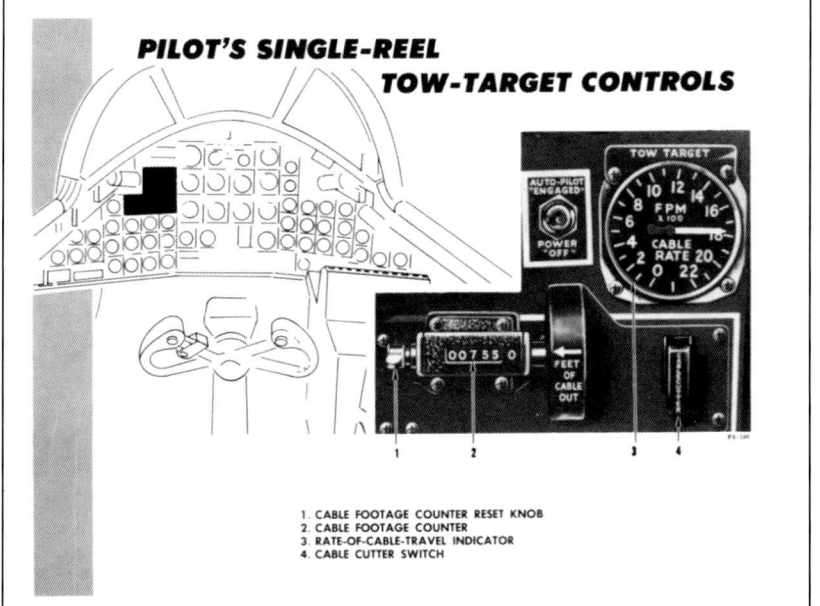

Photo 22: This image shows the pilot's single-reel tow target controls on the forward main instrument panel. (NMUSAF) Drawing 2: This image shows the pilot's hydraulic system controls and emergency cable cutter panel, just above the autopilot controller. (NMUSAF)

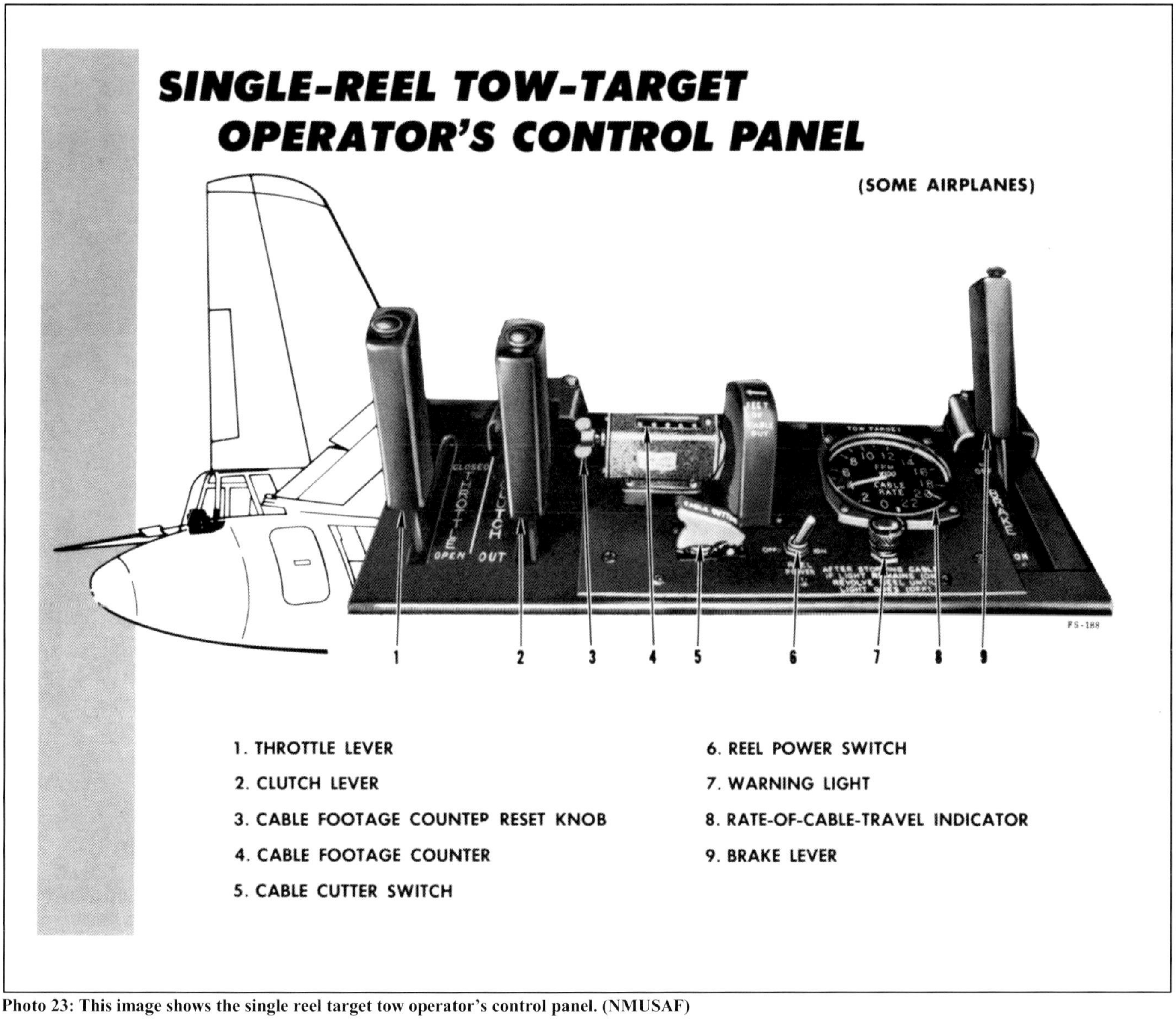

Photo 23: This image shows the single reel target tow operator's control panel. (NMUSAF)

Photo 24: TB-45A-1 7020 assigned to the 17th Target Tow Squadron at Yuma AFB, AZ was equipped with a dual reel system as the large external housing for the banners shows. The banners were the live-fire targets for the gun or rocket-armed interceptors. (B. Hostetter via David Menard)

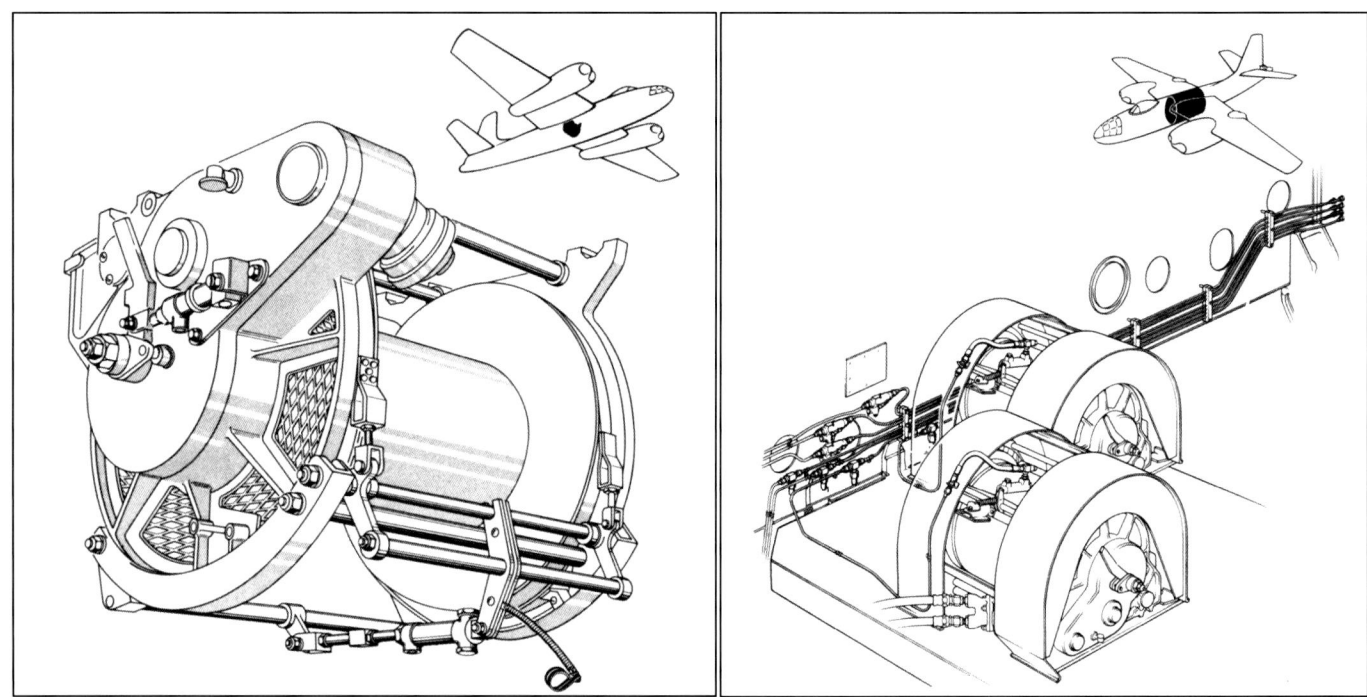

Drawing 3: This drawing shows a Mk 8 Mod 0 target tow reel capable of holding up to 6,000 feet of armored cable. One was mounted in the upper aft bomb bay on the single reel system. (NMUSAF) Drawing 4: This drawing shows the heavy tray that held both Mk 8 reels in the dual reel system and much of the hydraulic plumbing system used to power it from the two engine-powered pumps. (NMUSAF)

The TB-45As became the main teaching tool for interceptor units learning to defend against Soviet bombers from the Tupolev Tu-4 through the Tu-16 and the early Tu-95s. They also served as an important means to evaluate the proficiency of Air Defense Command (ADC) interceptors for that mission. Yuma AFB, AZ became the center of ADC evaluations with entire units of F-86D and F-89s flying south to conduct interception sorties against the TB-45s. Generally, the interceptors flew in four-ship formations with a precise number of non-explosive rounds in their machine gun magazines painted a particular color for each aircraft. The same applied to the non-explosive warheads of the rockets carried on each fighter, again a separate color for each aircraft in the flight.

Once on the range, the TB-45As would deploy their targets one at a time at an approximate altitude of 20,000 to 30,000 feet depending on the size of the target and weather. The interceptors would usually attack one at a time, two against each target. The first target would be retracted, then the second extended and the second pair of interceptors cleared to attack in order. Once all four aircraft had attacked the banners they would be retracted and the TB-45A would return to Yuma. Upon arrival, the banners would be removed from the target tugs and inspected by judges who would identify how many of each color of rounds or rockets hit the banners and identify scores for each color's aircraft. Scores were then computed and posted for the units to evaluate.

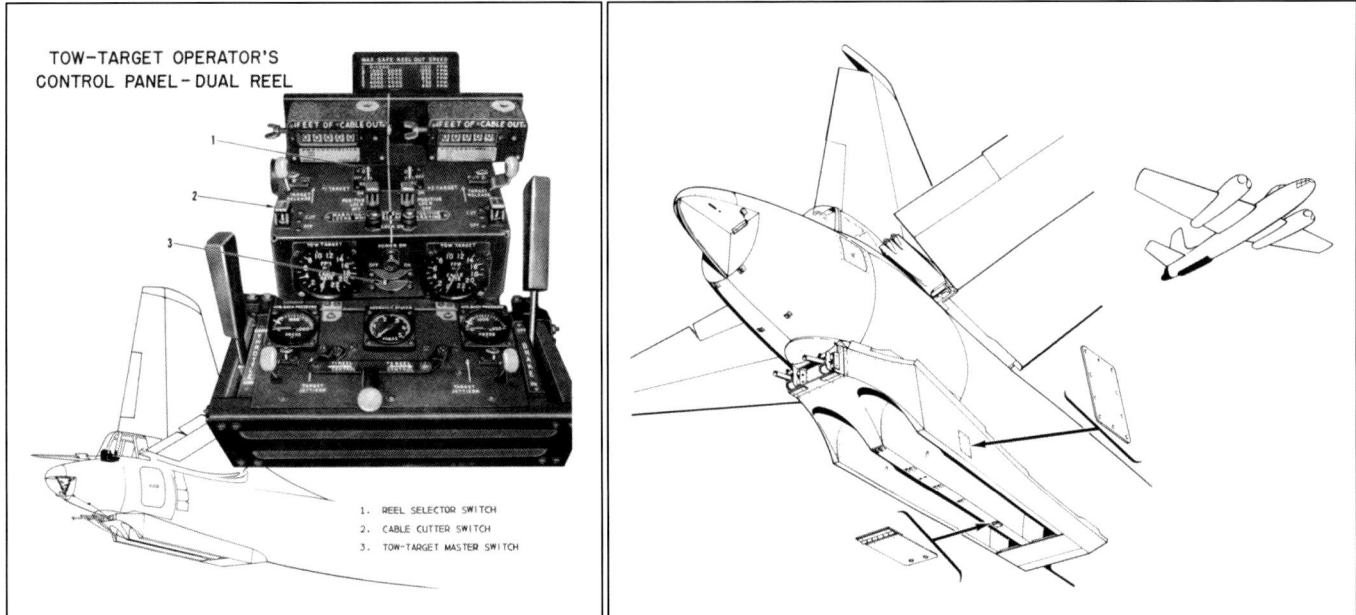

Photo 25: This image shows the dual reel control panel mounted in the target tow operator's station. (NMUSAF) **Drawing 5:** This drawing shows the dual reel system from below and the two banner protective depressions. The banners would be fully retracted for take-off and landing. They would be extended aft one at a time from these protective areas, then the cable attached to them would be reeled out from the pipes just aft of the covers. The banners would only be extended once over the range area, and then retracted before the tug left the range area (NMUSAF)

Photo 26: While not a great image, this photo shows TB-45A-5 7024 taxing out for another dual-reel target tug sortie at Yuma AFB. It appears its interceptors this day are to be ADC F-86Ds. The B-29 in the background is an interesting visitor to Yuma, as it imitates one of the common Soviet targets ADC trained against, the Tupolev Tu-4. (R. Wainwright via David Menard)

TB-45As served as target tugs for ADC until the late 1950s when they were replaced by Martin B-57E Canberra's. Only one TB-45A was lost in ADC service, in great contrast to those lost in TAC and United States Air Forces in Europe (USAFE).

Korea….

By mid-1950, other than test aircraft and target tugs, few B-45As were in service that involved or were preparing for direct combat roles. On 25 June 1950, the necessity for that to change grew massively. The People's Republic of North Korea Armed Forces launched a surprise attack across the 38th parallel dividing North and South Korea. By September most of the Korean Peninsula was under North Korean control except the very southeast portion of South Korea around Taegu and Pusan.

Other than Douglas B-26s and Republic F-84s that could be shipped by sea, little in the way of aerial reinforcements was available. Headquarters USAF immediately began plans to ferry B-45A-5s across the Pacific to Korea. But the planning quickly came to an end when the range of the B-45A-5 was determined to be 500 nautical miles too short to reach Hawaii, even with a full load of fuel in all available bomb bay tanks. And there were no means to modify the B-45A-5s to become aerial-refueling receivers. So, a combat mission for the Tornado would continue to wait.

An Atomic Mission....

In the summer of 1951, the Atomic Energy Commission notified the USAF that both the primary results of the 1948 Operation SANDSTONE tests had become operational weapons and that the Mk 5 and Mk 7 implosion atomic weapons would be stockpiled beginning in May 1952. Furthermore, the Mk 8 deep penetration weapon would be stockpiled beginning in April of 1952 to meet the need to destroy submerged and buried targets. This gave the service approximately 12 months to design, develop, plan, and manufacture the structures and equipment, and then carry out these plans to make the Republic F-84E and G, and the North American B-45A-5 aircraft atomic-capable. Furthermore, the service was to prepare and accomplish all requirements for the handling, servicing, and storage of the weapons overseas; and training the personnel to accomplish the maintenance, and combat delivery of these weapons. By April 1952, it was to have all modified aircraft and all personnel relocated to RAF Station Sculthorpe, Norfolk, UK.

Several motives were driving the USAF. First, the AEC had rapidly given the service a set of weapons that its top fighter-bombers and its sole light bomber could carry and deliver in combat in Europe if the Soviets and their allies contemplated a repeat of the North Korean adventures to secure control of Europe for themselves. Second, they now had a legitimate mission for a capable light bomber that could be in place as an atomic deterrent in less than a year. And finally, the USAF could aid NATO in establishing a critically needed air arm, to include an atomic aerial strike force to deter aggression against Western Europe.

As the only light bombardment aircraft in the inventory that could carry these new weapons, the B-45A-5 received a new lease on life and became an especially important element of defense and deterrence in Europe. The B-45A-5s were reassigned out of the 363rd Tactical Reconnaissance Wing on 11 March 1951 and the 47th Bombardment Wing was established on 12 March 1951, still at Langley AFB. The 84th and 85th Bombardment Squadrons were once again assigned under the 47th.

HQ USAF had already outlined a tentative plan to modify 60 B-45A-5s to carry atomic weapons. This provided three squadrons of 16 aircraft each and 12 attrition aircraft. In December 1950, five B-54A-5s were to be transferred to the Tactical Air Command to be configured with an "Auxiliary Bombing System" or the ability to carry atomic weapons, equipped with the AN/APQ-24 all-weather radar bombing system, the AN/APN-84 SHORAN, plus a M9C Norden Bomb Sight. In February, the number of TAC aircraft was increased to nine. Since final suitability and final configuration tests for the Mk 7 and Mk 8 were not complete, two Mk 5 capable B-45A-5s were assigned to the Special Weapons Command, and two to the Tactical Air Command.

BACKBREAKER....

In mid-1951 a program for the operational use of atomic equipped B-45A-5s was established and the aircraft modifications were to be completed under Operation BACKBREAKER which also included F-84E and G fighter-bombers. The program, directed by HQ USAF on 14 July 1951, was second in priority to the Strategic Air Command's Program ON TOP to equip SAC B-50s with Mk IV, 5, 6 and 8 atomic weapons. The first nine aircraft in BACKBREAKER were to be completed by 1 January 1952 and an additional 32 aircraft were added on 20 August 1951 to be completed by 1 April 1952. (One aircraft was lost due to a ramp fire at SBAMA so the total number was 40 aircraft by 1 April.) The additional 32 aircraft were the ones stored at the plant at Long Beach, CA.

Photo 27: B-45A-5s of the 84th Bombardment Squadron were imaged at Langley just a few days before their deployment to the UK began. (NARA II via Tommy Thomason)

The Program was complicated and extensive, shared between NAA and the SBAMA. New radars had to be installed, as well as electronic monitoring and control equipment for all three new weapons. In addition, there was a significant amount of GFE which was not installed on the production line but was required to make these aircraft combat capable. Given that three new bomb shackles were required, one for each weapon, much design, fabrication, and production work was required prior to their availability for installation in the aircraft. Finally, all aircraft due to be assigned overseas were to be fitted with new Emmerson A-6 mechanical turret, guided by an AN/APG-39 ranging radar. The turret was armed with two AN/M3 .50 caliber machine guns, each with 300 rounds of ammunition. Tie-ins between the AN/APG-30 and the turrets, the fuel-flow totalizer, and the extra fuel tanks were engineered and planned by North American and installed at Langley.

Each aircraft was newly equipped with the following: a reliable AN/APQ-24 radar bombing and navigation set; a Short Range Navigation system or SHORAN with a K-1A computer; an AN/APN-11 visual radar beacon; a Radar Bomb Scoring (RBS) training system; Emerson A-6 turret with an AN/APG-30 tail warning radar; an AN/ARC-27 Ultra High Frequency radio; a Type A-6 chaff dispenser with dispensing chute for easier maintenance; a new NAA fuel Totalizer; an AN/APX-6 Identification Friend or Foe set; two external 500-gallon Fletcher Aviation Corporation designed and manufactured fuel tanks under the engine nacelles (**Drawing 6**); and a standard 318-gallon upper aft bomb bay fuel tank regardless of what weapons were carried. Forty aircraft were modified by SBAMA and with kits produced by NAA and completed at Langley AFB, VA, by the end of April 1952. These aircraft would be assigned to the 84th and 85th Bombardment Squadrons (L) of the 47th Bombardment Wing (L) at Langley AFB by 1 May 1952.

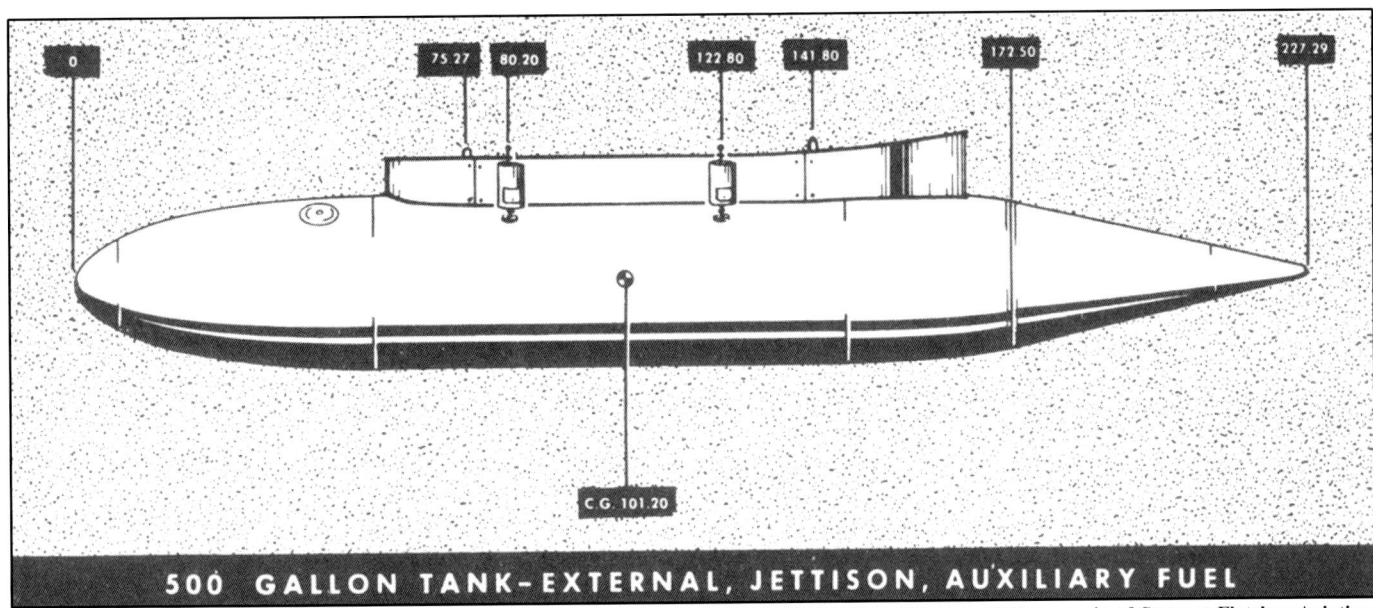

Drawing 6: The BACKBREAKER modifications increased the range of the BACKBREAKER B-45A-5s by fitting a pair of Sargent Fletcher Aviation 500-gallon JP-4 tanks, one under each nacelle. The tanks were 18.94 feet (227.29 inches) long, and 30.5 inches at their max diameter. They were suspended by 6-inch-deep pylons that were 9.5 feet (114 inches) long. Note that the taper of each end of the tank began at the ends of the pylons and the top of the pylon was concave to match the bottom of each nacelle. These tanks were among the first conversions installed in BACKBREAKER. (Courtesy Phillip Fridell)

SECOND CALL....

The two BACKBREAKER-modified squadrons became operational almost immediately, armed with the Mk 5 weapon. However as with all such complicated, fast-paced and self-reliant efforts, problems arose. In July 1952 HQ USAF initiated a second program, Project SECOND CALL, to increase the number of tactical light bombers based in Europe and to correct shortfalls in the capabilities of the B-45A-5s.

First, 15 additional B-45A-5s were to be reassigned from SAC upon delivery of SAC's new RB-47s. The BACKBREAKER configuration was to be the basis for the new aircraft but with some significant differences. The Mk 5 bomb rack was to be moved forward into the forward bomb bay along with the Mk 7 and Mk 8 racks. This would allow a removable 318-gallon and jettisonable 872-gallon tanks to always be carried in the aft bomb bay, adding approximately 270 nautical miles to the aircraft's range. Since the Mk 5 was the primary long-range weapon to be carried, this was extremely important to the 47th Bombardment Wing.

A new SHORAN, the AN/APN-84 system would be installed instead of the planned AN/APN-3 SHORAN. AN/APN-3 units were almost totally deployed to the Korean theater where they were compatible with the light and medium bombers as well as all of the turbojet fighter-bombers baser on the Peninsula and in Japan. To down-load the aircraft units and bring them and the ground-based equipment back to the States via overwater transport would be extremely expensive and carry a high risk of damage during shipment, primarily due to corrosion. A major reason for the use of the new gear was that SAC was planning to use the AN/APN-84 in their B-47 and later B-52 fleets. The cost of production for the aircraft units, ground units, spares, maintenance equipment and training for the BACKBREAKER/SECOND CALL fleet would be considerably less than recalling the AN/APN-3 gear from Korea, Finally, having the training and maintenance capabilities in the CONUS would make the installation of this new system into the B-57 Canberra considerably less expensive as well. The new General Electric A-5 fire control system was added to the turret control system, and the NAA Fuel Flow Totalizer system would be added to all aircraft as additional changes in BACKBREAKER/SECOND CALL.

The most important change was the fact that all 50 BACKBREAKER/SECOND CALL aircraft would be brought up to full SECOND CALL electronics and weapons standards. The modifications were carried out using kits provided by SBAMA and personnel assigned to the 49th Air Division in England, who had total responsibility for USAF atomic activities in the UK. Work began on 1 July 1953 and was not completed until March 1954. The delay was due to a major organizational change throughout Air Materiel Command. The 15 new airframes allowed a third squadron, the 86th Bombardment Squadron, to be assigned to the 47th Bombardment Wing on 23 March 1954. At this point the Wing was now fully capable of carrying and delivering each of the three atomic weapons assigned on targets throughout their area of operations in Europe. At last, after nearly five years of waiting, the B-45A-5s bombers had an extremely important operational mission in a vital area of US interest. During the SECOND CALL modification period, all combat-capable RB-45Cs were modified by SBAMA to carry a pair of the same Fletcher 500-gallon JP-4 tanks as the B-45A-5s when not carrying water/alcohol tanks.

The AN/APN-84 SHORAN system and the AN/APQ-24 radar system were key to bringing the B-45A-5 up to actual combat capability. The AN/APN-84 SHORAN was a new alternate bombing system supporting the AN/APQ-24 bomb/nav radar instead of the AN/APN-3 SHORAN. It relied on two ground-based beacons to which the aircraft sent repeating electronic signals. The beacons responded immediately to the attacking B-45As with replies that contained range and directional data. With this information the bomber could triangulate its position to a known preplanned point to release its weapon on the target. Radio Corporation of America was still fixing pressurization issues with some AN/APN-84 components that forced the aircraft to climb to at least 38,000 feet to use the system, costing considerable fuel and mission time while using SHORAN. It would be late 1955 until the pressurization issues were overcome. The range of the beacons, based on altitude was best at approximately 300 nautical miles, but it's most accurate range was approximately 150 nautical miles. Up to 20 bombers could operate off the pair of beacons at any given time.

Photo 28: At the time of this photo, B-45A-5 7072, assigned to the 85th Bombardment Squadron at Langley AFB, VA, the aircraft and its crews were spending quite a bit of time and flying hours on learning over-water navigation and emergency procedures for their upcoming mass overflight by two squadrons to the UK. (Steve Ginter)

Areas in the nose, bomb/nav station and aft fuselage were now occupied by the AN/APQ-24 as was intended in the original design of the B-45A. The antenna was under the nose transparency housed in a large fiberglass radome. The scope and bomb/nav system controls were centered in the bomb/nav station on and above a central table. Additional controls and equipment were located to the bomb/nav's immediate right-hand table and fuselage structure. Much of the large associated electronic equipment for the new system was in spaces throughout the aft fuselage which had been intended, but never used by a previous generation bomb/nav radar known as the AN/APN-23. Such was the state of the supply and equipment support for BACKBREAKER/SECOND CALL project, many of these GFE equipment installations did not exist or had been empty for many months, if not a few years. Non-existent items were manufactured by NAA, SBAMA, or Langley. Additional equipment and systems which supported the AN/APN-84 SHORAN included a new radar scope and view port, a new computer-capable airspeed indicator, a new radar control panel, weapon ballistics computer that would pass data and signals between the radar, the K-1 fire control system, and a new artificial pilot that would maintain constant or frequently changing headings, airspeeds, course corrections and altitudes as directed. The new SHORAN system required a host of similar systems such as analog computers, control panels, synchronizers, communication receivers and transmitters as well. The "special weapons", soon to be carried by these aircraft, had a host of highly classified and sensitive components which had to be re-designed to fit into areas, compartments and other locations they had never been intended for. Each weapon had its own set of monitoring and system control "T-Boxes" which had to be fitted and installed, then extensive wiring bundles designed, assembled, and installed between the bomb/nav station, the cockpit, and the bomb bay.

Each of these master systems, such as navigation, bombing, special weapons, communications, pressurization, emergency response and a myriad of others, had its own power generation and supply equipment, circuit breakers, voltage control and monitoring, modulation, amplification, junction and routing boxes, and the circuitry it took to connect, monitor, transmit data and directives between them. All this circuitry and that for other aircraft systems, such as the aircraft engines and pressurization, flight control, air-to-air and air-to-ground communications, had to be protected against any interference between any elements, large or small. Beyond these changes, a new A-6 chaff dispenser system that could jam Soviet designed-radars in multiple frequencies, and for the first time a twin-gun turret with radar ranging as well was installed.

Weapons and Alert Response....

The weapons used on the B-45A-5s were far from flexible and the 47th being the only tactical light-bomber atomic weapons unit in the USAF was far behind its training on weapons for tactical use, maintenance, and flight line operations to prepare the aircraft for atomic missions. First there were no schools in the Continental United States, or CONUS, or the "Zone of the Interior" to train air or maintenance crews on atomic weapons before they arrived at RAF Sculthorpe. Therefore, all personnel were effectively novices in all aspects of maintaining and delivering atomic weapons. Further, this problem existed throughout the chain of command. The 47th Bombardment Wing, the 20th and 81st Fighter-Bomber Wings were assigned to the 49th Air Division, who was assigned to the 3rd Air Force, responsible for all USAF atomic responsibilities in the UK, and the personnel were all neophytes. The result would be a period of some five years of extremely difficult and at times fruitless efforts to get the Wing even close to fully operational. Examples included complications of moving weapons to and from the secure storage areas, uploading and downloading weapons on the aircraft, and maintaining them on alert for a reasonable amounts of time. For the first year, weapons could only stand alert for periods of 36 to 48 hours. During this interval, a replacement was identified and put through long hours of functional inspections and operations check before being relocated to an alert hard-stand , and a new weapon uploaded. A team of 20 to 30 specialists were required to completely disassemble the down-loaded weapon for repair, reconditioning and replacement of items like batteries. The maintenance man-hours required for weapons care and inspections during normal up-loading cycles were high and these constraints alone reduced the number of aircraft that could be placed on ground-alert at any given time.

If a different model weapon were to be uploaded, the monitoring boxes or "T-boxes" would have to be installed to match that weapon, and settings reset for that specific weapon. This delayed the alert change-over time even further. Once the aircraft was fully armed, fueled, put on standby status for that mission, launching the aircraft was a lengthy process. The engines could not be started using cartridges to "spool up" the turbine shaft for ignition rapidly as the B-45A-5s and C-5s J47 was not equipped for cartridges. The warm-up time for navigation radar, SHORAN, pneumatics, and hydraulics had to be met. Another shortfall was the lack of a rapid communication system from 3rd Air Force to the 49th Air Division and on to the 47th Bombardment Wing. A line of communication to quickly alert the 47th to generate an atomic sortie, then notify the Wing of an actual order to take-off simply did not exist. Often that requirement was executed using Jeeps, bicycles and/or running. At best, even throughout 1956 and 1957, only about 50 to 60 percent of the wing bombers could be counted on to successfully launch an atomic strike against Soviet or Warsaw Pact forces. Few if any would be able to recover at locations from which they could be regenerated for a second atomic strike.

The initial BACKBREAKER modifications called for two bomb racks, one for a Mk 5 and one for a Mk 7 weapon, in the forward and middle bomb bay areas of the B-45A-5s. A third bomb rack to carry the Mk 8 weapon was to be installed in the aft bomb bay area. It became clear shortly after the modifications were made under BACKBREAKER that the aircraft could no longer carry either of the 872 gallon jettisonable fuel tanks it originally carried. During the planning for SECOND CALL, the 49th Air Division and the 47th Bombardment Wing made it clear to the Air Staff that the lack of both of these tanks would severely restrict the range of the aircraft and the ability to strike a great number of its most important targets in many of the NATO war-plans. Immediately, North American, SBAMA and the Boeing Aircraft Company, who had originally designed these atomic weapons racks, developed new plans that moved all three racks to locations in the forward bomb bay and returned the ability to carry the aft 872 gallon jettisonable tank in the aft bomb bay. These changes returned about a third of the B-45A-5's original range and put many targets back into the NATO war-plans. It was an example of the exceptional communication between USAF Headquarters, NATO and USAFE Headquarters, two major US aircraft manufacturers and a major USAF Depot that kept the B-45A a viable weapon for many years. With these SECOND CALL changes, all B-45A-5s could carry the total amount of the fuel of an original B-45C-1 including that held in the wingtip tanks.

In **Drawing 7**, the upper inboard-profile shows the location of the largest weapon carried, the MK 5. The U-2 Special Weapons Rack (**Drawing 18**) was angled slightly downward beneath the fuselage 318-gallon tank and the upper-forward 318-gallon removable tank. The Mk 5 also allowed the carriage of the aft 872-gallon jettisonable tank. The lower left drawing shows the location of the Mk 7 and its S-2 shackle and bomb adaptor (**Drawing 15**). The Mk 7 was the lightest of the weapons carried by the B-45A-5, therefore it could be carried closer to the limits of the center of gravity range of the aircraft. This location also gave the weapons crew plenty of room to remove and install the tail cone during installation of the pit. The major disadvantage of carrying the Mk 7 internally was that the size of the weapon forced the removal of both 872-gallon jettisonable tanks and the loss of 1,744-gallons of available fuel. The lower right drawing shows the primary shackling location of the heaviest weapon carried by the B-45A-5, the Mk 8 gun-type weapon. The drawing shows the most forward location of its D-7 shackle (**Drawing 20**) on the aircraft. The weapon's weight was driven by its fission initiation method. Its core was divided into two sections. One portion was propelled by an explosive through a cannon barrel into the remaining portion of the fissile material. If range was not a critical concern for the mission, the Mk 8 could be carried with the forward 872-gallon tank removed, and the aft 872-gallon tank installed for CG purposes. Monitoring CG was critical in this configuration and the 872-gallon aft bomb bay tank could not be used below about 600 gallons until the weapon was dropped. If range was an issue, then the Mk 8 would be mounted on a D-7 shackle where the Mk 7 was usually mounted. The forward 872-gallon tank was still removed, but the aft 872-gallon tank could have its entire load used prior to the release of the weapon and remain within CG limitations.

Photo 29: This photo shows B-45A-5 at RAF Sculthorpe shortly after their arrival in June of 1952. The aircraft was assigned to the 84th Bombardment Squadron and was refueling using one of the larger tankers available on the airfield. Refueling was done tank by tank and could take well over three hours depending upon the weather. (Author's Collection)

Many of the structural, arming, weapon monitoring components and power supplies were manufactured by the Boeing Aircraft Company. Most were similar regardless of which aircraft they would be mounted into. However, if changes to the aircraft structure or other components were required for their installation, the manufacturing company of the aircraft would modify or if necessary, re-construct any aircraft components to allow the Boeing weapon-related components to be installed.

Besides the newly modified B-45A-5s, a new model of B-45 was added to the family. Of the 15 new B-45A-5s in the SECOND CALL program, two were determined to be class 26, no longer airworthy, and one was lost to a second ramp fire at North American. Loses of the B-45A-5 SECOND CALL aircraft forced the conversion of four SAC RB-45C airframes to SECOND CALL configuration. The aircraft were brought up to full SECOND CALL standards with complete electronics and weapons criteria. The first, **8035**, was converted in April 1954, and other three new aircraft, **8013**, **8014** and **8023** were converted in August of 1954.

Three of the new aircraft, RB-45Cs **8013**, **8014**, and **8023** were fitted with Emerson A-6 turrets as used BACKBREAKER/SECOND CALL B-45A-5s. **8035** would be an exception, and was fitted with a Bell M-7 turret fitted on all other armed RB-45Cs. All four aircraft were redesignated B-45C-5s and assigned to the 47th Bombardment Wing at RAF Sculthorpe. The full fuselage interior illustrations of the B-45A-5s and B-45C-5s (**Drawings 7 and 8**) modified by BACKBREAKER/SECOND CALL Programs covers the installation and details the Emerson A-6 radar ranged self-defense turret, the Bell M-7 visually aimed turret, the Gunner's Stations; and the installation and details of the A-6 Chaff Distribution System.

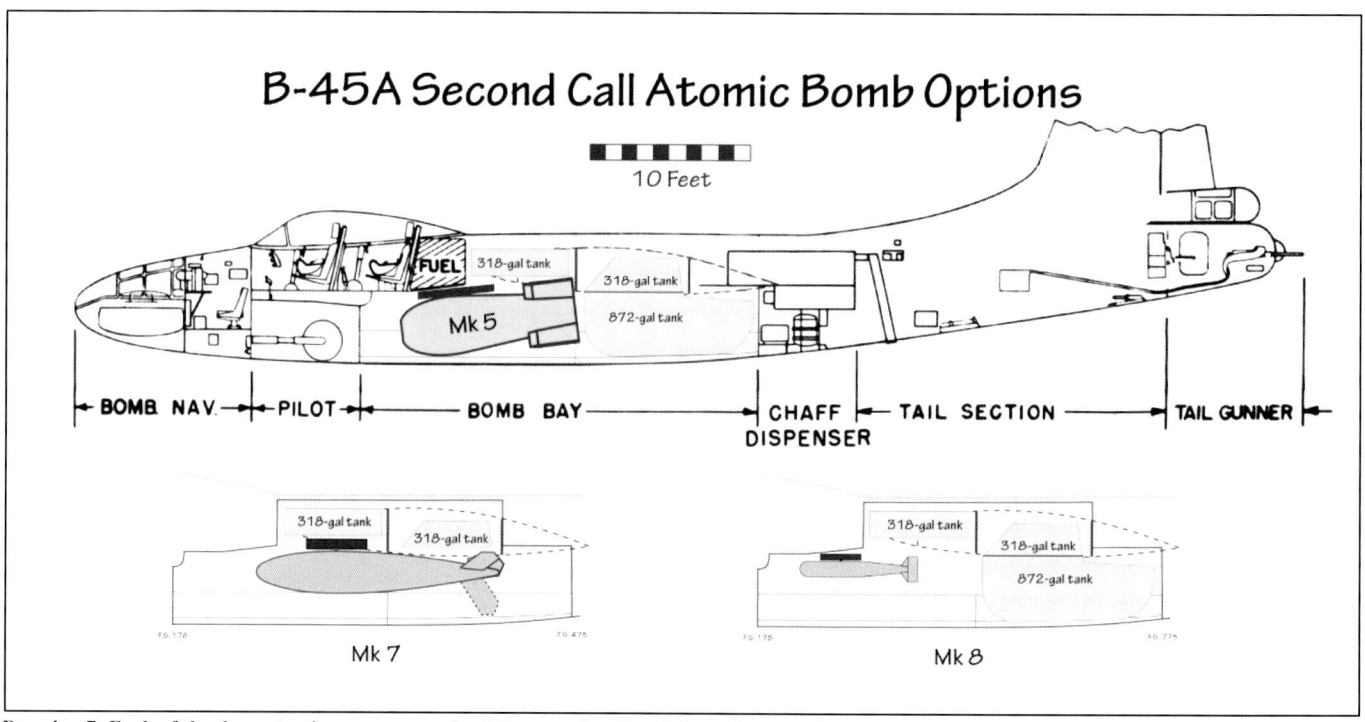

Drawing 7: Each of the three atomic weapons was loaded on a unique bomb-bay adapter and rack that had to consider the size and shape of the weapons as well as its weight and the quantity of fuel in the aft bomb bay tank when installed. This maintained the aircraft's center-of-gravity within limits from take-off, weapon delivery, through mission recovery. This drawing shows the locations of the A-7 chaff dispenser and Emerson A-6 turret installations following the BAACKBREAKER/SECOND CALL modifications. (Tommy Thomason)

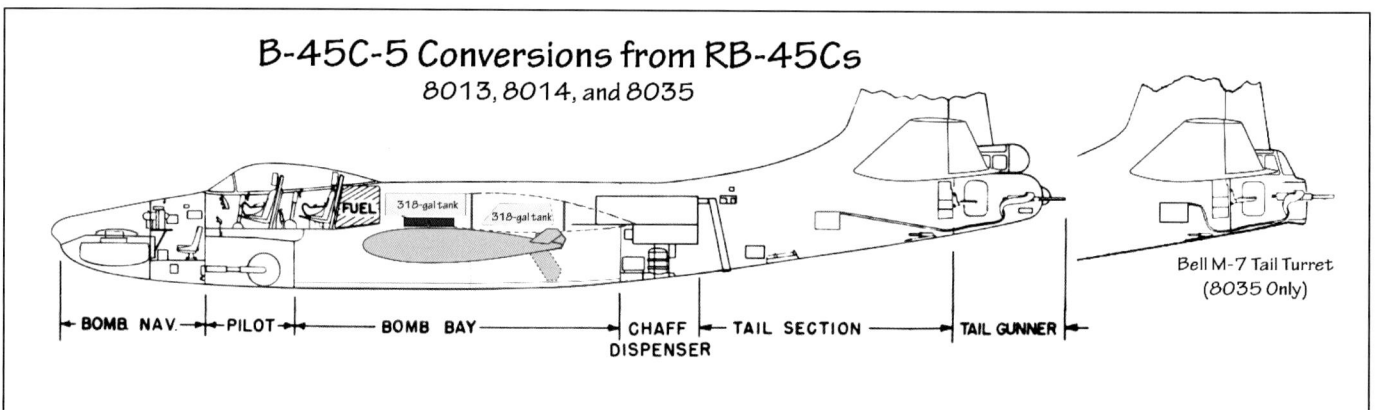

Drawing 8: All camera gear and photographic support systems were removed from the RB-45Cs. The AN/APN-84 SHORAN and the AN/APQ-24 bomb/nav radar were installed along with Emerson A-6 radar ranging tail turret fitted to 8013 and 8014. 8035 retained its Bell M-7 turret installed when the 19th TRS was reassigned to Europe in 1954. Full atomic weapons capacities were fitted and the ground single-point-refueling and in-flight-aerial-refueling capability was retained. 8035 was identical in all respects to 8013, 8014 and 8023 except for the Bell M-7 turret fitted. This turret had its ammo boxes fitted farther aft with shorter ammo belts to the turret and no ranging radar. (Tommy Thomason)

Photo 30: The only external indication that 8014 was a B-45C-5 was its serial number and the tip tanks. Other than that, 8014 looked like all BACKBREAKER/SECOND CALL B-45A-5s. B-45C-5 8014 was assigned to the 86th Bombardment Squadron. Note that the aircraft's tail turret was armed, which was quite common on all USAF aircraft stationed in the UK at the time. 8014 was the first B-45C-5 delivered to the 47th Bombardment Wing on 10 August 1954 and internally was identical to a SECOND CALL B45A-5. The aircraft previously never had a turret fitted even though it flew combat missions in the Far East; and therefore, carried an Emerson A-6 instead of a Bell M-7 when selected for transfer to Europe. (Courtesy Paul Minert)

Photo 31: B-45C-5 8035 was photographed on the ramp at RAF Sculthorpe during a summer airshow in 1955. 8035 was assigned to the 84th Bombardment Squadron and carried its squadron insignia on the outboard sides of its drop tanks. It was different from the other trio of B-45C-5s in the 86th in that this aircraft was fitted with a Bell M-7 turret, as were all 19th Tactical Reconnaissance Squadron RB-45Cs. (Big Bird Aviation)

Photo 32: B-45A-5 7082 assigned to the 84th Tactical Bombardment Squadron is shown on static display at RAF Sculthorpe in the late 1950s. The image shows the late ECM antennae. (Paul Minert)

During the research for this book, it was discovered that the SECOND CALL B-45A-5s and B-45C-5s had the capability to carry the Mk 7 weapon externally. One Mk 7 could be carried under either engine nacelle, with a single 500-gallon Fletcher tank carried under the opposite nacelle for balance **(Drawing 9)**. The information was contained in the Standard Aircraft Characteristics document for the B-45A (BACKBREAKER) Tornado, dated 25 January 1957. That date would have meant the information was from SECOND CALL-modified aircraft. The first detail was contained in the Landing and Performance – Typical Mission section which displayed data for four different sorties. Mission Three, labeled EXTERNAL STORES, called for an external load of 1,700 lbs., or the weight of a single MK 7, and a fuel weight of 40,500 lbs. Combat radius was listed as 795 nautical miles, an increase of 391 nautical miles. In the section on Formula NOTES for computing range, airspeed, altitude and so on, it states: "Under nacelle tanks and droppable bomb bay tanks are dropped when empty except in Mission III when one under-nacelle tank is carried to target to balance drag of the external load under the other nacelle". (Thanks to Tommy Thomason for finding this information.)

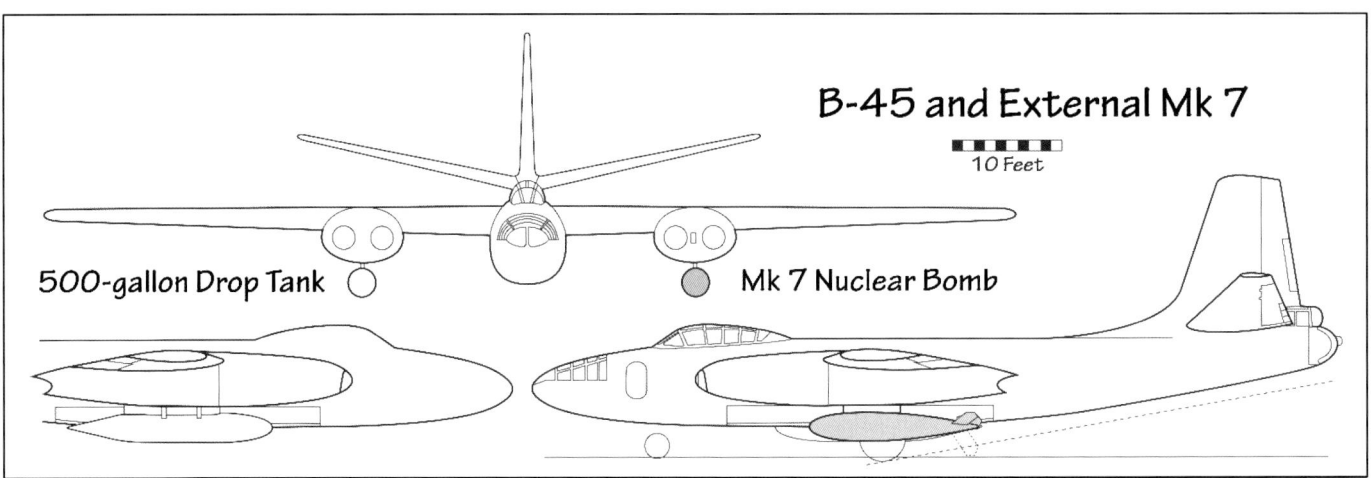

Drawing 9: This drawing shows what the external Mk 7 carried under the port nacelle and its balance drag of a 500-gallon external tank under the opposite nacelle would have looked like. Due to the approximately 25% increase in mission radius, it appears that was the reason for modifying an unknown number of SECOND CALL B-45A-5 and B-45C-5 fleet with the monitoring, arming and release electronics to carry a MK 7 externally. (Tommy Thomason)

Targeting was applied to major transportation nodes for rail and highway links, airfields, and armor casernes (barracks), as well as to destroy Soviet and Warsaw Pact forces likely to pass through the Fulda Gap with the intent to attack and destroy allied forces in West Germany. The list of targets that was short however despite what was known about the enemy. By 1955 the wing had some 55 aircraft in three squadrons, with a launch estimate of only 30 aircraft and no restrike capability.

Throughout its service, the SECOND CALL B-45A-5s were never able to plan or conduct a rapid reaction, rapid start, and taxi-to-take-off with armed-weapons, as SAC would pioneer and practiced for decades in just a few more years. Therefore, generation and mission launch was a slow process that relied on critical early-warning to be successful.

In summary, the entire fuselage was totally re-designed, re-configured, re-fitted or re-equipped. What is truly astonishing was that in many, if not most cases, this was the first use on this equipment, especially special weapons, in a multi-engine turbojet aircraft. Further, these projects were completed from scratch on over 54 previously non-combat-capable aircraft between 14 July 1951 and 30 August 1954. And finally, three, now combat-capable atomic-armed squadrons, personnel and their families were moved overseas.

Photo 33: B-45A-5 7046 of the 85th Bombardment Squadron was photographed at an airshow at Prestwick, Scotland in 1956. Prestwick was familiar with USAF aircraft, being a common refueling stop for aircraft crossing the Atlantic. (Author's Collection)

Internal Modifications....

The following illustrations show the B-45A-5s and B-45C-5s of the 47th were reworked by the BACKBREAKER/SECOND CALL Program. First are illustrations of the forward fuselage modifications and changes:

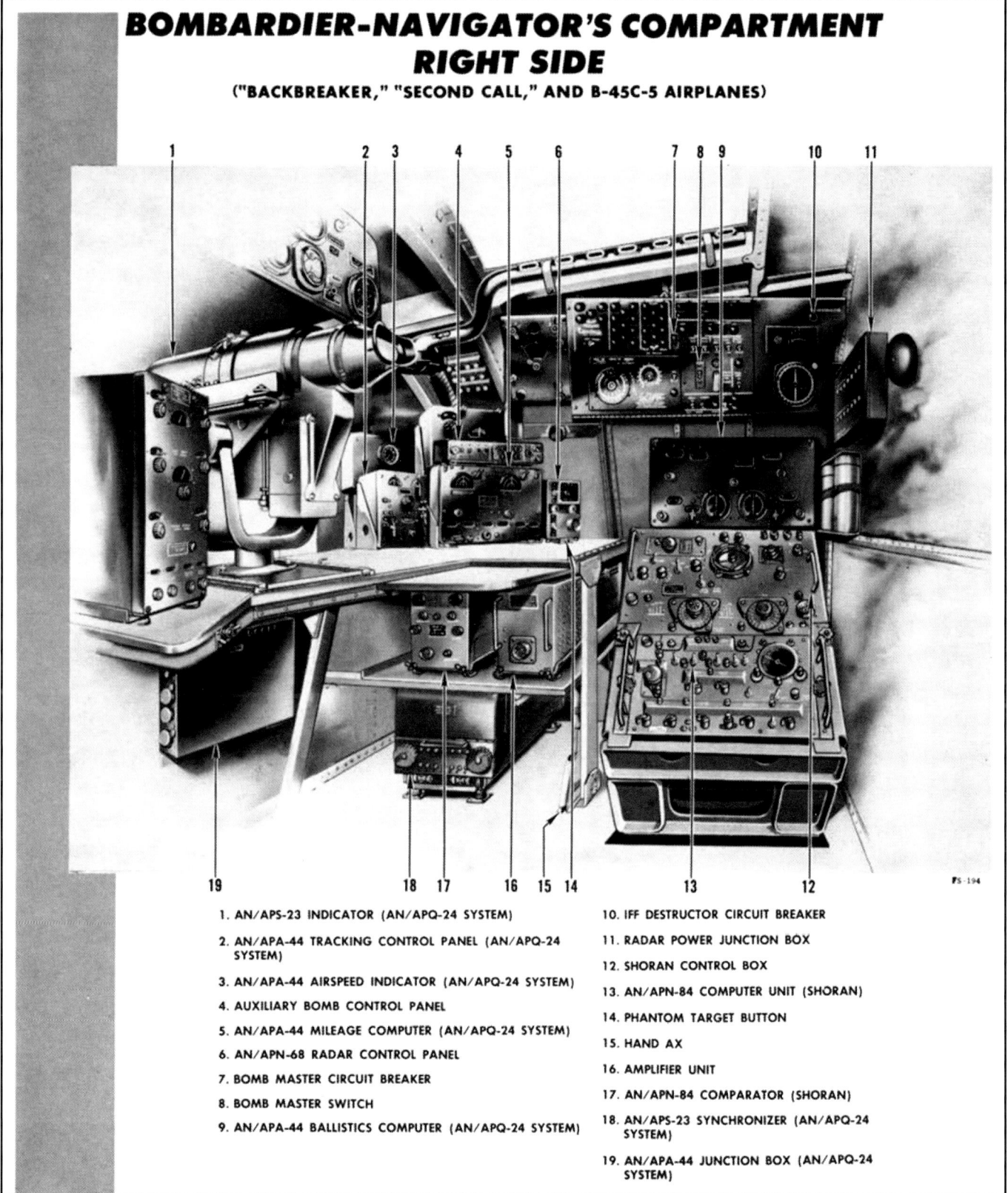

Photo 34: BACKBREAKER / SECOND CALL modifications for the installation of the SHORAN system and equipment related to the AN/APQ-24 Bomb/Nav radar took up what spare space there had been in the Bomb/Nav Station. (NMUSAF)

BOMBARDIER-NAVIGATOR'S COMPARTMENT
LEFT SIDE
("BACKBREAKER," "SECOND CALL," AND B-45C-5 AIRPLANES)

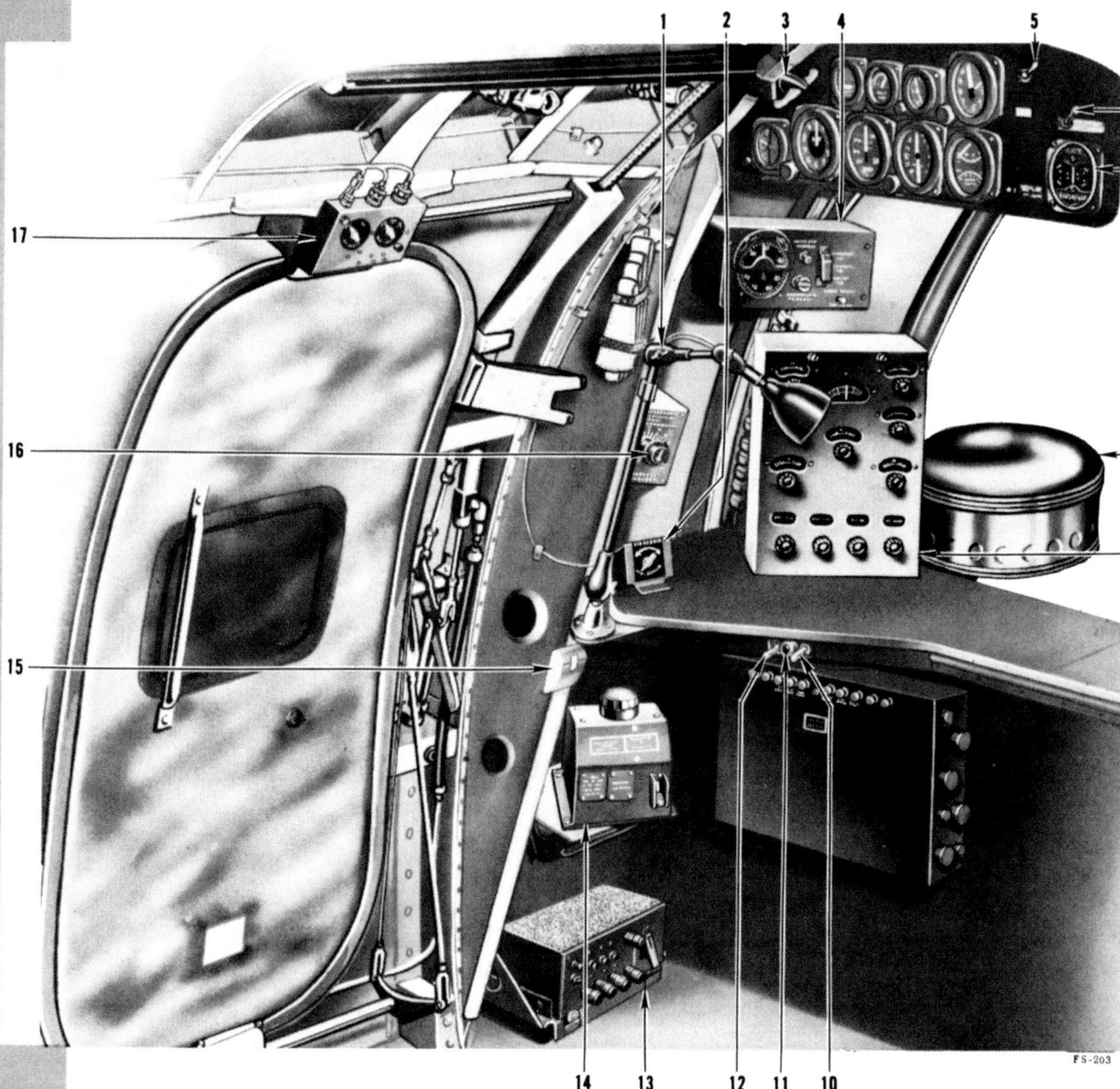

1. CHART TABLE SPOTLIGHT
2. CHART TABLE SPOTLIGHT RHEOSTAT KNOB
3. MANUAL EMERGENCY AUXILIARY BOMB RELEASE HANDLE
4. RADAR PRESSURIZATION CONTROL HANDLE
5. CABIN PRESSURE SWITCH
6. SHORAN SENSITIVITY SWITCH
7. POSITION DEVIATION INDICATOR (SHORAN)
8. AN/APS-23 RECEIVER-TRANSMITTER (AN/APQ-24 SYSTEM)
9. AN/APS-23 CONTROL PANEL (APQ-24 SYSTEM)
10. RADAR BOMB SCORING-TONE SWITCH
11. RADAR BOMB SCORING-TONE LIGHT
12. RADAR BOMB SCORING-TONE MASTER SWITCH
13. T-24 TEST BOX STOWED (SPECIAL STORE SYSTEM)
14. AUTOPILOT CONTROLLER
15. ASH TRAY
16. AUTOTRANSFORMER KNOB (AN/APQ-24 SYSTEM)
17. CAMERA CONTROL PANEL

Photo 35: New equipment including four new components for the AN/APQ-24 Bomb/Nav Radar and for monitoring the atomic weapon carried by the aircraft were added to the port side of the Bomb/Nav station. Note that the crew entry and exit door now had a window allowing the crew to ensure the escape area near the door was free from fire in ground emergencies. (NMUSAF)

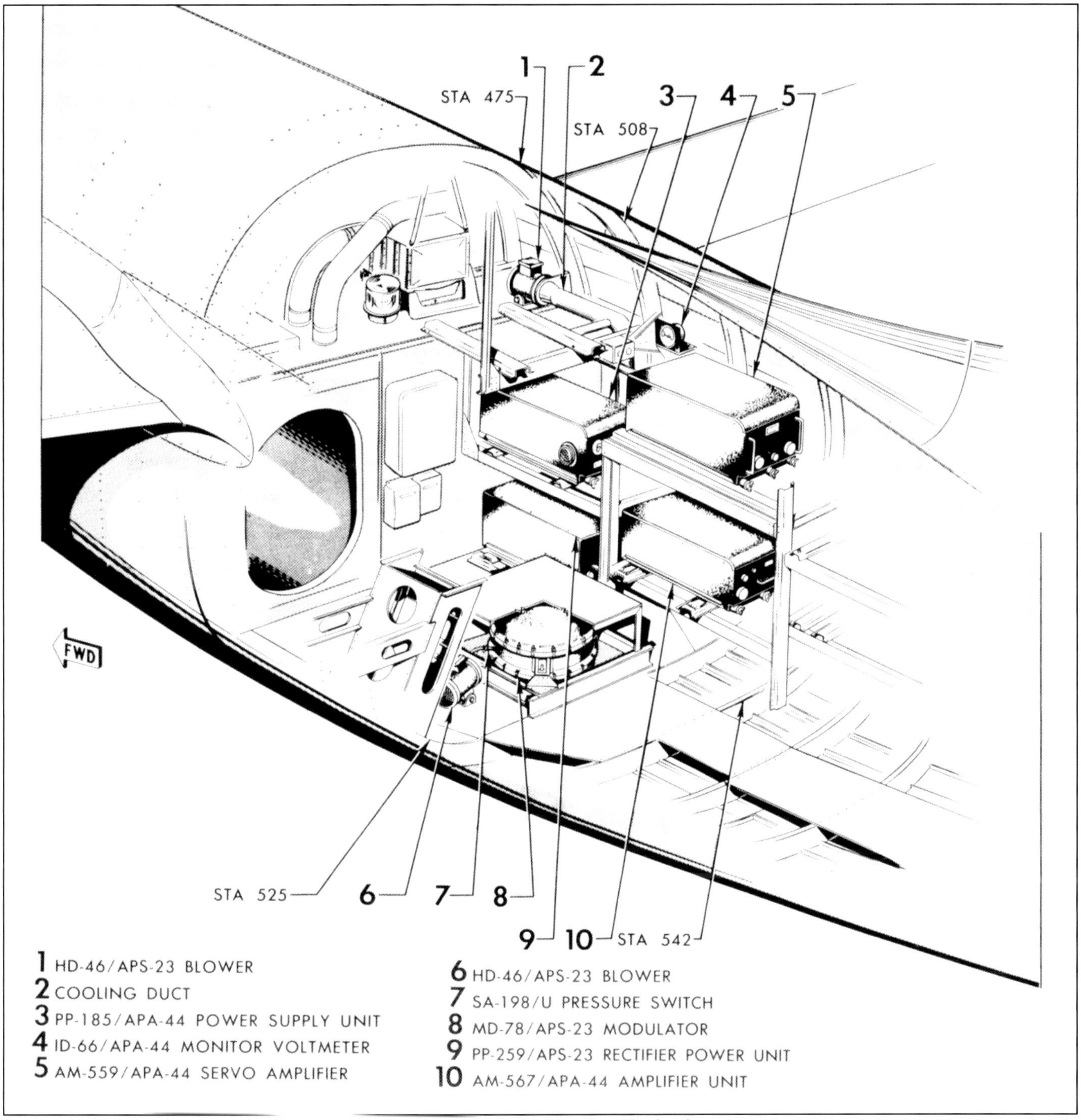

1. HD-46/APS-23 BLOWER
2. COOLING DUCT
3. PP-185/APA-44 POWER SUPPLY UNIT
4. ID-66/APA-44 MONITOR VOLTMETER
5. AM-559/APA-44 SERVO AMPLIFIER
6. HD-46/APS-23 BLOWER
7. SA-198/U PRESSURE SWITCH
8. MD-78/APS-23 MODULATOR
9. PP-259/APS-23 RECTIFIER POWER UNIT
10. AM-567/APA-44 AMPLIFIER UNIT

Drawing 10: This drawing shows the aft unpressurized portion of the B-45A-5 fuselage. The hatch on the port side of the forward bulkhead, open in this drawing, leads to and from the aft end of the unpressurized bomb bay. (NMUSAF)

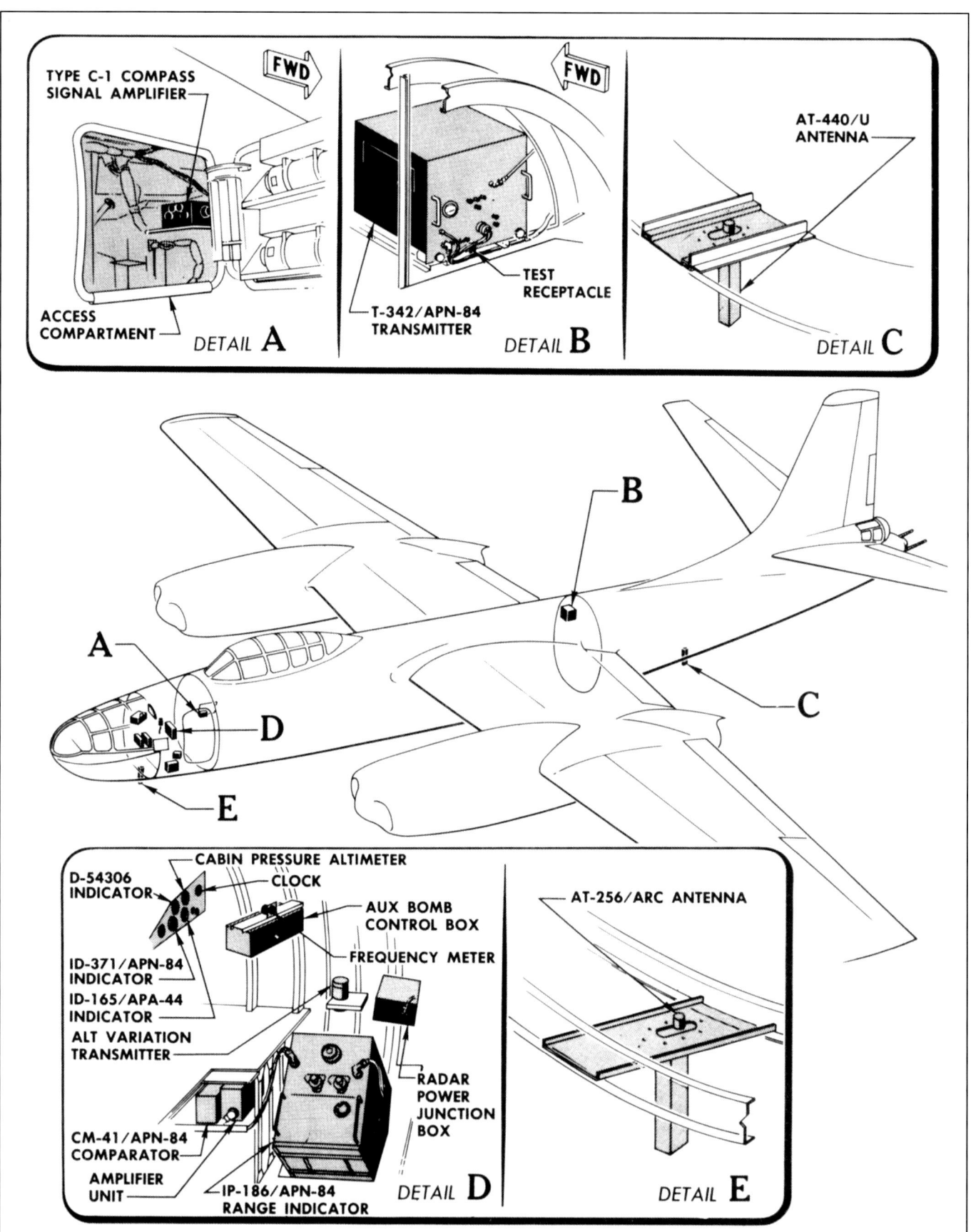

Drawing 11: This drawing shows the location of some of the beacon transmitters and receivers of the AN/APN-84 SHORAN system. The location of the SHORAN antennas were shown in Details C and E and if seen from the outside, were a good indication that aircraft was equipped with SHORAN. The AN/APN-84 SHORAN and AN/APQ-24 installation also increased the empty weight of the aircraft. The BACKBREAKER/SECOND CALL aircraft were nearly 8,000 pounds heavier than the original production B-45A-1s and A-5s. (NMUSAF)

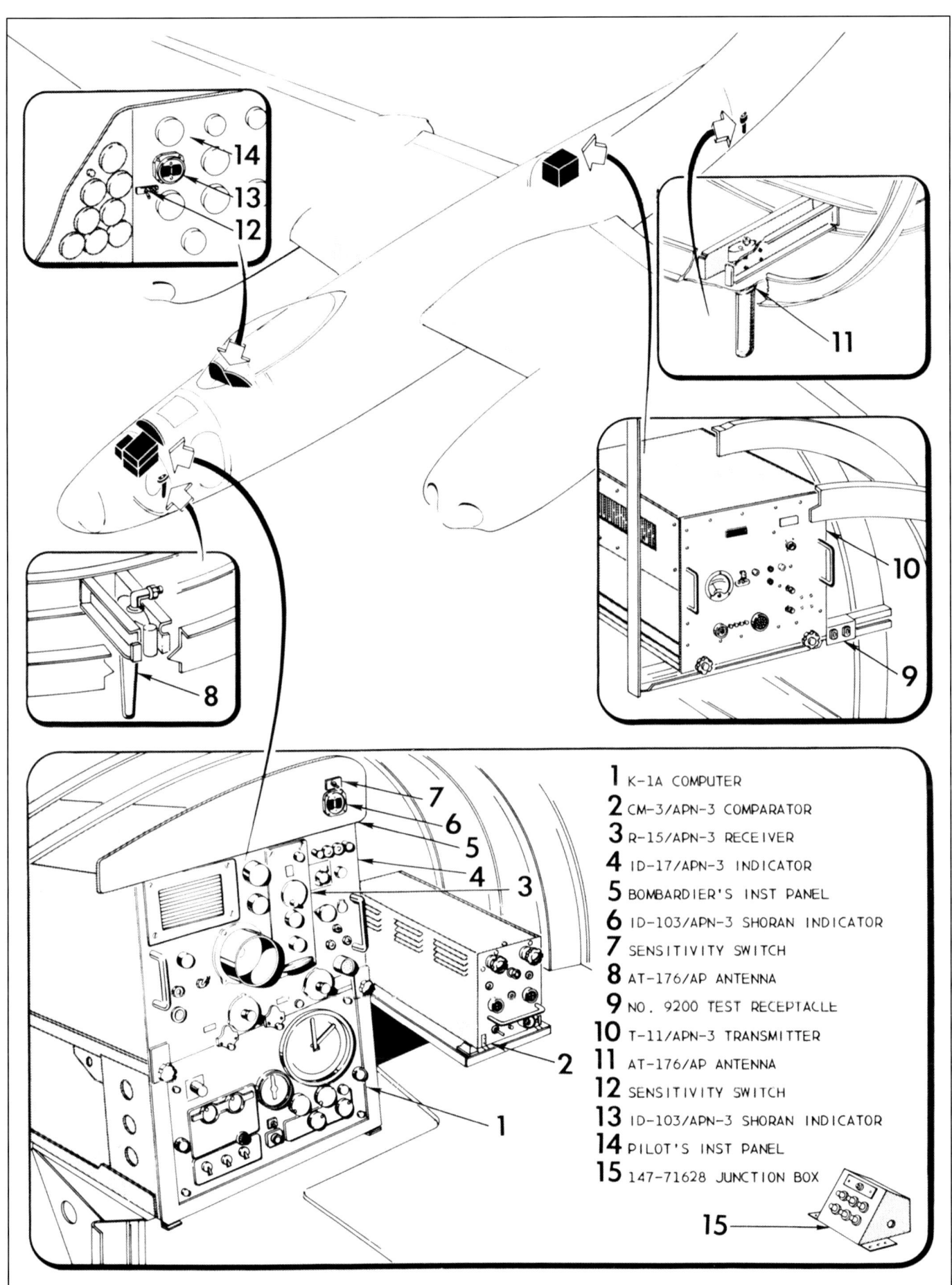

Drawing 12: This drawing shows the details of the controls on the front panels of the AN/APN-84 SHORAN gear on board the **BACKBREAKER/SECOND CALL** aircraft. (NMUSAF)

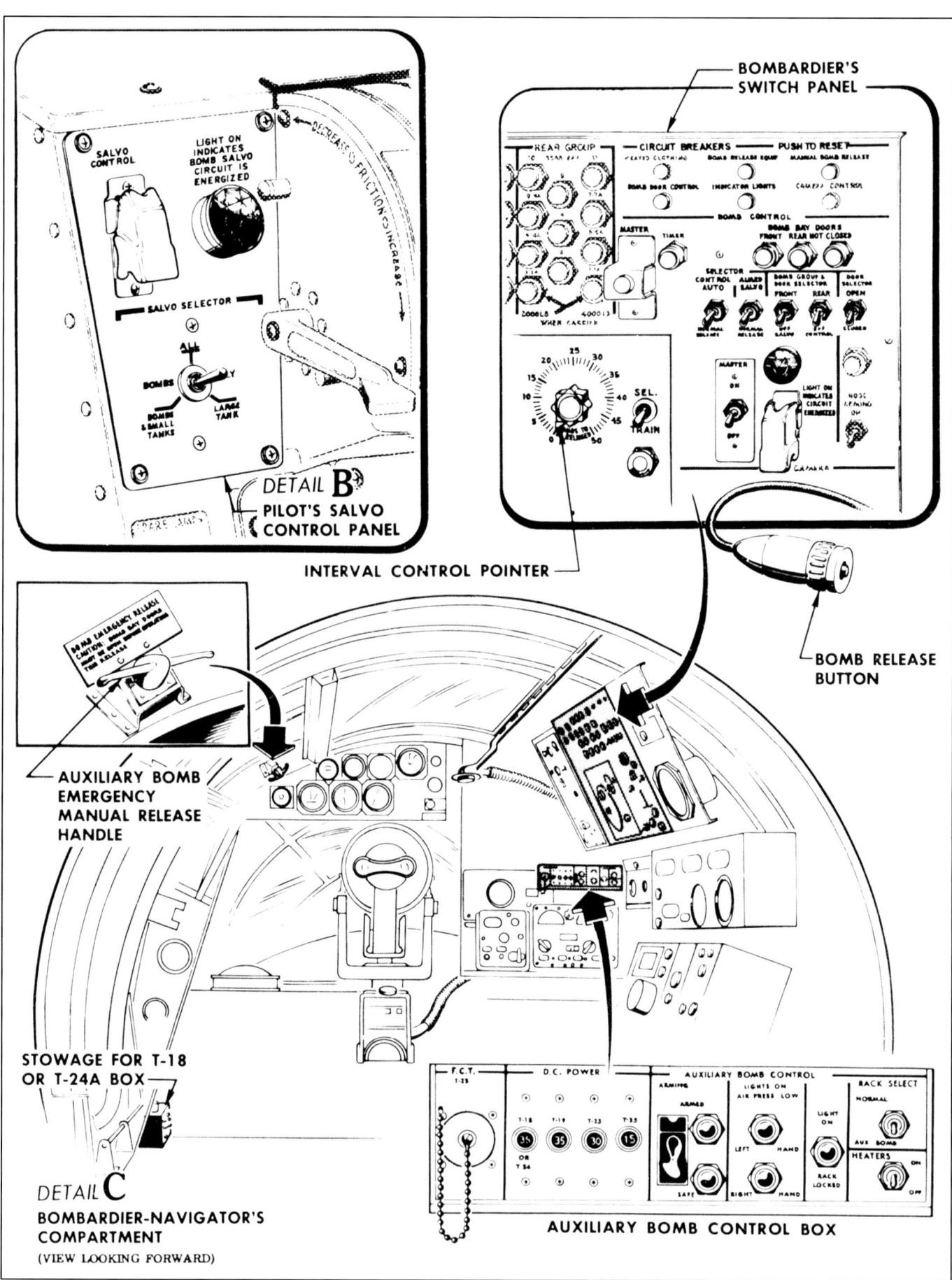

Drawing 13: This drawing shows changes required to the bombardier/navigation station as a result of carrying atomic weapons. The presence of equipment and controls related to atomic weapons was generally referred to as the Auxiliary Bomb System, so information related to the weapons could be presented in an unclassified manner. (NMUSAF)

The second section of the forward fuselage illustrations of the B-45A-5s and B-45C-5s modified by BACKREAKER/SECOND CALL, cover the pilot's and co-pilots stations and systems related to their stations.

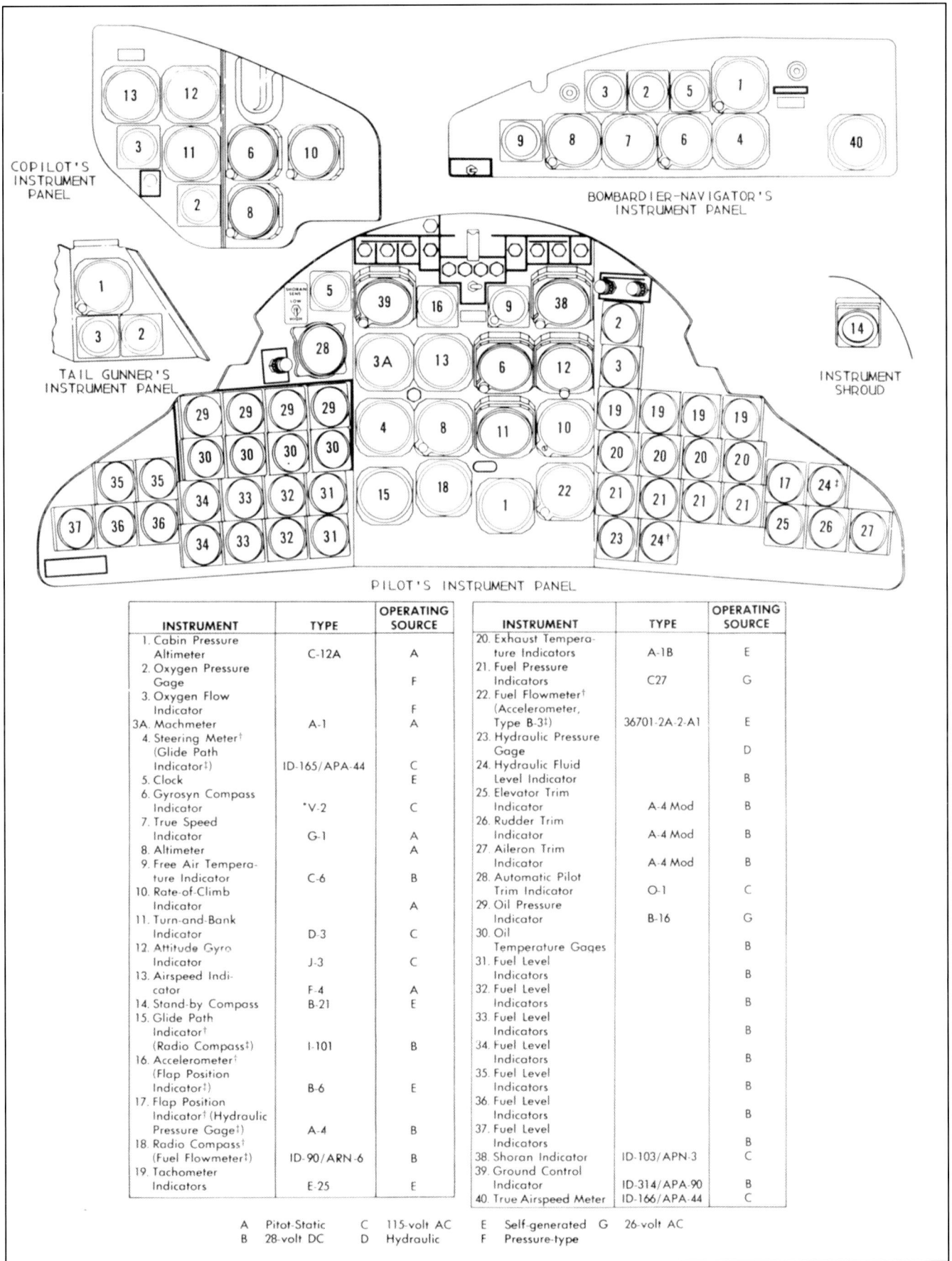

Drawing 14: This drawing lists the instruments, their type, and power sources for all the crew stations on the BACKBREAKER/SECOND CALL-modified aircraft. (NMUSAF)

PILOT'S INDICATORS
("BACKBREAKER," "SECOND CALL," AND B-45C-5 AIRPLANES)

1. SHORAN SENSITIVITY SWITCH
2. CLOCK
3. AN/APA-90 COMMAND INDICATOR (AN/APW-11 RADAR)
4. ACCELEROMETER
5. FLAP POSITION INDICATOR
6. POSITION DEVIATION INDICATOR (SHORAN)
7. FUEL FLOWMETER AND TOTALIZER INDICATOR
8. CABIN PRESSURE ALTIMETER
9. RADIO COMPASS INDICATOR
10. GLIDE SLOPE INDICATOR

PILOT'S STATION—RIGHT SIDE

1. HEAT AND VENT OUTLET
2. LIAISON RADIO CONTROL PANEL (PROVISION FOR AN/ARC-8 SET)
3. PNEUMATIC SYSTEM BLEEDER VALVE *
4. COMMAND RADIO CONTROL PANEL
5. RADIO COMPASS CONTROL PANEL
6. INTERPHONE AND MIXER SWITCH PANEL
7. INSTRUMENT APPROACH CONTROL PANEL
8. FLUORESCENT LIGHT
9. CONTROL LOCK LEVER
10. OXYGEN REGULATOR
11. PNEUMATIC SYSTEM BLEEDER VALVE
12. HEATED-CLOTHING PANEL
13. EMERGENCY BRAKE HYDRAULIC SYSTEM PRESSURE GAGE
14. EMERGENCY HYDRAULIC SELECTOR VALVE HANDLES
15. CIRCUIT-BREAKER PANELS
16. RADIO-FREQUENCY CARD †
17. CANOPY EMERGENCY RELEASE HANDLE
18. STAND-BY COMPASS LIGHT RHEOSTAT KNOB †
19. CABIN AIR KNOB
20. LANDING GEAR LOCK EMERGENCY RELEASE KNOB
21. WING TIP TANK MECHANICAL RELEASE HANDLE ‡
22. LANDING GEAR HANDLE
23. CABIN TEMPERATURE MANUAL OVERRIDE SWITCH §
24. HYDRAULIC MAIN SYSTEM PRESSURE CONTROL HANDLE
25. AUTOPILOT ENGAGING HANDLE
26. WINDSHIELD WIPER SWITCH

* Airplanes not changed by T.O.
† B-45C-5 Airplanes
‡ B-45C Airplanes
§ Some Airplanes

Photo 36: This image shows the various key indicators on the pilot's instrument panel. The primary instrument is the Instrument Landing System Indicator (item 10) that allowed safe landings down to ceilings of 200 feet and visibility of ¼ mile. (NMUSAF) Photo 37: This image shows the new Ultra High Frequency (UHF) radio communication set used for aircraft on a strike mission to clearly communicate over short distance with one another. UHF was clear with little static or other interference and became the standard frequencies of nearly all military communications. (NMUSAF)

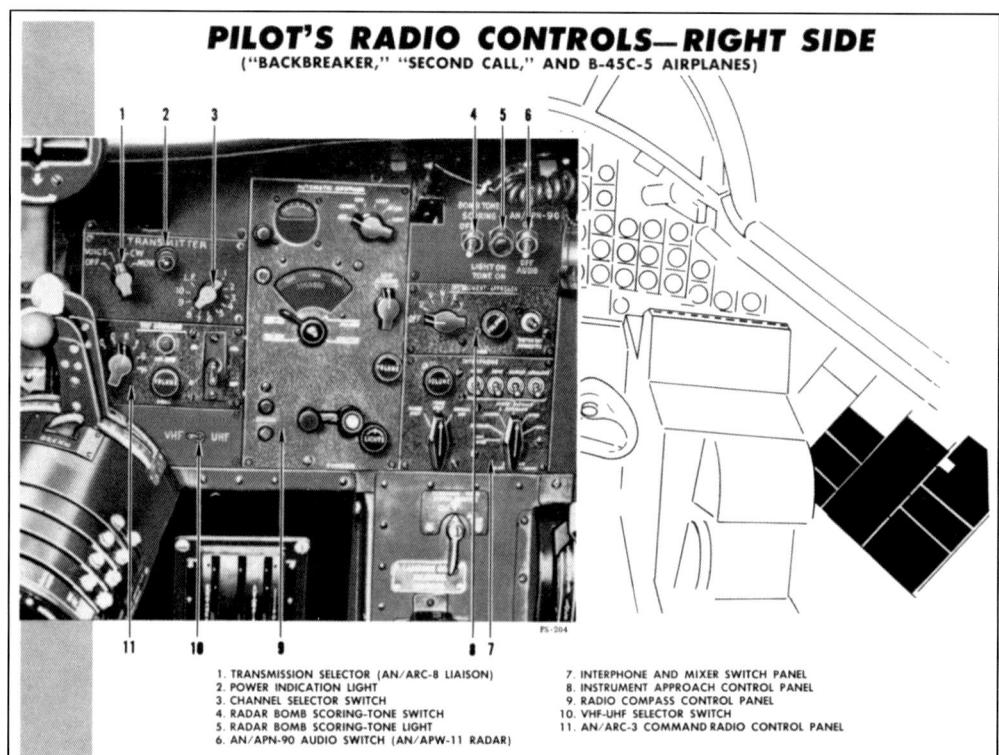

PILOT'S RADIO CONTROLS—RIGHT SIDE
("BACKBREAKER," "SECOND CALL," AND B-45C-5 AIRPLANES)

1. TRANSMISSION SELECTOR (AN/ARC-8 LIAISON)
2. POWER INDICATION LIGHT
3. CHANNEL SELECTOR SWITCH
4. RADAR BOMB SCORING-TONE SWITCH
5. RADAR BOMB SCORING-TONE LIGHT
6. AN/APN-90 AUDIO SWITCH (AN/APW-11 RADAR)
7. INTERPHONE AND MIXER SWITCH PANEL
8. INSTRUMENT APPROACH CONTROL PANEL
9. RADIO COMPASS CONTROL PANEL
10. VHF-UHF SELECTOR SWITCH
11. AN/ARC-3 COMMAND RADIO CONTROL PANEL

The ILS system had entered service just at end of WW II and there were still few airfields that were equipped with it, but RAF Sculthorpe was one. It was radio-based, with one transmitter broadcasting a constant "localizer" signal that was in effect an extension of the centerline of the runway. The other transmitter sent a "glide slope" signal that provided the aircraft with a constant rate of descent indication. Deviation from either the centerline extension or the glide path was shown by the changing position of two bars, one horizontal and one vertical from two fixed indicators on the ILS indicator.

Photo 38: This image shows the pilot's radio control panel and controls that handled the aircraft's use of a ground-based radar directed bombing system that was used for tens of thousands of all-weather and night bombing sorties during the South East Asian War beginning ten to twelve years later. (NMUSAF)

COPILOT-RADIO OPERATOR'S RADIO CONTROLS—
RIGHT SIDE
(SOME AIRPLANES)

1. CODE TRANSMISSION KEY
2. AN/ARC-8 LIAISON RECEIVER POWER SWITCH
3. COPILOT'S CHECK LIST
4. FREQUENCY BAND SELECTOR SWITCH
5. INTERPHONE CONTROL PANEL
6. INTERPHONE AMPLIFIER AUXILIARY GAIN KNOB
7. COPILOT-RADIO OPERATOR'S CHAFF DISPENSER CONTROL PANEL
8. RADIO CIRCUIT-BREAKER PANEL
9. LIAISON RADIO MONITOR SWITCH
10. LIAISON RECEIVER TUNING KNOB
11. LIAISON RECEIVER VOLUME KNOB
12. LIAISON RECEPTION SWITCH
13. AN/ARC-8 LIAISON TRANSMISSION SELECTOR SWITCH
14. LIAISON TRANSMITTER POWER SWITCH
15. LIAISON TRANSMITTER CHANNEL SELECTOR SWITCH
16. LIAISON TRANSMITTER TRANSFER SWITCH

Photo 39: This image shows the co-pilots starboard-side radio panels. One of the co-pilot's main duties was to handle, tune and communicate via the aircraft's radios throughout the mission. (NMUSAF)

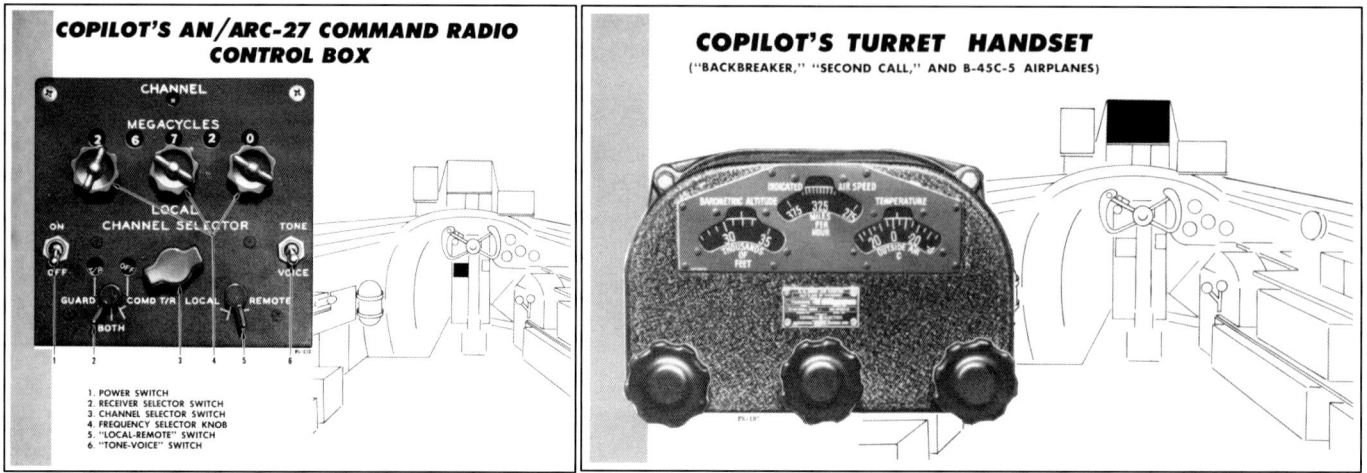

Photo 40: This image shows the co-pilot's radio control box for the "Command" set which was used if the aircraft's crew were the combat leaders of the mission. It allowed them and only them to direct all aircraft in that mission. (NMUSAF) Photo 41: This was the co-pilot's turret handset that allowed him to provide data to the turret operator's gunnery computer. Barometric altitude, indicated air speed in miles per hour, and outside air temperature in degrees Celsius, were passed to the turret gunner's system through this instrument. (NMUSAF)

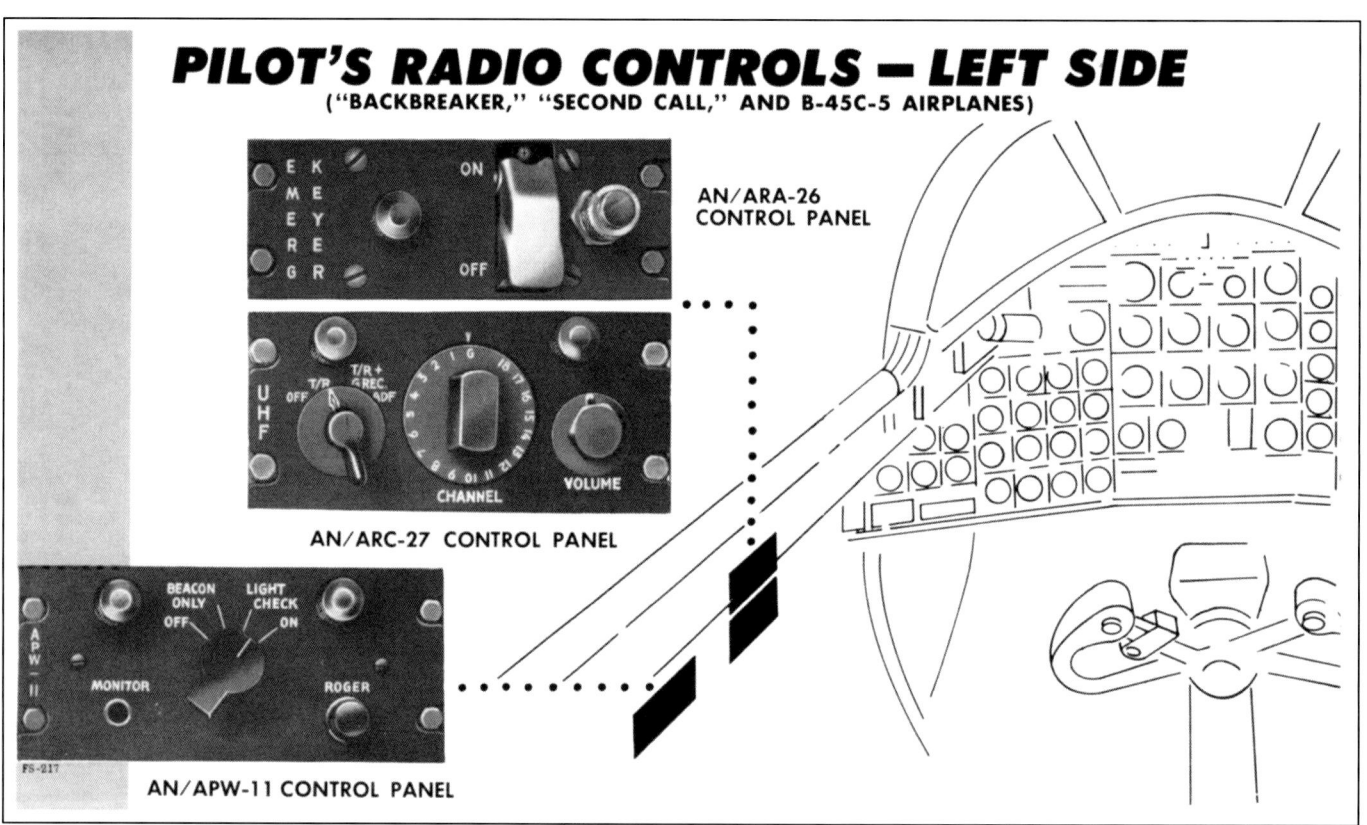

Photo 42: This image shows the pilot's radio panels along the port side of the cockpit, just below the canopy frame. (NMUSAF)

The third section of the fuselage interior illustrations of the B-45A-5s and B-45C-5s modified by BACKBREAKER/SECOND CALL, cover the modifications and additions to the forward, center, and aft Bomb Bay areas of the aircraft.

The following drawings show the three bomb shackles used in the B-45A-5 and B-45C-5 bomb bays. They are shown in their fore to aft locations.

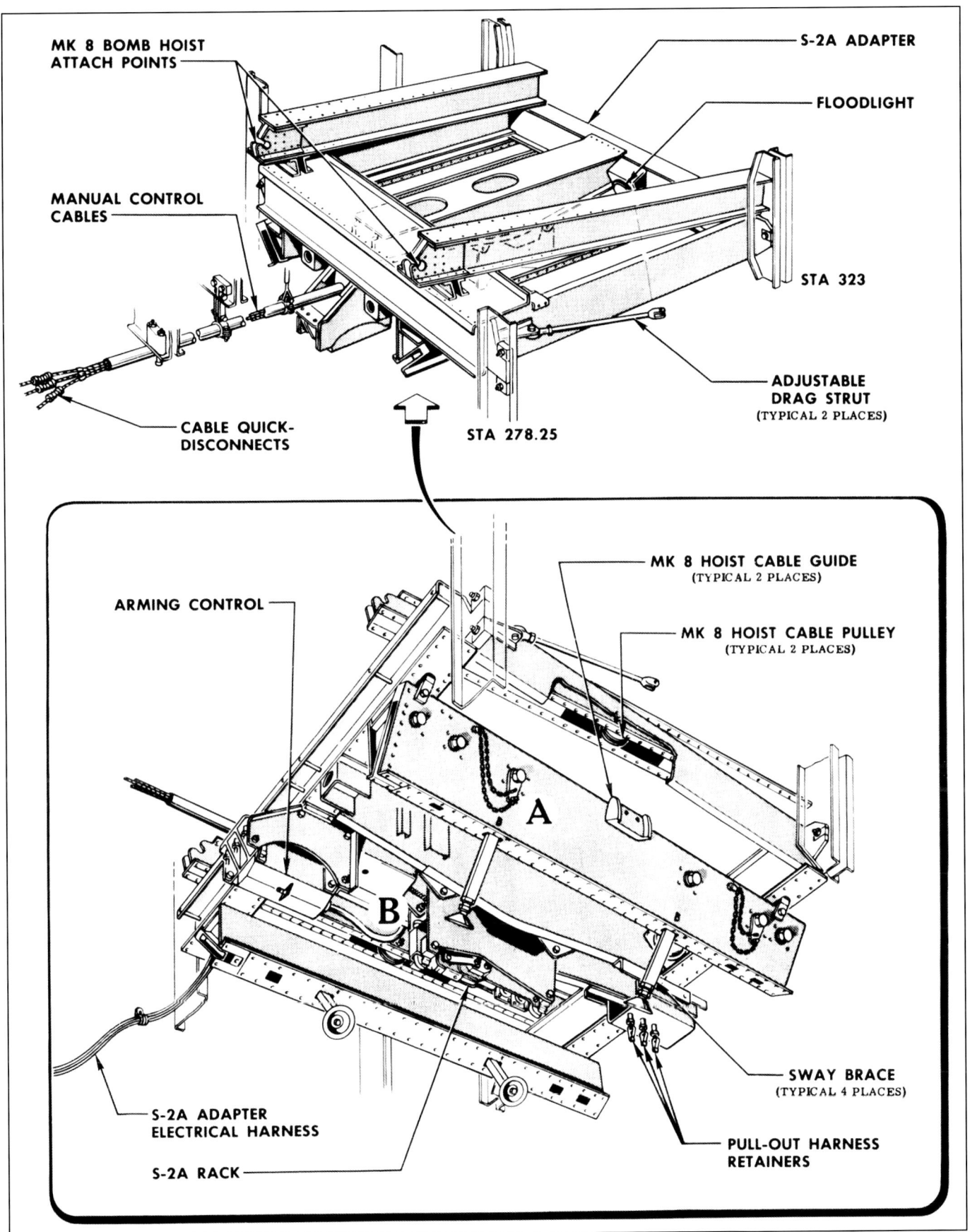

Drawing 15: This drawing shows both surfaces of the S-2 Shackle for the Mk 7 weapon. The upper surface drawing shows the heavy structure that attached to the aircraft's vertical mounting attachments and the manual control cables from the cockpit to the shackle. The lower portion of the drawing shows the four sway braces which secured the weapon and the hooks that suspended the weapon at Detail B. Detail information is shown in Drawing 16. (NMUSAF)

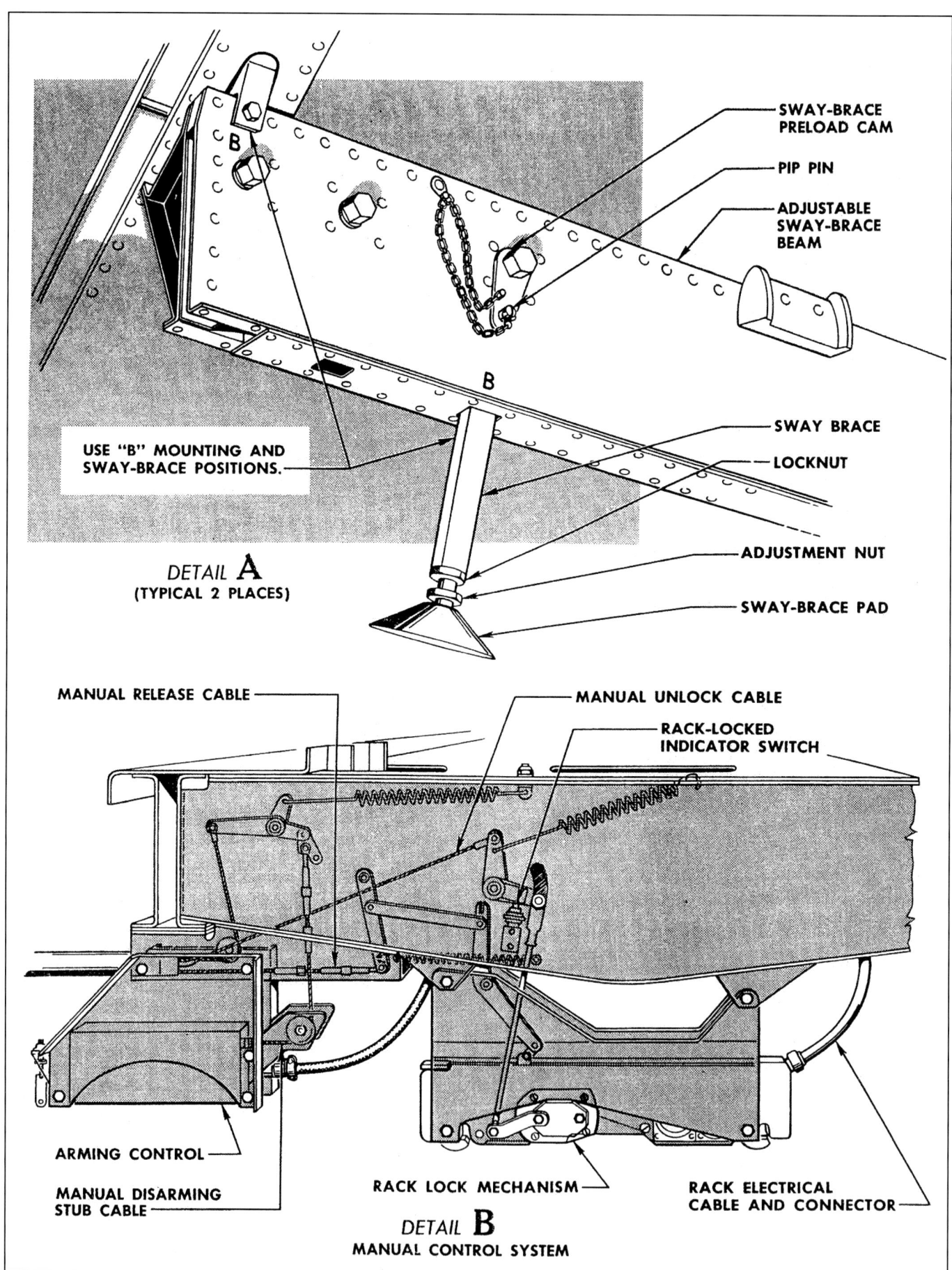

Drawing 16: This drawing shows details of the lower portion of the S-2 Shackle. Detail A covers the sway braces that secured the weapon during flight. Detail B covers the manual release system and the hooks that suspended the weapon and released it on automatic or manual command from the cockpit. (NMUSAF)

Photo 43: This image shows a mockup of the U-2 Rack designed for the Mk 5 weapon, looking forward. The sway braces and release rack are the key details shown, as well as the heavy structure needed to support the rack and its weapon in the bomb bay. (Courtesy The Boeing Aircraft Company)

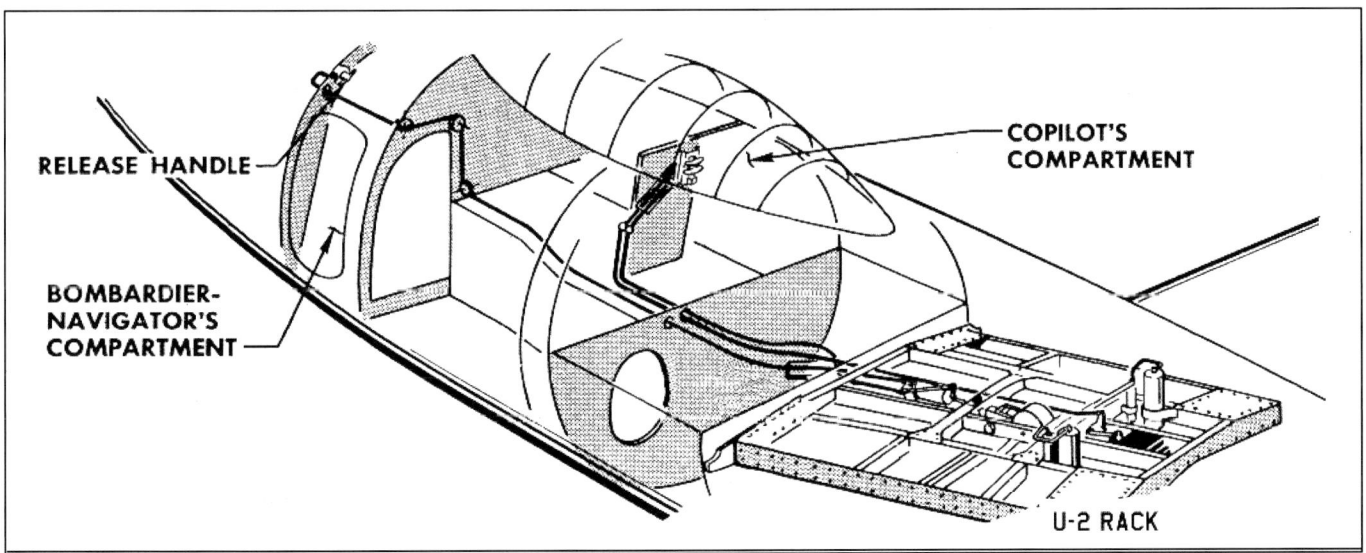

Drawing 17: This drawing shows the location of the U-2 special weapons rack (Drawings 18 and 19) for the Mk 5 weapon and its sloping angle in the forward Bomb Bay. (NMUSAF)

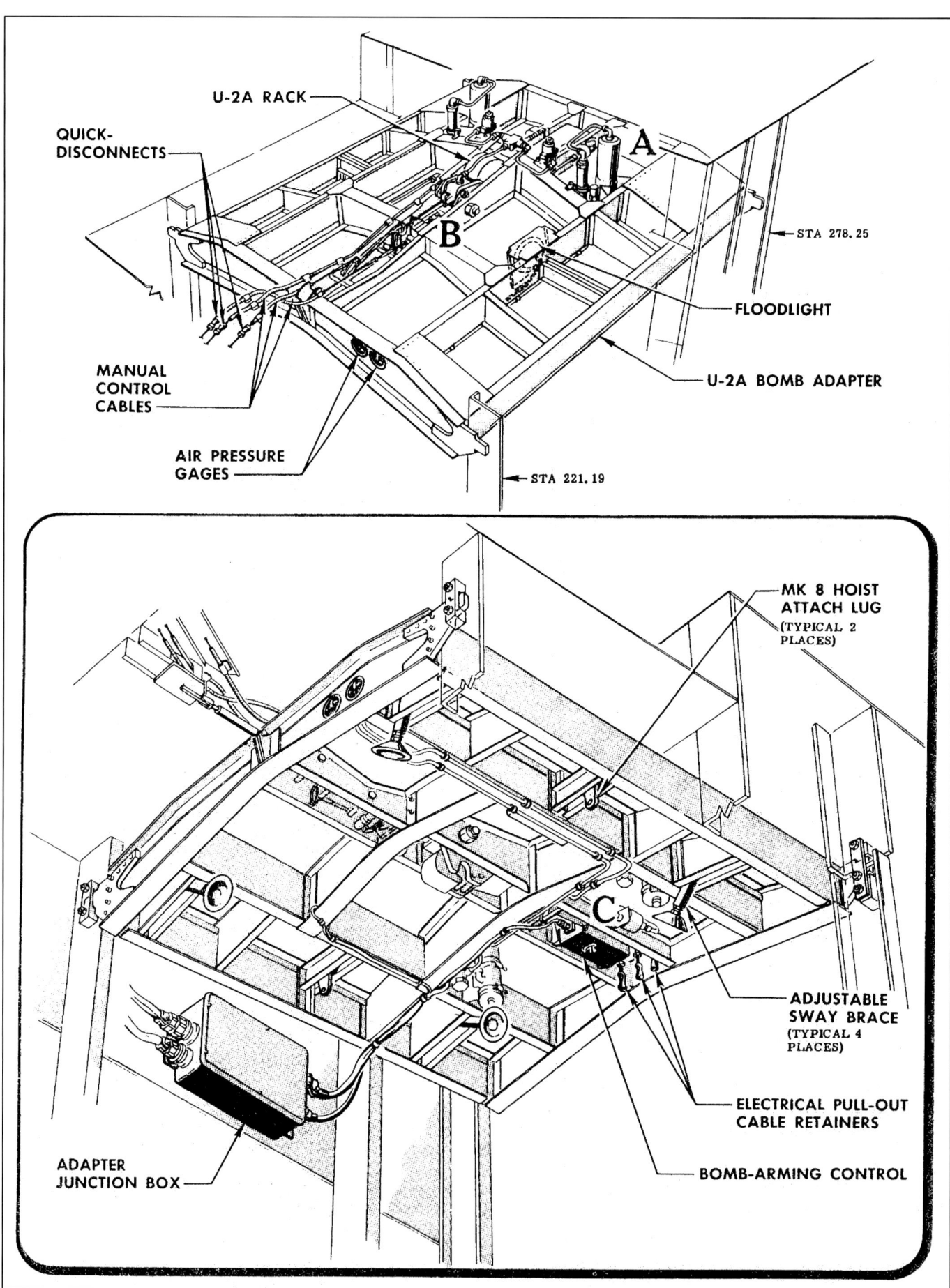

Drawing 18: This drawing shows the U-2 rack for the largest weapon carried by the B-45A-5 and B-45C-5, the Mk 5. The upper surface is on the top portion of the drawing and the lower surface is on the bottom portion. Details of the rack are on Drawing 19. The heavy hook which suspended the weapon is located in the center of the lower side of the rack. (NMUSAF)

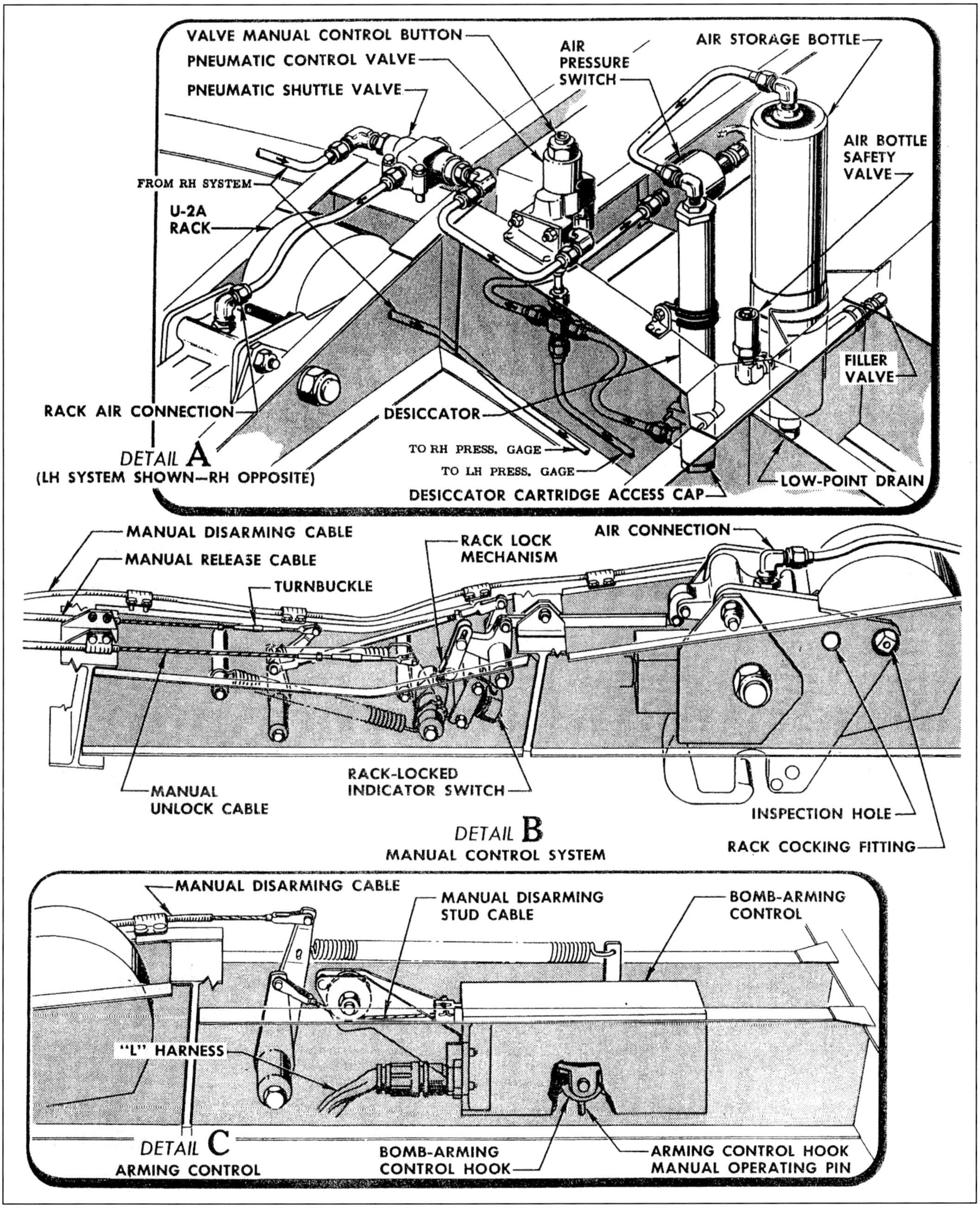

Drawing 19: These are the U-2 rack details. Detail A covers details of the pneumatic system on the upper surface, and an upper view of the hook that suspended the weapon. Detail B shows the manual control and release system on the lower surface of the rack. The suspension hook is at the aft end of the main structure in the Drawing. Detail C covers the arming controls for the Mk 5 when loaded, also on the lower surface of the rack. (NMUSAF)

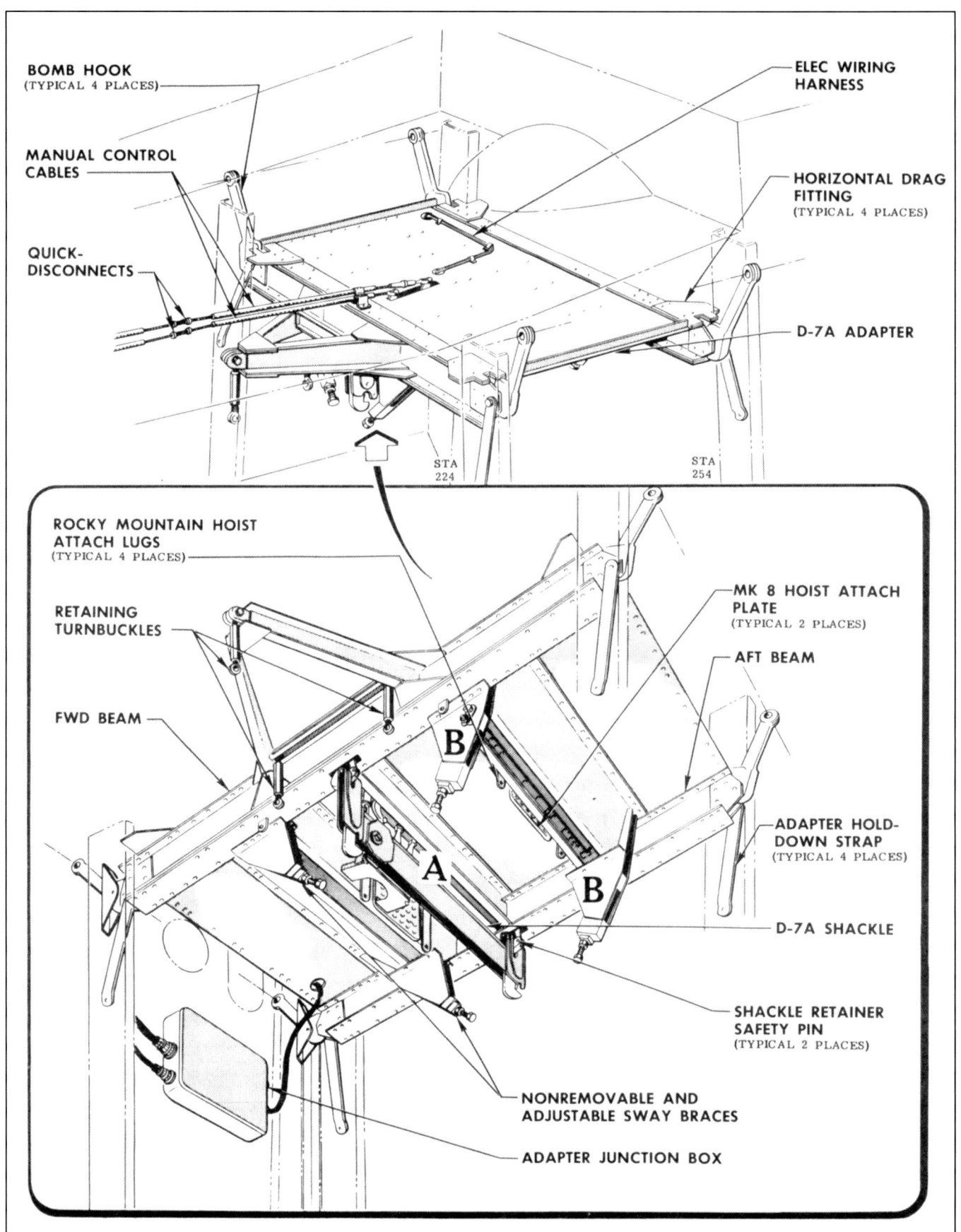

Drawing 20: This image shows the upper surface (on top) and lower surface (below) details of the D-7 shackle for the Mk 8 weapon. Details A, B, and C are located on Drawing 21.

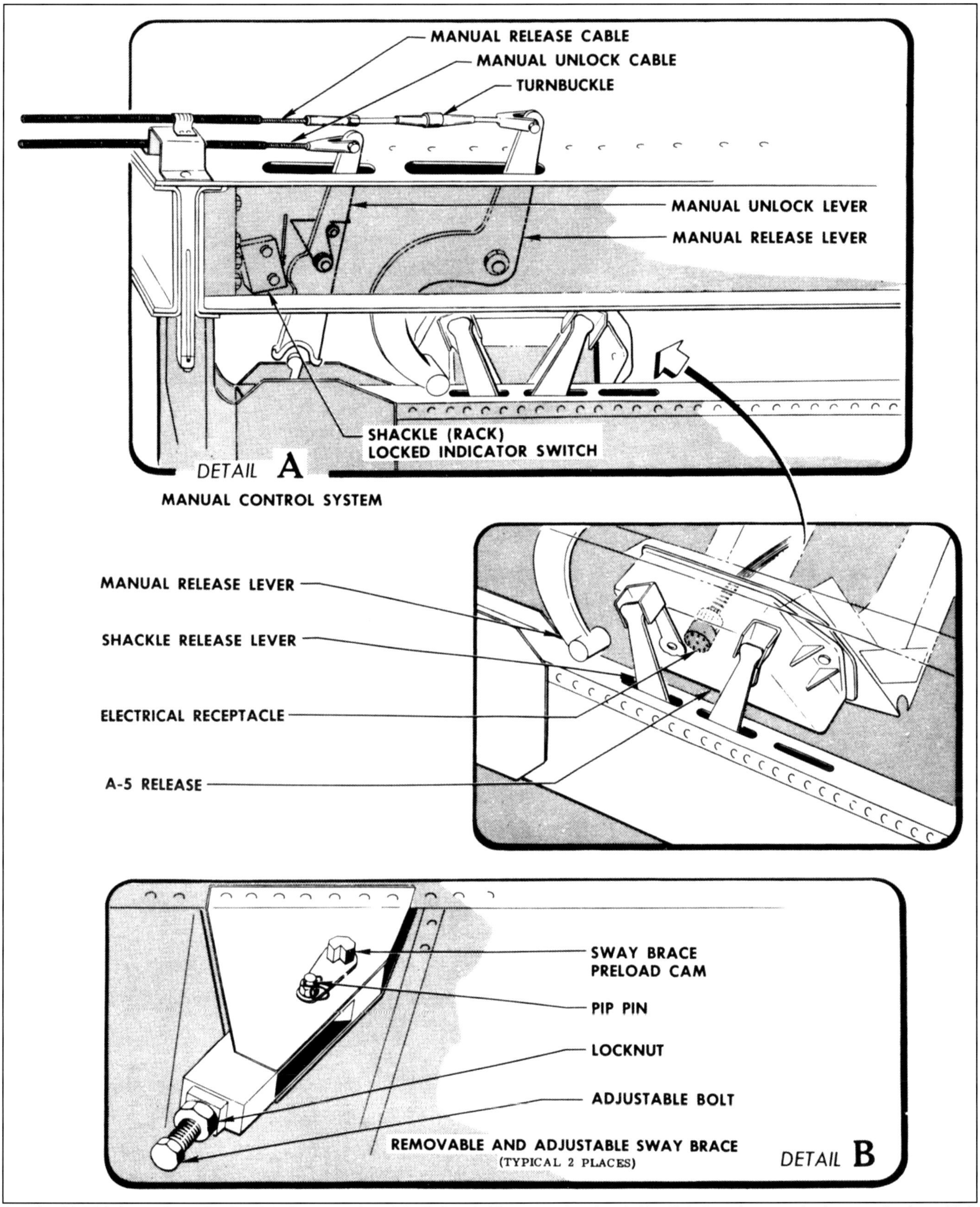

Drawing 21: This image shows details of the lower portion of the D-7 rack. The upper drawings in Detail A show the manual release mechanism with its cable from the cockpit which pushes the manual lever that pushes the two shackle release levers closed to release the weapon. Detail B shows one of the two adjustable sway braces that steady and hold the Mk 8 weapon firm in flight. (NMUSAF)

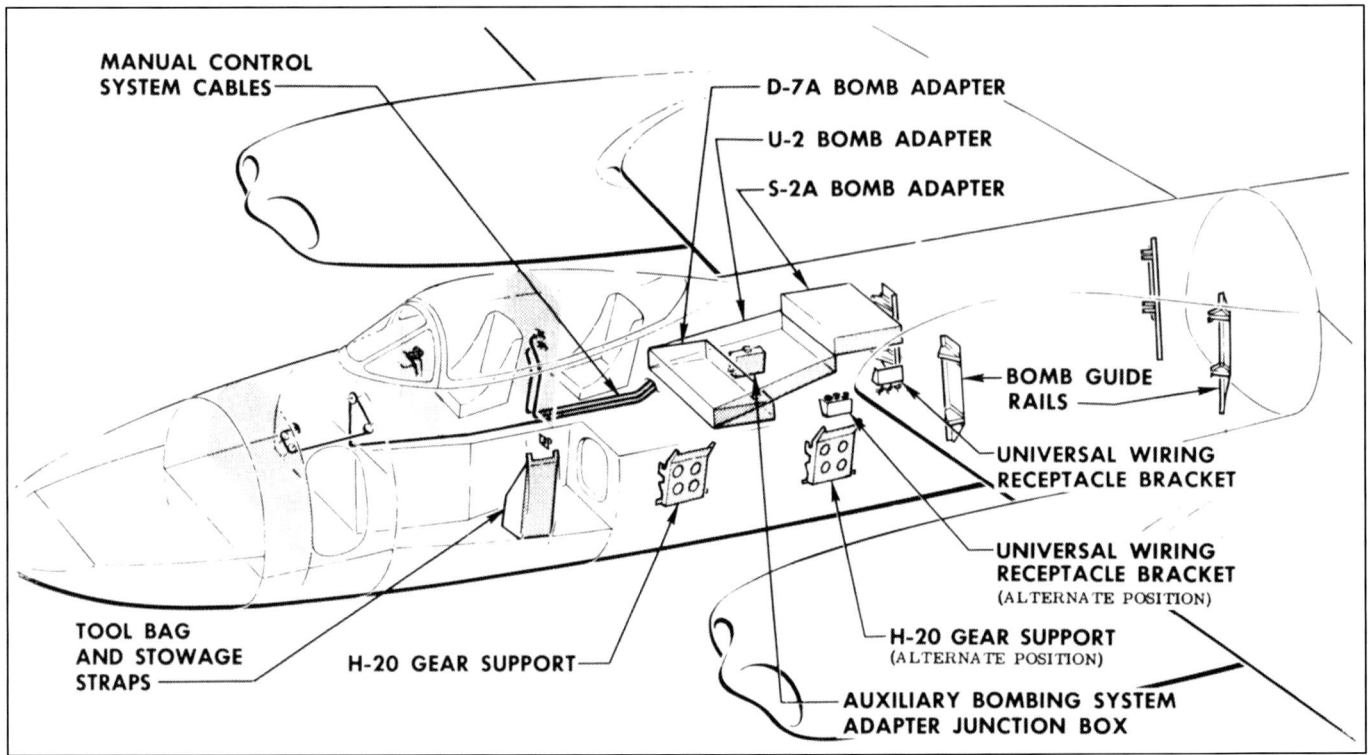

Drawing 22: This image shows the order the weapons racks are mounted in the bomb bay and some additional details. The H-20 gear was part of the weapon arming equipment. (NMUSAF)

The fourth section of fuselage interior and turret exterior illustrations covers the Emerson A-6 turret and the A-6 chaff dispenser equipment installations:

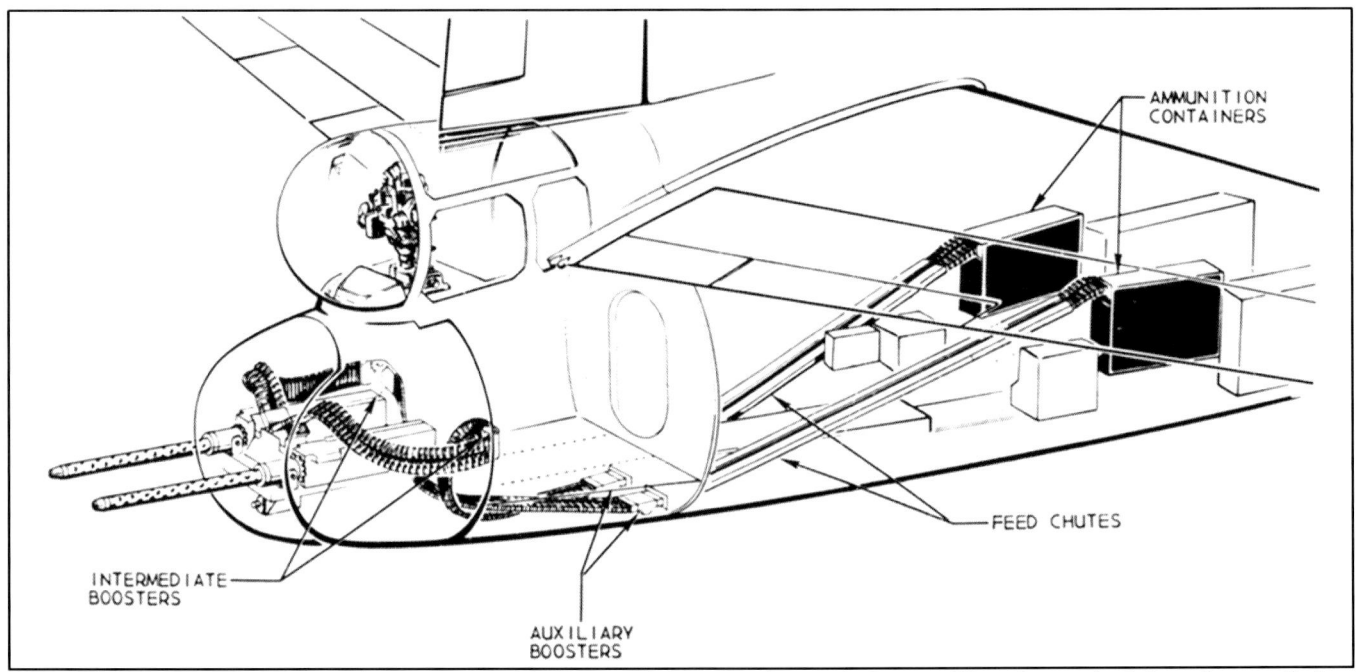

Drawing 23: This drawing shows the major components of the Emerson A-6 turret and ammo feed system. The radar ranging information was displayed in the turret aiming assembly on a shelf aft of the Gunner's Station, where the azimuth of the target was also displayed. To the right, were two 300 round ammo boxes, one for each weapon, and the two lengthy ammo-belts. These belts and boxes were in a non-pressurized portion of the aircraft, with the belts running beneath the pressurized gunner's station and directly into the non-pressurized turret. The twin AN/M3 .50 caliber machine guns road on a combined rack and were fired together from the gunner's station being trained in azimuth by sight and in range by an AN/APG-30 radar unit. (NMUSAF)

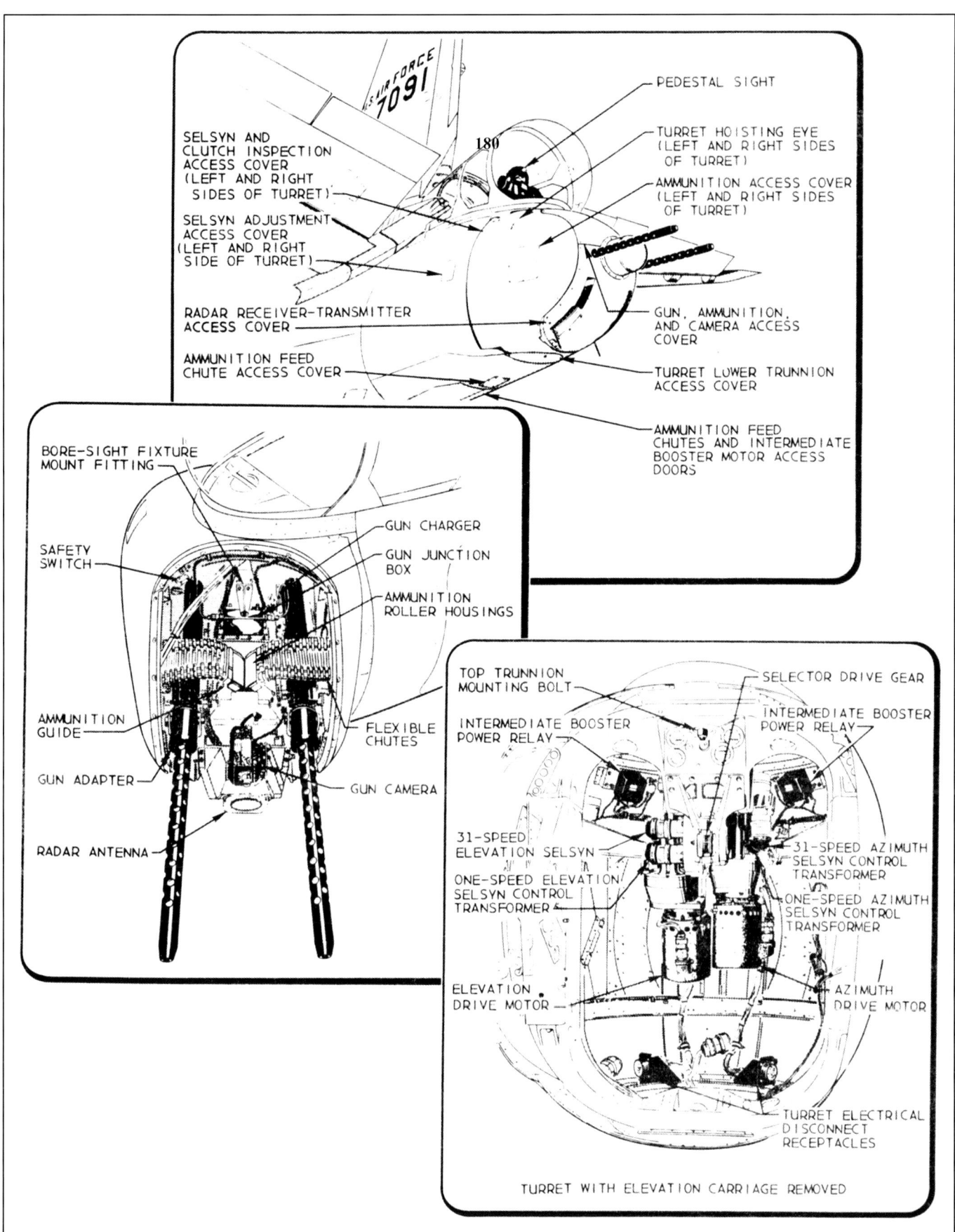

Drawing 24: This illustration shows the exterior detail and access covers and panels of the Emerson A-6. It also shows detail within the turret itself with the elevation carriage removed. Note the AN/APG-30 radome between M3 .50 caliber weapons. (NMUSAF)

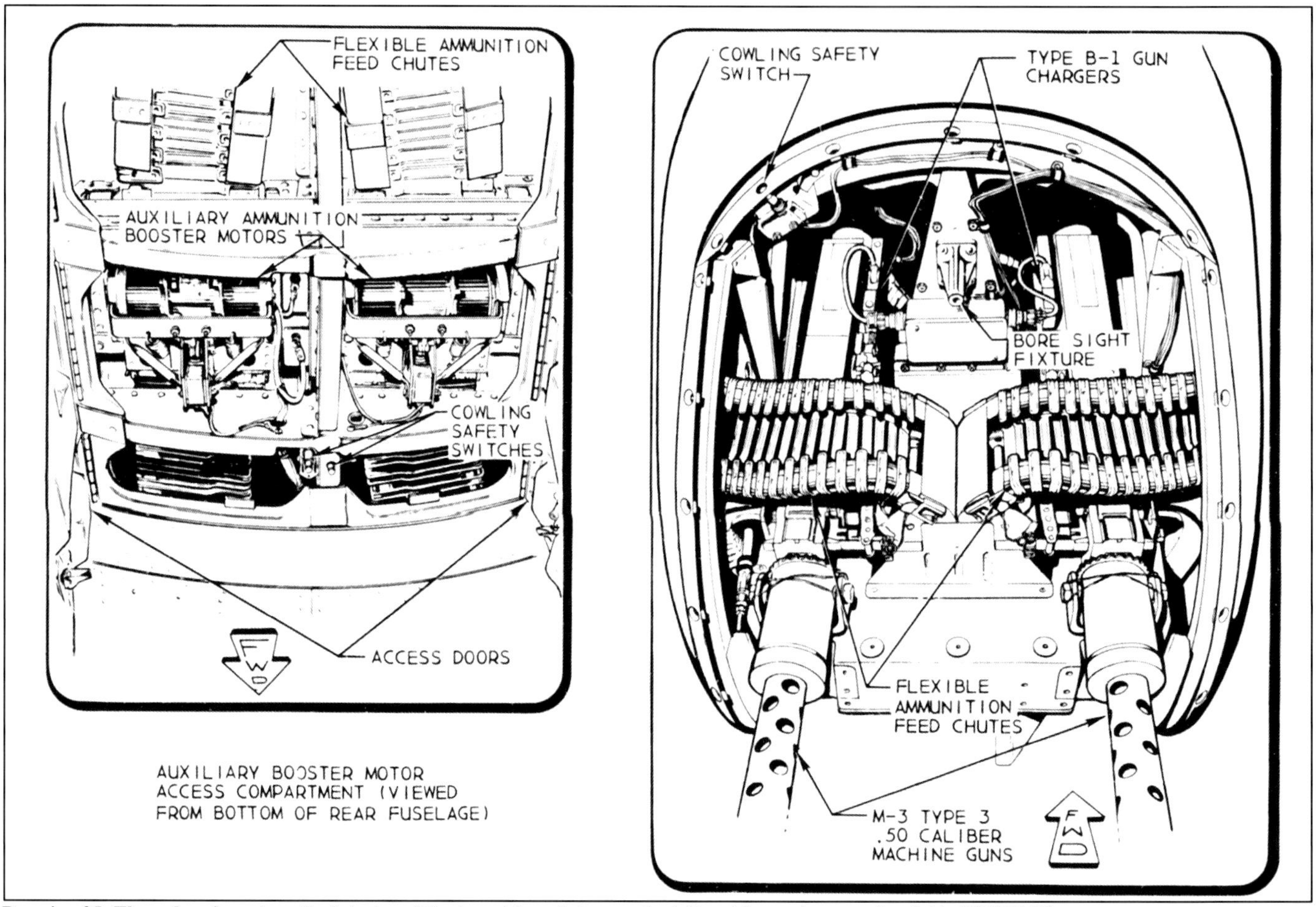

Drawing 25: These drawings show the interior of the turret from below on the left and from above on the right. (NMUSAF)

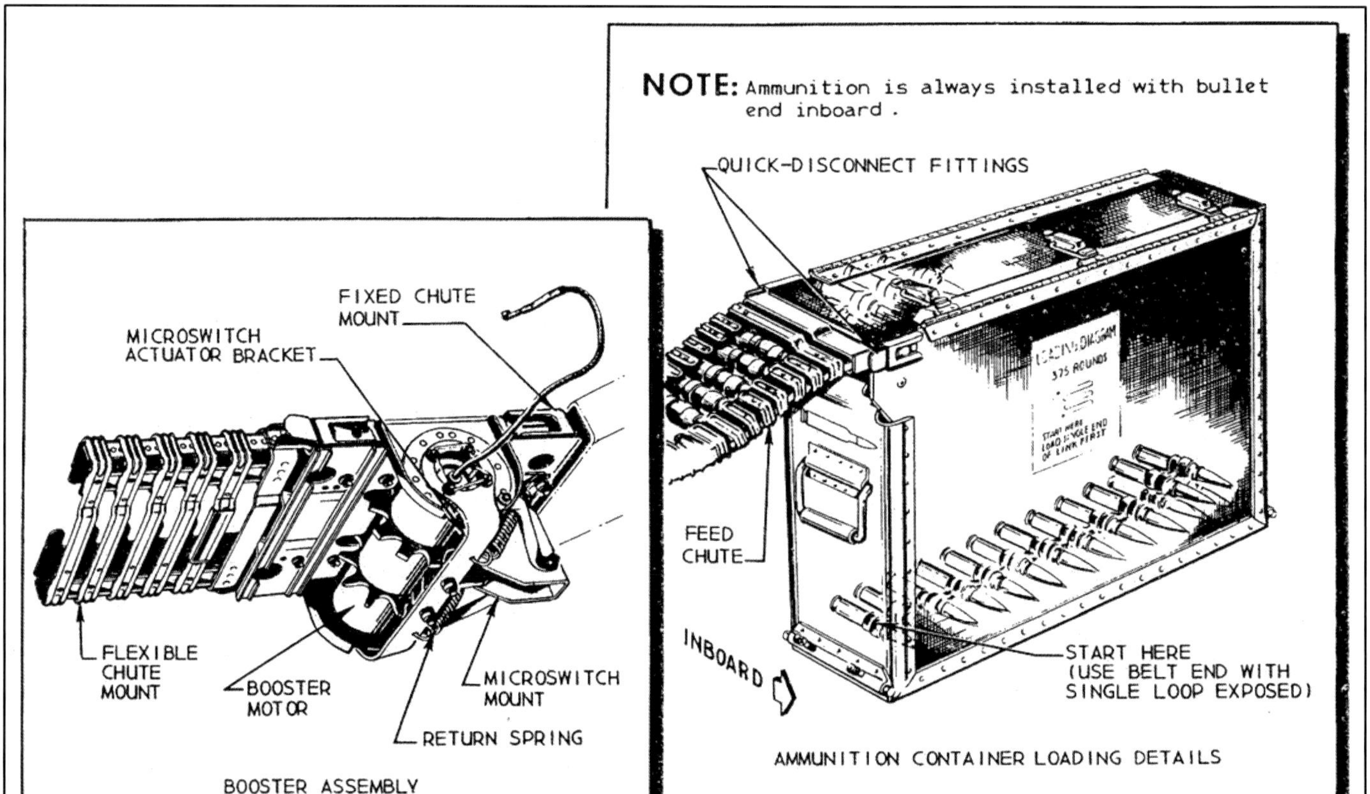

Drawing 26: This drawing shows details of the ammo boxes and booster motors which moved the shells and their casings through the long chute from the ammo boxes to the turret. (NMUSAF)

Photo 44: This image shows the starboard side of an Emerson A-6 with the new triangular late 1950s Electronic Counter Measures (ECM) antenna just forward of the turret on the aft fuselage. Images of the ECM equipment were rarely seen. (NMUSAF) **Photo 45:** This photo shows the port side of the A-6 turret at night giving a good perspective of how large the aft dome was on the turret glazing. (Steve Ginter)

One of the lessons learned in the original Emerson A-1 through A-3 turret design and glazing was that it was extremely narrow, making it difficult for the gunner to clear the aircraft to each side, or to scan for enemy aircraft in nearby airspace. One of the major improvements in the Emerson A-6 turret was a new, greatly expanded, and heightened glazing. The glazing grew in height and width as it extended aft. It ended in a large semi-circular dome which provided a large unobstructed area through which to aim the pedestal gun sight through.

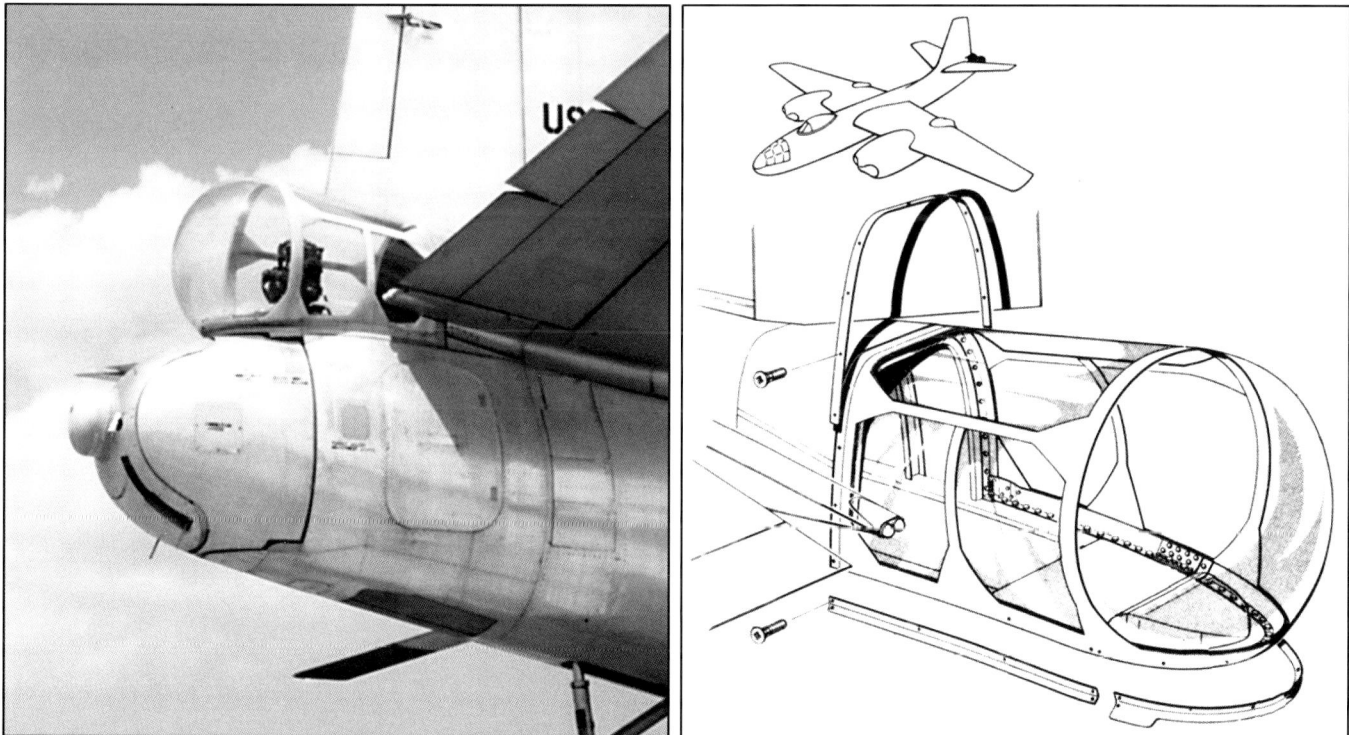

Photo 46: This image shows a clear shot of the starboard side exterior of an Emerson A-6 turret and the AN/APG-30 radar antenna. The gunner's access and escape hatch is shown, as well as the air deflector just in front of the hatch to aid in-flight parachute escape. Just below the hatch were the two five-sided arming hatches through which the ammo belts were fed to the turrets and forward to both ammo boxes in the aft fuselage. (NMUSAF) **Drawing 27:** This image of an A-6 turret shows the crop made on the lower rudder edge to allow the new expanded canopy for the A-6 turret. One issue never resolved with the B-45 tail turrets were the orange and white position lights mounted on the fixed portion of the inboard horizontal stabilizer. At night they severely restricted sight in their direction. (NMUSAF)

TAIL CABIN–RIGHT SIDE
("BACKBREAKER," "SECOND CALL" AIRPLANES)

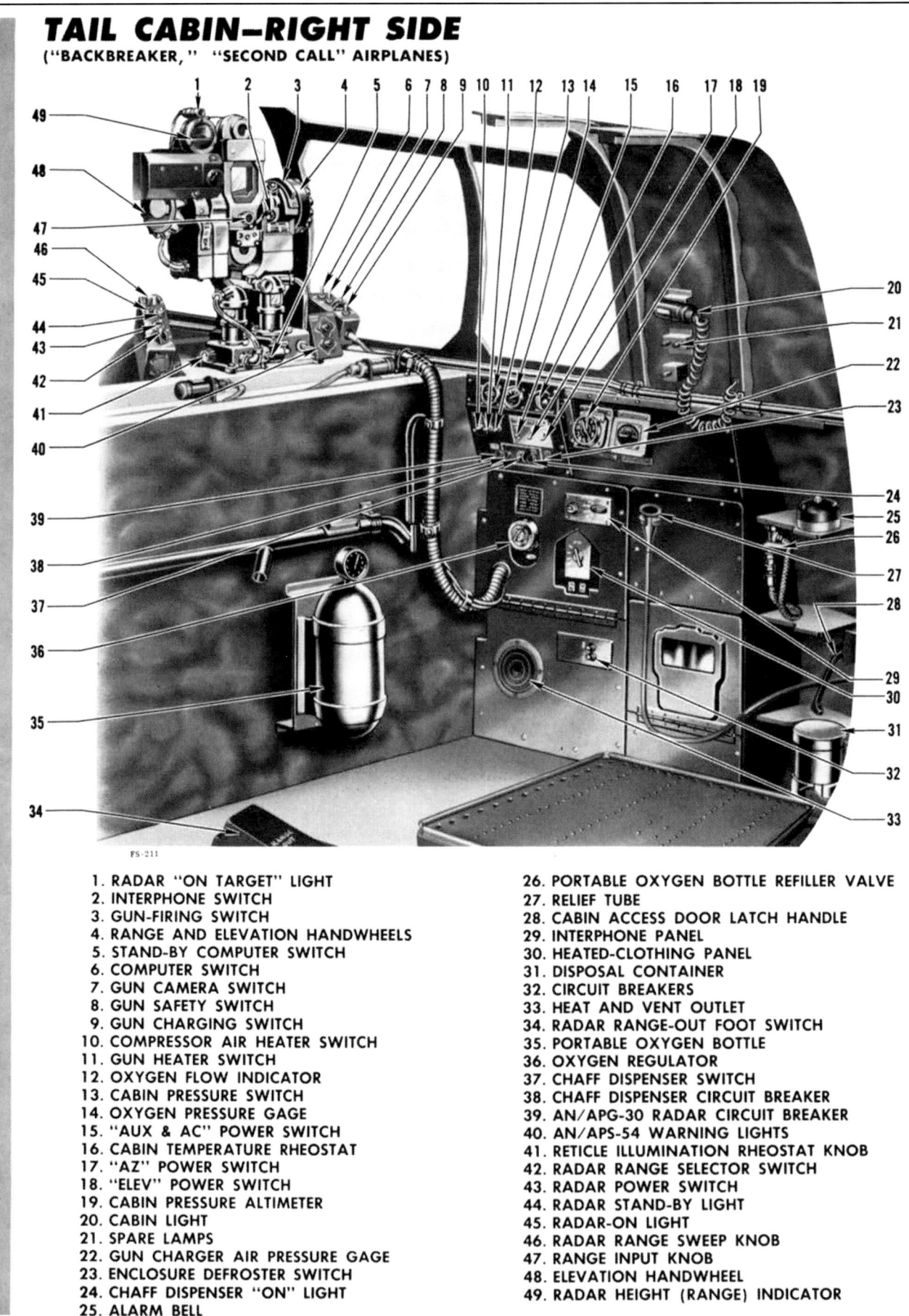

1. RADAR "ON TARGET" LIGHT
2. INTERPHONE SWITCH
3. GUN-FIRING SWITCH
4. RANGE AND ELEVATION HANDWHEELS
5. STAND-BY COMPUTER SWITCH
6. COMPUTER SWITCH
7. GUN CAMERA SWITCH
8. GUN SAFETY SWITCH
9. GUN CHARGING SWITCH
10. COMPRESSOR AIR HEATER SWITCH
11. GUN HEATER SWITCH
12. OXYGEN FLOW INDICATOR
13. CABIN PRESSURE SWITCH
14. OXYGEN PRESSURE GAGE
15. "AUX & AC" POWER SWITCH
16. CABIN TEMPERATURE RHEOSTAT
17. "AZ" POWER SWITCH
18. "ELEV" POWER SWITCH
19. CABIN PRESSURE ALTIMETER
20. CABIN LIGHT
21. SPARE LAMPS
22. GUN CHARGER AIR PRESSURE GAGE
23. ENCLOSURE DEFROSTER SWITCH
24. CHAFF DISPENSER "ON" LIGHT
25. ALARM BELL
26. PORTABLE OXYGEN BOTTLE REFILLER VALVE
27. RELIEF TUBE
28. CABIN ACCESS DOOR LATCH HANDLE
29. INTERPHONE PANEL
30. HEATED-CLOTHING PANEL
31. DISPOSAL CONTAINER
32. CIRCUIT BREAKERS
33. HEAT AND VENT OUTLET
34. RADAR RANGE-OUT FOOT SWITCH
35. PORTABLE OXYGEN BOTTLE
36. OXYGEN REGULATOR
37. CHAFF DISPENSER SWITCH
38. CHAFF DISPENSER CIRCUIT BREAKER
39. AN/APG-30 RADAR CIRCUIT BREAKER
40. AN/APS-54 WARNING LIGHTS
41. RETICLE ILLUMINATION RHEOSTAT KNOB
42. RADAR RANGE SELECTOR SWITCH
43. RADAR POWER SWITCH
44. RADAR STAND-BY LIGHT
45. RADAR-ON LIGHT
46. RADAR RANGE SWEEP KNOB
47. RANGE INPUT KNOB
48. ELEVATION HANDWHEEL
49. RADAR HEIGHT (RANGE) INDICATOR

Photo 47: This image shows details of the Gunner's Station looking to Port and Aft. (NMUSAF)

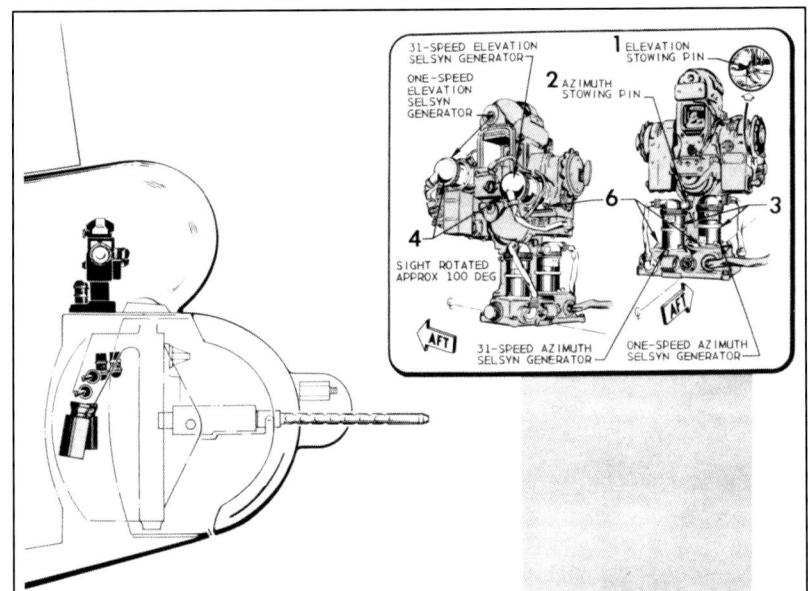

Drawing 28: This drawing shows a side view of the Emerson the A-6 and the AN/APG-30 Ranging Radar. The turret image shows the portions of the turret which rotated in the horizontal plane and the pressurized portion of the Gunner's Station immediately in front of it. (NMUSAF)

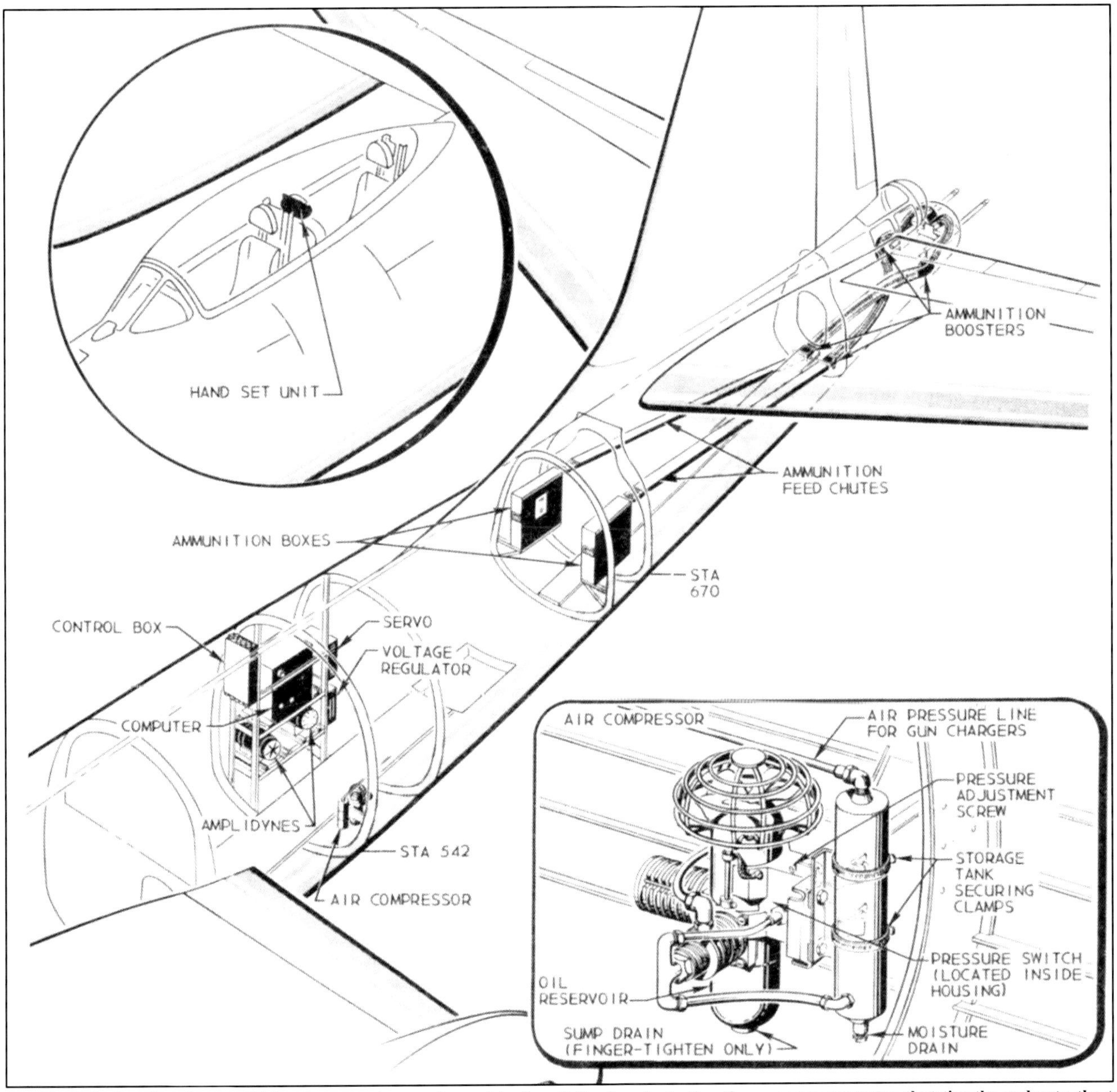

Drawing 29: This drawing shows additional gun system equipment in the aft fuselage including the gunnery computer that ties the radar to the turret movement mechanism; the gun charging compressed air system; and gun oiling system. (NMUSAF)

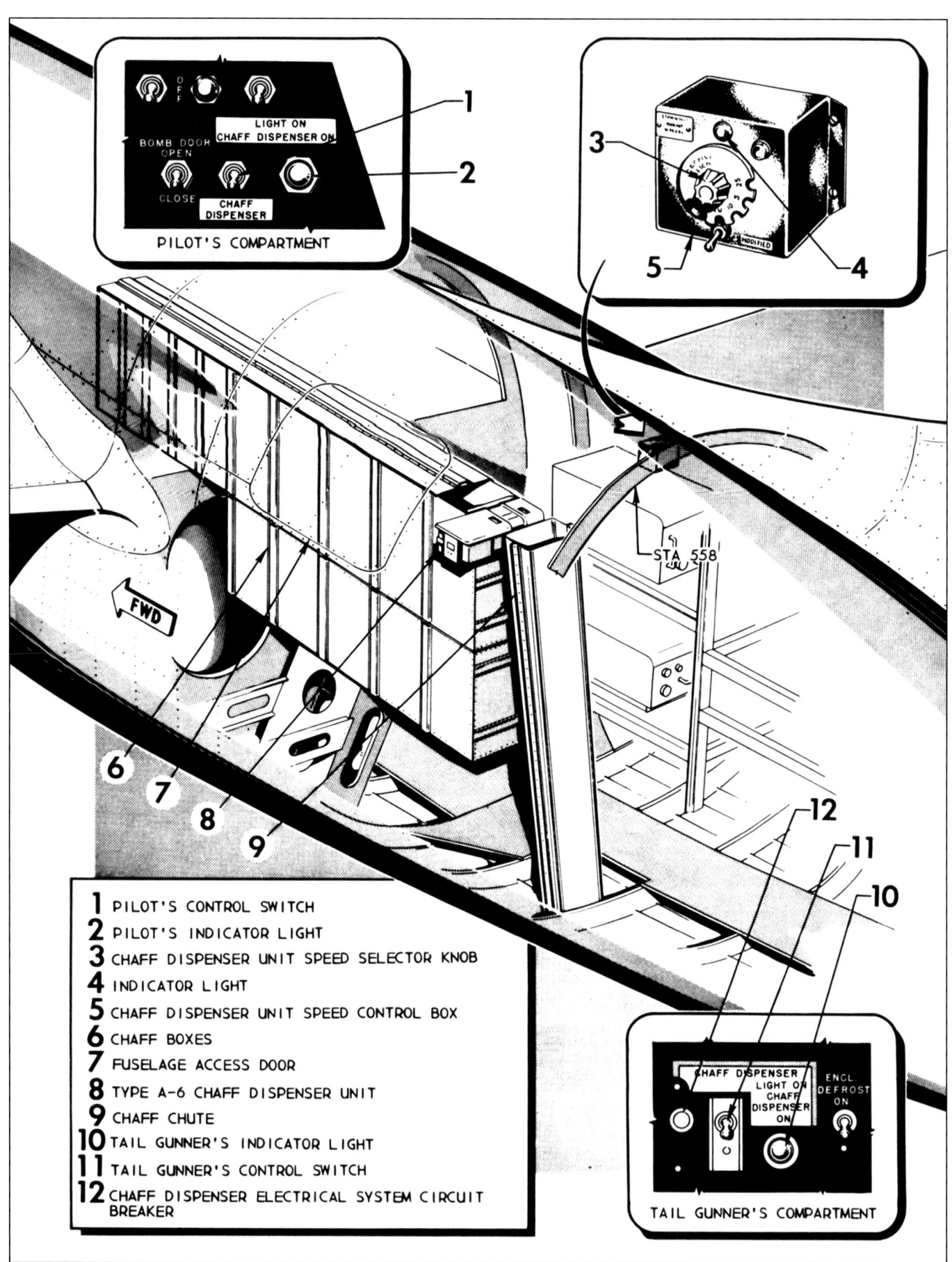

Drawing 30: This drawing shows the location of the A-6 Chaff Dispensing system in the aft unpressurized section of the fuselage. The key elements were the chaff boxes (item 6), the chaff cutter (item 8) which cut the chaff to the proper length to jam specific radar frequencies, and the chaff chute (item 9) developed and manufactured by SBAMA to make dispensing and maintenance much more efficient. (NMUSAF)

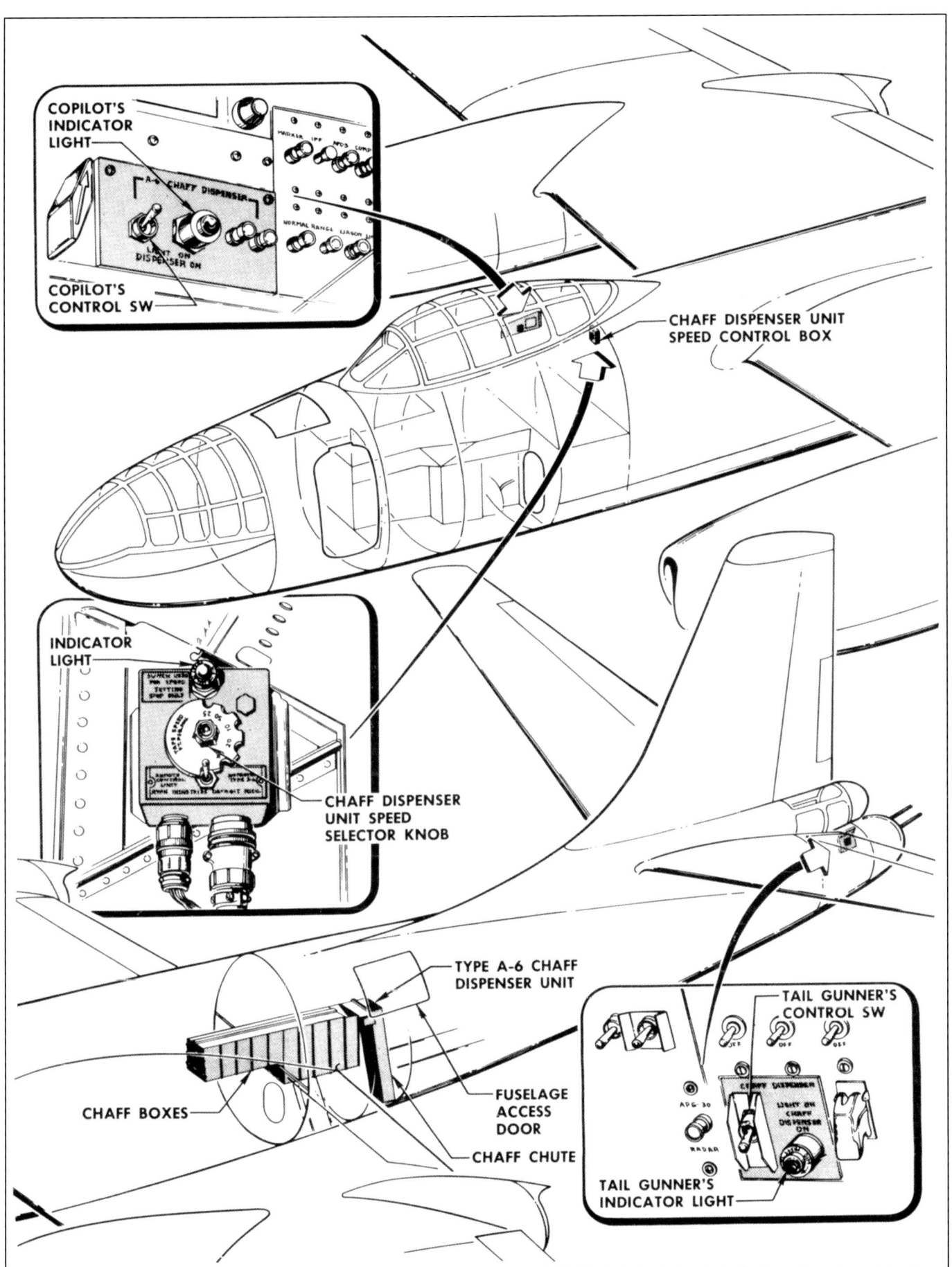

Drawing 31: This drawing shows a view of the chaff system looking to starboard and slightly aft. It also shows the locations of chaff system controls in the co-pilot's and gunner's stations. (NMUSAF)

Photo 48: This image shows details of the Chaff Dispenser Speed Control Box in the Co-pilot's station and on the aft bulkhead in the unpressurized aft fuselage. (NMUSAF)

UK And European Operations....

The 40 BACKBREAKER Aircraft were deployed to RAF Station Sculthorpe the first week of June 1952, only two months later than HQ USAF plans. Their route was from Langley to RCAF Goose Bay, Labrador, then across portions of Greenland, the North Atlantic and on to the Icelandic base Naval Air Facility Keflavik, Iceland. Their last leg was from NAF Keflavik, Iceland non-stop to RAF Sculthorpe in Norfolk, East Anglia, the UK, supported by KB-29P in-flight-refueling tankers. All aircraft made the crossing safely and with no incidents.

Photo 49: A B-45A-5 of the 84th Bombardment Squadron is refueled at Goose AB in Central Labrador on their way to the UK the first week of June 1952. It is probable that the personnel and equipment belonged to the USAF unit stationed there to handle the heavy flow of combat and Military Air Transport System (MATS) aircraft on their way to bases throughout Europe via Goose Bay and Keflavik, Iceland. Note the NCO on the wing refueling tank-by-tank was a USAF Master Sargent. (Craig Kaston via Steve Ginter)

The 47th relied on USAF units permanently stationed at locations such as RCAF Goose Bay or Goose AB as it was called by Air Force units transiting through. These units would provide minor maintenance to the aircraft, refuel, and replenish oxygen systems on board before their next take off. The base also serviced Military Air Transport Service (MATS) aircraft transitioning through with 47th Bombardments Wing personnel transiting to the UK and supporting the Wing's aircraft as they passed through RCAF Goose Bay, and NAF Keflavik, Iceland on their way to

RAF Sculthorpe. Refueling operations of a single B-45A-5 would take about three and a half hours because each tank had to be fueled individually.

Photo 50: B-45A-5s of the 85th and 84th Bombardment Squadrons in the distance remain overnight (RON) at RCAF Goose Bay, Labrador on the Wing's North Atlantic crossing to the United Kingdom in June 1952.

Photo 51: B-45A-5 7078 of the 85th Bombardment Squadron, 47th Bombardment Wing begins its rotation on take-off at Keflavik, Iceland Airport on the last leg to the UK, destination RAF Sculthorpe. All the B-45As on the deployment carried the new 500-gallon Fletcher Aviation drop tanks. Weather was a constant concern across the North Atlantic. For example, even in relatively good weather conditions, 45 knot crosswinds at NAF Keflavik were a common occurrence for aircraft transiting through. (Steve Ginter)

Photo 52: The first four B-45A-5s of the 47th Bombardment Wing arrive in a diamond formation across the windsock at RAF Station (later RAF) Sculthorpe on 1 June 1952. Sculthorpe would be the European B-45A's home until 1958. These four aircraft were assigned to the 84th Bombardment Squadron. (Author's Collection)

From June of 1952 through mid-1954 the 47th Bombardment Wing lived on a short string. First was the training shortfall in atomic operations and maintenance mentioned above. Second were a series of logistical shortfalls that caused serious operational issues in every aspect of the Wing's operations from electronics to power plants to instruments in the crew stations. Third was the lack of practice ranges for the AN/APN-84 SHORAN and the AN/APQ-24 Bomb /NAV system, dispersal bases and alternate recovery airfields. Fourth was secure or even dependable open lines of communications between unit areas and various higher headquarters locations. This was due to local communist-leaning labor parties that tended to keep what communications lines that did exist, inoperable. Nevertheless, Wing, Air Division and Numbered Air Force exercises became a regular and critical aspect of wing life in a never-ending effort to find resolutions to these and dozens more serious operational hurdles.

By late 1953, logistics was still a crucial problem for the Wing. The wholesale reorganization of Air Materiel Command that had caused the serious delay of Project SECOND CALL, continued to significantly delay
33the completion, and return of aircraft sent to SBAMA for Inspection and Repair as Necessary (IRAN) depot work. Crews were often delayed at Norton AFB, CA for over two months before they could obtain a serviceable B-45 to return home or even "dead-head" on a MATS aircraft going eastbound. The most serious issue that this problem caused however, was a severe lack of crews on-site at Sculthorpe to meet the Wing's Emergency War Order crew requirements.

Night missions, weapons loading, navigation and bombing practice sorties using radar and SHORAN steadily improved with early 1954 SHORAN effectiveness rates around 81%. This improvement was due to consistent practice, improved location of fuel and maintenance servicing units, the stand-by readiness of air and ground crews, and the faster loading rate of weapons. Exercise tasks began to vary greatly including striking unfamiliar targets in poor weather or at night in one-shot probability attacks that raised that proficiency rate to 77%.

Photo 53: B-45A-5 7069 of the 85th Bombardment Squadron was photographed at Norton AFB, San Bernardino, CA in 1955 just after completing an IRAN at SBAMA. The aircraft commander and his crew would return 7069 to RAF Sculthorpe in about four days flying time. SBAMA provided truly outstanding support to the B-45 fleet throughout the service life of the aircraft, despite significant disruptions such as the reorganization of AMC beginning in 1953. Their depot maintenance following the retrieval of 33 B-45As in 1950 and 1951 and the installation of the SHORAN AN/APN-84 and AN/APQ-24 radar system made the aircraft a truly combat-capable aircraft. (Courtesy Paul Minert)

Photo 54: B-45A-5 7072 of the 85th Bombardment Squadron was imaged inflight over the English countryside on a sortie out of RAF Sculthorpe in late 1954. Note: The serious issue of cracking Plexiglas, especially in the nose, was widely spread across the fleet. It was first reported to SBAMA, North American and AMC in March, 1950 and several fixes had already been attempted. By 1955 and 1956 the entire Plexiglas nose on the majority of B-45As was being replaced by an all-metal skin over the metal framework inside the bomb/nav station. (Steve Ginter Collection)

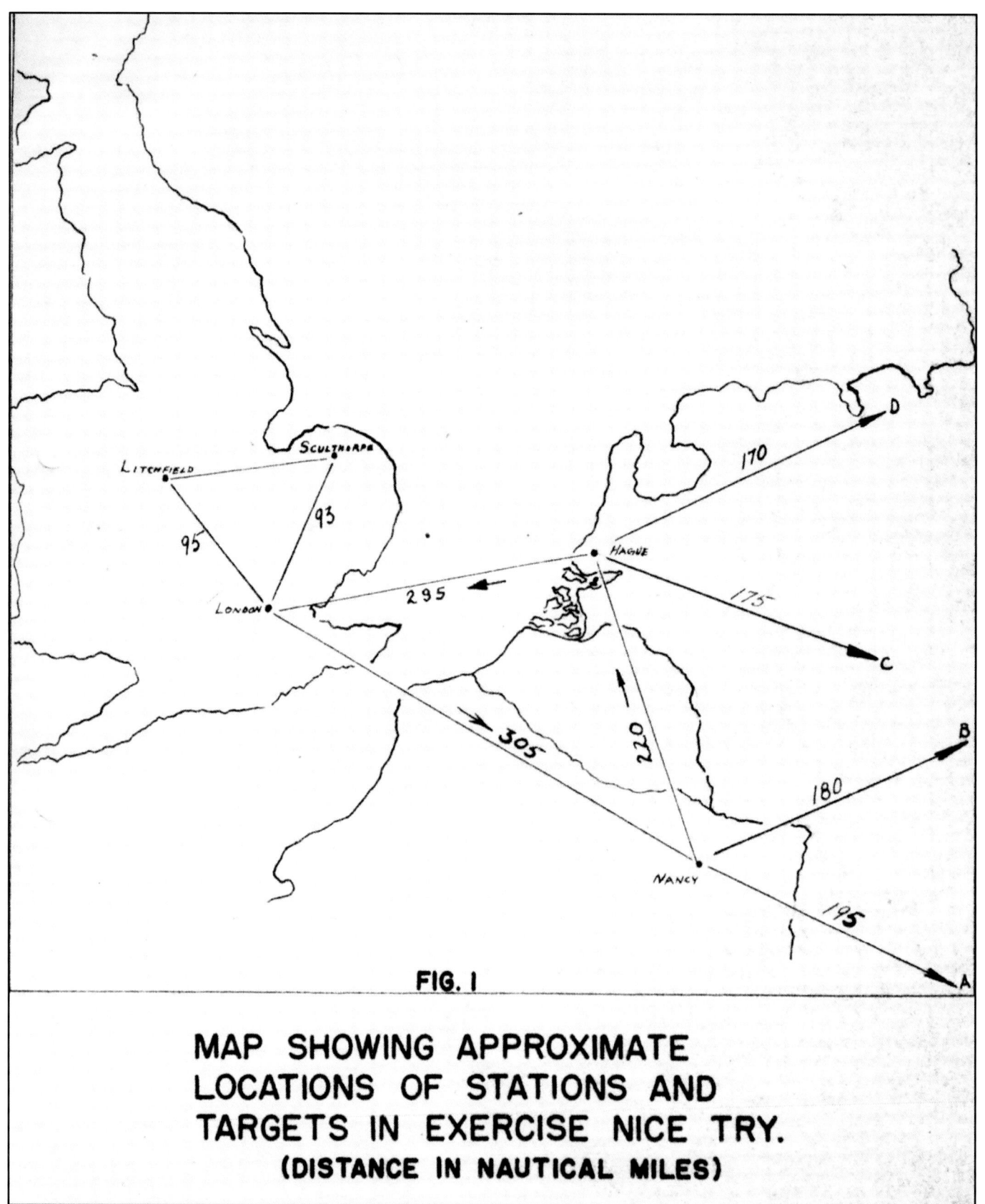

Drawing 32: SHORAN stations in this drawing of an exercise of the 47th Bombardment Squadron between 22 June and 2 July 1954 were located at RAF Litchfield, RAF Sculthorpe, London, Hague in the Netherlands, and Nancy in France. Additional unnamed targets in The Federal Republic of Germany were A, B, C and D. The max distances shown between some stations and targets were near the max range of AN/APN-84 System, such as between London and Nancy, at 305 miles. It took two stations locked on by an aircraft to triangulate and chart a course to a targeted weapon release point. For example, it would require The Hague and Nancy to be locked on by the aircraft seeking to bomb targets A through D. (HQ AFMC/HO via Tony Landis)

Photo 55: A pair of 85th Bombardment Squadron B-45A-5s, 7067 and 7065, with 500-gallon Fletcher Tanks installed, fly formation on a long-range training sortie over the UK in 1954. (Norm Taylor Collection via Steve Ginter)

The third method of bombardment was the use of the bomber's radar to navigate the aircraft to a predetermined release point based on radar returns of known and well-located points that gave good radar returns; then factoring in altitude, winds, and ballistics of the weapon. A ground-based radar, the AN/MSQ-1, would then track the aircraft to its release point and using the same basic information and data, determine how close the aircraft came to releasing the weapon at the correct point and where the weapon would have impacted. A beacon on the aircraft, the APW-11A, insured the aircraft could receive the ground-based radar signal and return it accurately to the radar site. The aircraft flew at altitudes between 38,000 and 40,000 feet and at 390 Knots Indicated Airspeed or KIAS.

The site, operated by SAC near London, was located 110 nautical miles from the target. The bomber departed RAF Sculthorpe and flew a planned course until picked up by the AN/MSQ-1 radar which then directed it to the Initial Point (IP) for the bomb run against the target near London. An exercise in August of 1954 involved 100 successful missions. 22 % simulated strikes hit within 900 feet of the target and 36% fell within 600-feet of the target, a simulated 5,000-foot long runway. These accuracies were better than those achieved in combat conditions in Korea by MSQ-1 directed sorties.

In the second half of 1954, several problems were significantly improving, only to be replaced by some new ones of equal seriousness. First the issue of deployment locations and aircraft dispersal was improving with particularly good cooperation and support from the RAF. More and more exercises with good efficiency ratings were being conducted to simulated short notice atomic strikes on planned and newly assigned targets. Secondly, mathematical experts arrived from the States that developed and implemented new methods of much more accurate geographic location computations for SHORAN beacons and target locations that significantly improved accuracy of SHORAN and MSQ-1 ground-based radar strikes. Mission ratings and crew confidence in both systems significantly improved.

In mid to late 1954 the first batch of BACKBREAKER/SECOND CALL B-45A-5s began arriving with the updated SHORAN system now used on the B-47, the AN/APN-84. While 15 aircraft were requested by USAFE for the 47th Bombardment Wing, only 12 aircraft were available at first. This group of 15 aircraft provided a new squadron to the wing, the 86th Bombardment Squadron, and a total of 55 B-45A-5s and B-45C-5s at RAF Sculthorpe.

A new exercise was added to the recurring schedule. It began in May of 1955 and consisted of long-range simulated strike mission against targets along the borders of Soviet and Warsaw Pact controlled territory in Eastern Europe and East Germany. No overflights were conducted in denied airspace, but the Allied aircraft were certainly on Soviet and Warsaw Pact radars. The missions were a great morale and confidence boosters for crews in the 47th.

Two continuing problems remained into 1955. The first was training in the Zone of the Interior (ZI) for personnel in atomic weapons specialties and SHORAN and AN/APQ-24 maintenance. Personnel arriving in the UK unqualified forced the 47th to devote considerable effort on the part of its most experienced personnel to this problem, instead of issues of dispersed maintenance, dealing with new aircraft with a new SHORAN system and excessive transfer rates of personnel out of the 47th to new atomic-assigned units sent to the UK and France. Logistics remained the second major problem. Turbulence from the 1953/54 reorganization of AMC remained a serious issue that effected basic supply and depot maintenance for engines, atomic weapons, bombing and navigation systems as well as specialty tools and equipment needed to handle atomic-armed aircraft.

Despite these issues, tests and exercises with the RAF dealing with Air Defense capabilities continued with great benefit to the USAF and 47th Bombardment Wing as well. Further, by the middle of 1955 the Wing's Dispersal Plan was firm with significant assistance from the RAF. Now most exercises practiced some element of the plan, including pre-attack dispersal and post-attack recovery at locations across the UK. The Wing had long ago learned that dispersal was the only means by which a multi-aircraft strike could be conducted with any reasonable probability of a significant number of aircraft becoming airborne and clear of the area before the Warsaw Pact/Soviet Forces delivered their first strikes.

1955 began well but serious problems in the Wing's combat capacity began to grow throughout the year. Logistics was still a major issue and the Wing was forced to rely on its War Reserve Supplies to keep its assigned aircraft airworthy. By Regulation, War Reserve Supplies were not supposed to be used except under combat conditions to execute combat sorties. However, if the Wing were to maintain a 70% Combat Readiness Rate of its assigned aircraft, it had no choice. Aircrew, ground crew, maintenance and handling personnel, communications, radar and SHORAN maintenance, and atomic weapons and monitoring maintenance were major shortfalls in every specialty. CONUS training was still being built and it would be some time before overseas units around the world would begin to see a long-term solution to the training issue.

In mid-1955 HQ USAF designated the 47th Bombardment Wing as a Tactical Bombardment Wing, written as 47th Bombardment Wing (T). It meant that the unit became an asset of USAFE and was under USAFE's sole direction in peace time and in war. Another organizational change occurred in mid-1955 when all support activities were grouped together and a "support wing" was established at each Air Division base taking command of the base, freeing the flying units to focus on their mission of atomic strike.

Photo 56: B-45A-5 7030 was one of the initial B-45s assigned to Europe as part of the 84th Bombardment Squadron which also was designated a "Tactical" unit in mid-1955. (Norm Taylor via Steve Ginter.)

Photo 57: B-45C-5 8014 was photographed on the ramp at RAF Sculthorpe with mostly empty fuselage tanks, hence the aircraft's nose-high posture. Note that this newer auxiliary-power cart is fitted to provide simultaneous start on both engines 1 and 2. (Courtesy of the RAF Sculthorpe Heritage Centre vie Ian Brown)

Photo 58: B-45A-5 7057 of the 86th Bombardment Squadron had just arrived at RAF Sculthorpe and was photographed there in late 1954 as part of the third squadron stationed on the base. (NMUSAF via the Author)

In March 1956, a new Emergency War Plan was established which revised the order in which targets were to be prioritized and attacked. Target attack priorities were based on the targets' importance in winning elements of the air war. Fighter bomber units such as the 20th and 81st Fighter-Bomber Wings (now flying Republic F-84F-45s) would attack in single aircraft strikes against targets in "bands" laid out from the eastern boundaries of East Germany toward the West. The B-45A-5a of the 47th Bombardment Wing (T) would now be used to strike targets at night and in bad weather using their AN/APN-42 radar, and SHORAN, and the AN/MSQ-1 ground-based radar.

Photo 59: This image shows another image of a B-45A-5 on the ramp at RAF Sculthorpe on a cold post snowy departure on a training mission. The image shows good detail of the nose area and detail of the unit emblems and its trim paint done in red. (NMUSAF)

One of the more serious issues in mid to late 1956 was the continuing reduction in the availability of crews and aircraft for assignment to meet Emergency War Plan mission requirements. By late 1956, only 24 of the Wing's 50 of so aircraft could be assigned to alert missions. This was due to supply and logistics issues, crew and maintenance turnovers and replacement training, and the non-availability of runways or of a way to support dispersal needs. Squadrons were routinely transferring in mass to RAF bases such as RAF Alconbury, RAF Burtonwood, RAF Molesworth, and shared facilities with the runway at RAF Sculthorpe. The work included moving ground support equipment, spares including auxiliary power plants, weapons, petroleum, oil, and lubricants and finding suitable locations for them to be properly and securely placed at dispersal locations. Further, personnel requirements such as working to get billeting, housing, and messing locations and services up to minimum standards were also needed. Communications still remained a serious problem and a rapid means of passing Emergency War Orders was lacking throughout 1956.

Photo 60: In late 1954 after the arrival of the Wings' third combat Squadron, the 86th Bombardment Squadron, Flight International Magazine conducted a number of inflight photo sessions with B-45A-5 7047 of the 86th Bombardment Squadron producing some beautiful medium and high-altitude images of the B-45A. When produced in California, all B-45As had their anti-glare panels on the engine nacelles and the upper nose painted matte dark green, similar to today's U.S. Federal Standard 595a No. 34079. During all aircraft's first deport maintenance or IRAN at SBAMA that color was changed to Matte Black, F.S. 595a No. 37038. The color remained the same for the remainder of the aircraft's life. (Courtesy Flight International Magazine)

Photo: 61: Flight International caught B-45A-5 7047 of the 86th Bombardment Squadron at medium altitude over the snow-covered Norfolk, East Anglian farm lands in early 1955. (Courtesy Flight International Magazine)

Photo 62: B-45A-5 0895 of the 84th Bombardment Squadron was photographed taxying for take-off on an extremely cold day in the winter of 1954. (Coutesy of the RAF Sculthorpe Heriage Center via Ian Brown)

Photo 63: B-45A-5 7047 was imaged at high altitude revealing great detail of the underside of the aircraft including the Chaff dispenser chute developed by SBAMA aft of the bomb bay and beneath the fuselage national insignia. (Courtesy Flight International Magazine)

By the end of 1956, a survivable location for a remote Operations Center had yet to be found and a temporary facility in London was in use. Therefore, there remained no survivable backup for the facility in the UK. Only approximately 8 of the 50 or so assigned aircraft could participate in all-weather and night bombing exercises by early 1957 due to logistics issues. Further, Identification Friend or Foe (IFF) on US and RAF radar systems remained unworkable due to needed repair to equipment installed on either nation's radars. One positive factor in operations support were frequent overseas exercises to Scandinavia, France, Spain, North Africa, Italy, Greece, and Turkey, which were now routine and successful. Exercises around the Mediterranean were extremely important as a month's flying there was the equivalent to three months flying in the UK and northern Europe.

Photo 64: One of the unspoken missions of the 47th Wing was to have one of their aircraft at airshows, fly-ins, and static displays in the UK and across Western, Northern, and Southern Europe as much as possible. The goal was to allow eastern-agents see the aircraft (except the inside and under controlled conditions) and let them know this aircraft was real, operational, and a legitimate threat to their forces. If there was to be a flying exhibition, a second aircraft was dispatched as well. Here B-45A-5 7055 of the 86th Tactical Bombardment Squadron was imaged on display at RAF Sculthorpe in 1956. (Courtesy Paul Minert)

Another favorable result of the exercise routine and frequency was that response time for Emergency War Order alerts had decreased from three hours to approximately 1.5 hours. This gave the Wing a bare minimum of time to get their aircraft off the ground before the first Soviet or Warsaw Pact strikes arrived. In general terms however, given the type and fragility of the weapons the Wing was equipped with, preparing an aircraft for a strike mission, or "generating" it, only occurred if there was an intelligence indication that the Soviets or Warsaw Pact were preparing some portion or all of their forces for an offensive ground or air strike against NATO. The weapons could not be left on the aircraft for extended period (less than three days or so) without considerable rehabilitation and replacement of many components. Further, the risk of damage or accidental repair rose considerably when the weapons were moved, loaded and then downloaded from an aircraft. Also, as mentioned earlier, even with the weapon up-loaded it was still a time-consuming process to get the aircraft airborne and locate the dispersal location to avoid destruction in the first wave of attacks from Eastern Europe.

Photo 65: B-45A-5 7083 of the 84th Tactical Bombardment Squadron was photographed in late 1956 at an airshow at RAF Sculthorpe parked next to an RAF Blackburn Beverly C.1. It was lost on 20 Sept 1957 in a crash near RAF West Raynham, just a short distance south of RAF Sculthorpe. Note the large number of additional fiberglass reinforcements on the nose to secure the Plexiglas nose from cracks. (Steve Ginter)

In late 1956 the 47th Bombardment Wing (T) was notified that it would begin aircraft transition in mid-1957. The Wing would be equipped with the bombardment version of the Douglas B-66 Destroyer and retain the same roles as an atomic bomber for USAFE, but would be equipped with Hydrogen or Thermonuclear weapons. It is interesting that this new mount was constrained by many of the same issues as the B-45, under-powered engines, control issues, and less performance in terms of altitude and airspeed than would be needed in this new role. The B-45A-5s began leaving RAF Sculthorpe and the other temporary operating locations in May of 1957 and all were relocated by January of 1959. Their new role after relocation and disposal was to become fire-fighter and aircraft accident response airframes at USAFE and SAC air bases throughout the UK, France, Germany, Spain, Italy, and Morocco.

Photo 66: This image captures the fate of some 48 atomic capable B-45A-s assigned to the 47th Tactical Bombardment Wing. 7056 of the 86th Tactical Bombardment Squadron was ultimately destroyed in fire response training at RAF Alconbury mid-1958. (Craig Kaston via Steve Ginter)

Engine Testing and Hollywood....

Despite the number of B-45A-5s that were committed to USAF Test Programs, and operational missions such as Target Tow Squadrons and the BACKBREAKER/SECOND CALL Project, there still a few B-45As that were available to be bailed to aircraft turbojet engine companies for flight test work.

Photo 67: 7096 was bailed to the Wright Aeronautical Corporation between August 1952 and April 1957 for flight test work which involved bringing the Armstrong Siddeley Sapphire turbojet into a production version known as the J65 to be used on the Martin B-57 Canberra and Republic F-84F Thunderstreak. (NMUSAF)

Photo 68: B-45A-5 7049 was severely damaged in an accident at Langley AFB, VA on 8 October 1951 and the U.S. Navy received permission to obtain the airframe and arrange for its repair to airworthy status. It became a U.S. Navy flight test aircraft painted in US Navy overall glossy Sea Blue, FS 595a 15042. It was bailed to the Westinghouse Aviation Gas Turbine Division at Olathe Naval Air Station south of Kansas City, MO. The aircraft tested the Westinghouse J40 turbojet intended to power the Douglas XA3D-1 Skywarrior, Douglas XF4D-1 Skyray, Grumman XF10F-1 Jaguar, Donnell F3H Demon, and the NAA X-10. But the engine was cancelled by the Navy in 1952 due to extremely poor performance. 7049 also flew from the Texas Engineering and Manufacturing Company (TEMCO) airfield known as Hensley Field on the Dallas Naval Air Station just east of Grand Prairie and west and north of Mountain Creek Lake. The field also became home to the Vought Aircraft Company producing the Vought Crusader. (NMUSAF)

Photo 69: This image was taken in the mid-1950s at Westinghouse's very wet test facility at Olathe Naval Air Station,6 south of Kansas City. At this time, 7049 was likely flight testing the Rolls Royce Avon 15 stage compressor turbojet for possible use in several U.S. projects. While the engine became one of the most successful Rolls Royce engines ever produced it saw little use in the U.S. And yes, mid-western thunderstorms do get that dark sometimes! (Craig Kaston)

Photo 70: Throughout the 1950s an unknown B-45A-5 served as an airborne camera platform for several Hollywood studios. This image was taken during the production of the RKO Radio Pictures film *Jet Pilot,* staring John Wayne and Janet Leigh. It is highly likely that the individual on the aircraft's entrance ladder is William H. Clothier, the cinematographer of the movie, and the person holding the entry door is Joseph von Sternberg, the film's Director. The airborne footage of the various U.S. Air Force aircraft including the X-1 and B-36 was truly outstanding. The forward crew entry door was modified with a larger optically-perfect glass window, for cameras such as the one in the bomb/nav station. The door is also marked with 43 camera sorties recorded. (NMUSAF)

CHAPTER 7 – THE RB-45C IN SERVICE

This Chapter relies heavily on unit histories of the 91st Strategic Reconnaissance Wing and its subordinates; Far East Air Force reconnaissance units flying RB-45s; the 363rd Tactical Reconnaissance Wing; and the 19th Tactical Reconnaissance Squadron. Further, it draws heavily from the outstanding research and authorship of the late John Fredriksen and his book *"The B-45 Tornado"* published by McFarland. I recommend it to anyone with an interest in the B-45 and its history. For clarity this Chapter is divided into three parts with Part 1 covering RB-45C operations in the Far East during and after the Korean War. Part 2 covers the operations of the 91st SRW in the CONUS and overseas in the UK from late 1950 through 1953. And finally Part 3 covers the operations of the 19th Tactical Reconnaissance Squadron based in the United Kingdom from April 1954 through July 1957.

PART 1: RB-45C OPERATIONS IN THE FAR EAST

War Begins....

The RB-45C was not due to be declared Initially Operationally Capable unto mid-1951. North American was to start delivering fully equipped and operational RB-45Cs in late 1950 for aircrew and maintenance operational training. All related plans for putting the RB-45C into service were thrown out on 25 June 1950 when the Democratic People's Republic of North Korea, the DPRK military, invaded the Republic of South Korea (ROK) without warning or allied intelligence alert.

The 91st Strategic Reconnaissance Squadron (SRS), a Far East Air Force (FEAF) unit based at Yokota Air Base (AB), Japan, hosted the few "strategic" reconnaissance assets which were available in the Far East. At war's start, these included a few RF-80 Shooting Stars photo recon jets, and a small number of RB-29A Electronic Intelligence (ELINT) and Photo Recce platforms. There was an immediate need for assets that could penetrate North Korean airspace at will to monitor what resources the DPRK might be preparing to move south to fully occupy the ROK.

As FEAF had no assets capable of accomplishing these missions, the Pentagon turned to the SAC who had on order the USAF's first turbojet-powered reconnaissance platform, the RB-45C. The 91st Strategic Reconnaissance Wing (SRW) was due to receive its first RB-45C in August, 1950 at Barksdale AFB, LA with 12 more due by the end of the year. SAC was directed to transfer the first three RB-45Cs scheduled for delivery to Continental Air Command (CAC) at Langley AFB, VA for assignment to the 363rd Tactical Reconnaissance Wing (TRW). While some SAC crews had been flying B-45A-1s on loan from the 47th Bombardment Group (BG) at Barksdale, the most experienced B-45A crews were assigned to the 363rd TRW's 84th and 85th Bombardment Squadrons (BS).

The First RB-45Cs over Korea....

RB-45Cs **8013**, **8014** and **8015** were delivered to Langley instead of Barksdale on 25 July 1950, and assigned to Detachment (Det) 4149A, 84th Bomb Sq. (known simply as "Det A"), 363rd TRW. Volunteers from the 84th and 85th Bomb Squadrons immediately began intensive training on their new mounts, and ferried the three RB-45Cs to Barksdale AFB for a 30-day crash course on photo reconnaissance in late August. Emphasis was given to the bombardier-navigators who would now become reconnaissance-navigators and camera operators at the same time. The crews did have one advantage in preparing for the long overwater delivery flight. All were well experienced on the B-45A-5, equipped with the same J-47 engines used on the RB-45C.

All three aircraft were flown from Barksdale AFB to McClellan AFB, CA in mid-September 1950 as the jumping off point for the Pacific crossing. AMC, SAC, CAC, Air Transport Command, NAA, GE, and a host of contractors, and suppliers who made extraordinary logistical efforts to ensure supplies, spares, consumables, technicians and advisors were in place at Yokota AB before the first aircraft arrived. While this of course meant that the 91st SRW, and to some degree the 47th BG went short, it insured an incredible sortie and mission success rate for Det A. It would become an achievement that spanned three years of fighting using only 7 airframes.

8013 (**Photo 1**), **8014** (**Photo 2**) and **8015** (**Photo 3**) departed McClellan AFB, CA on 20 September 1950, and after stops at Hickam AFB, HI and Midway Island Naval Air Facility (NAF), arrived at Yokota AB, Japan on or about 28 September. There they came under FEAF tasking for photomapping, bomb damage assessment, and target data collection. One supply issue that had not been solved was camera parts, and a shortage of Land Camera components delayed the first combat mission until 2 November 1950. After the Allies pushed closer to the DPRK/PRC border during October and November 1950, the People's Republic of Chinese (PRC) invaded in December 1950. Nearly daily photo reconnaissance missions over the Yalu River and the border with Manchuria became the daily routine.

With the PRC invasion, the mission grew to include the ability to penetrate Chinese airspace to determine the size and capability of their forces, especially air resources that might be committed in an attempt to upset the Allied air superiority over the Peninsula. In early 1951 it became apparent that it might not be PRC assets that were the most significant threats, but forces from the Soviet Union. Thereafter, the area of interest expended west and south from the DPRK and Manchuria to the area bounded by Peking and Shanghai, including installations in the Soviet Far East and transportation links between the central USSR to and through the PRC.

The combat nature of the assignment was driven home first by the confirmed presence of MiG-15s in the theater on 1 November and Det A's first combat loss 32 days after their first combat sortie. **8015** was shot down by 4 MiG-15s of the 523rd Fighter Aviation Regiment 45 nautical miles east of Andong, North Korea on 4 December. The co-pilot, Capt. Jules Young, and navigator, Capt. James Pucucci, were killed in the attack, while the pilot and Detachment Commander, Capt. Charles McDonough, was injured and captured. Following severe interrogation and limited medical care, he died a short time later in captivity. A fourth individual, a Headquarters USAF senior intelligence officer, Colonel John Lovell, was also on board and successfully bailed out. He was captured but was quickly allowed to fall into the hands of an enraged mob who murdered him. His presence and purpose onboard 8015 remains a mystery to this day.

Photo 1: RB-45C, 8013, the first B-45C in the Pacific theater, is shown in November, 1950, at Yokota AB, Japan. (NMUSAF)

Photo 2: RB-45C, 8014 at Yokota AB, October, 1950 showing the arrival markings of the RB-45Cs as assigned to Det A. (Keith Davidson)

This incident added emphasis, as if any was needed, to Det A's efforts at self-defense armament which had started almost from the first day in theater. They asked for tail turrets at best or at least tail warning radar that could fill the blind spot behind and below the aircraft. But the wait for a satisfactory solution would be long. In the meantime, in-house attempts were made at fitting a tail warning scanner, but these failed due to airframe and electrical interference. The incident also brought home another lesson, that scheduling near identical times-on-target, no matter how short or infrequent was deadly. It would be a lesson that had to be re-learned a decade and a half later over Southeast Asia.

By October 1950, HQ USAF was already working to exploit the potential of the new jet-powered recce platform for "overflights" of North Korea's neighbors, the PRC, Manchuria, and the USSR. The initial Washington reaction was cold, but those in the Theater recognized the importance of such flights, particularly after the Chinese invasion of North Korea. It is likely that the combined efforts of General George Stratemeyer, Commander, FEAF, and General Emmett "Rosie" O'Donnell, commander, 5th AF Bomber Command, were the spark plugs behind the first sustained jet aircraft clandestine "strategic" reconnaissance program. On 1 February 1951, Det A came under the operational control of 5th AF Bomber Command and flew its first Top Secret overflight mission on 3 February 1951, likely over airfields in Manchuria.

Photo 3: Believed to be RB-45C, 8015, *Timely Exposure*, in November 1950 at Yokota AB, Japan. The aircraft was lost to MiG-15s on 4 December 1950. (Keith Davidson)

Det A Becomes Det 2, 91st SRS, Strategic Air Command…

The next step in formalizing the growing importance of overflights in the RB-45's future came on 23 March 1951 when Det 4149A became a SAC asset assigned to the 91st SRW and re-designated 91st SRS, Det 2 (simply Det 2), at Yokota AB. While tasking of the RB-45s was still shared with FEAF, SAC now had a more direct role in their use in support of SAC's strategic tasking. The first visible change was the markings of the two RB-45Cs at Yokota. Both had gained nose art and now they carried full 91st SRW markings as well. **8013** had been named "*State Side*" (**photos 4 and 7**), and **8014** was now "*Split Vertical*" (**photo 5 and 6**).

Photo 4: RB-45C 8013, *State Side*, 91st SRS, Det 2, Yokota AB, Japan in mid-1951 shows the 91st SRW markings they gained after March, 1951. Note the 91st AREFS KB-29P in the background. (Keith Davidson)

Both aircraft gained the 91st SRW Square "I" on the vertical fin in matte black, shark's teeth between the forward oblique camera bay fairing and the radome, and on the nose of both tip tanks. Mission marks, shown as vertical cameras, magazines and lens, appeared in matte black on the port side, the tallies of which had increased rapidly as tasking for the pair grew significantly in the New Year. Working for a new

Major Command also brought a complete changeover in aircrew. Capt. Stacey Naftel became the new Det commander, with 1st Lieutenant James McGrath as his deputy.

Photo 5: RB-45C 8014, *Split Vertical*, 91st SRS Det 2, Yokota AB prepares to start engines in fall of 1951. (Keith Davidson)

Photo 6: Capt. Stacey Naftel, Det 2's first Commander, and 1st Lieutenant James McGrath, his deputy, pose in the forward crew hatch of RB-45C 8014, *Split Vertical*, in April, 1950 at Yokota. (Keith Davidson)

Photo 7: RB-45C 8013 at Yokota, mid 1951. The new shark's teeth between the camera bay cover and the radome is clearly visible along with the *State Side* nose Art. The mission marks had been recorded in just a little over 6 months of operations. (Keith Davidson via the Author)

Photo 8: Ground crew fire up the auxiliary power units (APU) for use in starting engines 3 and 4. Each engine required its own APU for startup, as the aircraft did not have internal APUs or large enough batteries. Note the security on the ramp at Yokota was high as both ground personnel were armored with .45 caliber weapons. ((Keith Davidson) Photo 9: RB-45C 8013, *State Side*, was the first Tornado to arrive at Yokota at the start of the Korean War. Here in the spring of 1951 it is having film cassettes for the K-22 forward oblique camera reloaded. The next step in the progress in this photo is replacing the single piece cover for the K-22 camera bay and its Cyclops-style camera and lens cover. (Keith Davidson)

April and May 1951 were orientation months for the new SAC crews, with missions flown only in support of FEAF, south of the Manchurian border. June 1951 brought significant FEAF restrictions on unescorted recce missions into *MiG Alley* due to the lack of 5th Fighter Command F-86 escort sorties. June also brought the first attempted overflight mission by SAC crews into the PRC west of Shanghai. However it was aborted due to a significant air defense fighter response detected by ELINT assets. July brought the second overflight attempt, a successful one across North Korea, into the PRC at Port Arthur and on to Darian. About 30 minutes into the penetration near another target, Harbin, at least 7 MiGs intercepted the RB-45C at 35,000 feet and it took nearly 29 minutes of evasive maneuvers and near red-line airspeeds to save the crew and aircraft. Radar imagery was taken of Harbin and the RB-45C exited the PRC near Kimch'aek, and the Sea of Japan. It made a successful but near-minimums landing at Yokota at the end of the 5 hour and 50 minute mission. The mission provided the FEAF's first proof that the PRC had night interception capability using ground-based radar. The airborne ECM war in Korea began in earnest after this night.

Det 2 completed 17 successful recce sorties over North Korea in July. The 15 July mission was the first recorded sortied to use IFR under combat conditions. Three KB-29P tankers had been deployed to Yokota AB by the 91sr SRW to test the combat potential of IFR. A KB-29P (**photo 8**) of the 91st Air Refueling Squadron (AREFS) off-loaded approximately 8,000 gallons or 54,700 pounds of Jet Propulsion 1 (JP-1) into the RB-45, nearly doubling its time over enemy territory. Interestingly, this highly classified mission was reported on the front page of a leading Tokyo newspaper just a few days later, having been released by HQ FEAF without telling the 91st SRS.

Photo 10: 91st ARS KB-29P 484107 refuels 91st SRW RB-45C 8031 over the US in 1951. The 91st ARS was activated at Barksdale AFB, LA in April, 1950, moving to Lockbourne AFB along with its parent unit, the 91st SRW, in September, 1951. This image was likely taken in the first half of 1951 over the CONUS. (USAF)

In August 1951, a new Det 2 Commander, Capt. Robert Christensen arrived; however Capt. Naftel had one last overflight to fly. This mission was flown into Soviet territory, a low altitude, high speed (800 feet at 440 KIAS) all-camera imagery sortie over Vladivostok Harbor on or about 8 August. Having descended through a solid overcast over the harbor from the north, they were able to completely surprise all defenses and received no anti-aircraft fire. Once past the harbor they immediately entered the overcast once more and climbed to 16,000 feet for the return trip to Yokota.

August 25th saw the next overflight mission, an imagery and radar scope photography sortie over the port city of Shanghai, commanded by Lt. Robert Hempen. Two lengthy high-altitude passes were completed without opposition. The aircraft had been refueled on the outbound leg in flight, but his return flight was through heavy monsoon rains and wind. Unable to air refuel again, and with Yokota well below instrument minimums for ceiling and visibility, the aircraft recovered at Haneda airport, about 14 miles south of downtown Tokyo. Its longest runway, at 6,700 ft. was long enough, but just barely. Minimum fuel onboard was required for a safe departure and return to Yokota a few days later.

Conflicting Priorities…

Radar scope imagery (actual photos of the aircraft's radar scope at pre-set intervals) was an important goal of all overflight missions, and the source of a fair amount of conflict between SAC and FEAF that would fester for at least another year to 18 months. FEAF, who was fighting a war in real time, naturally needed as many RB-45 missions as possible, dedicated to their needs. The focus had to be on North Korea and the immediate PRC and Soviet neighborhoods. SAC, who owned the aircraft, had a different perspective. Their priority was to obtain as much intelligence as possible on Soviet and PRC strategic threats against US, Taiwanese, and Japanese Forces. As the probability of success increased, so too did the deterrent effect upon the Soviets and Chinese.

At the top of SAC's intelligence priorities were radar scope images of what these targets would look like, at least the ones they could overfly. The B-29s and B-50s of the time (and the B-36 just around the corner) were the only bombers with the range and load-carrying capacity to conduct atomic strikes within the Soviet Union and the PRC. Given the known vulnerabilities of these bombers, missions would have to be flown at high altitude, at night and in bad weather if at all possible. That meant the bombers would depend on their onboard radars to find and strike the targets. This in turn meant that crews would have to have some means of studying for and practicing their missions ahead of time to have any reasonable chance of success.

For targets that <u>could</u> be overflown, actual radar scope images could be readily available. Analysts would amass large amounts of data about what these targets and the surrounding topography looked like on radar, and information about construction, layout, and proximity to lines of communication. Further, if there was high terrain in the target area, such as hills or mountains, it might reveal how this landscape might "shadow" or hide the target on certain attack headings. (The lack of information about what areas of the North Korean mountains caused severe radar "shadows" was already a problem for RB-45s and B-29s using radar for navigation.) Not only would these images give data on target radar "returns", but also provided a reasonably good idea of how the topography of the target area appeared on a scope. Finally radar images of this topography was an important element in developing accurate navigation charts for these areas, many of which had never been aerially imaged in daylight, let alone covered by radar.

While these radar scope images were very important, they might not properly present the target from a particular distance, heading or altitude required for a particular mission. Specific weapons to be used, enemy defenses, and winds aloft, might force the planners and crews to attack from totally new headings and altitudes. The data amassed from actual overflights could then be combined with other information such as commercial or travel maps, physical descriptions from human intelligence sources, and commercial surveys. This material could be combined to complete a hand-drawn illustration of a radar scope and its returns which portrayed a "prediction" of what specific targets would look like on radar from any given heading and altitude. These "radar artwork predictions" could be photographed to produce prints that looked very close to what a scope would look like for crew target study.

For targets that, as yet, could not be overflown, the problem was far more difficult and solutions much more a product of educated guesswork than analysis based on facts in hand. Charts and imagery, photographic or radar, was effectively non-existent in 1951, save those few of the central Soviet Union made by the German Luftwaffe known as GX imagery taken before, or early in World War II. Data from overflown targets could be extrapolated to a degree if any information about the facilities and locales in the eastern Soviet Union was available. SAC in particular made several efforts to mine various libraries in the US and elsewhere for maps, decryptions, images, surveys, and the like to begin the great challenge of producing accurate charts for the entire Soviet Union and PRC. Topography had to be estimated as did radar return sizes and shapes. Similarities could then be combined to provide at least a preliminary assessment of the target and local topography appearance on radar in the form of radar artwork predictions. It was indeed an exercise in deduction, but it was far better than nothing.

For years, in fact into the 1980s, the use of hand drawn radar artwork predictions continued until replaced by computer and satellite data-based methods. The problem of accurate and available navigation charts which could be readily updated lasted for decades as well. The issues of resource allocation and prioritization were not fully resolved until SAC had been re-equipped with the RB-47 and both FEAF and USAFE had tactical reconnaissance aircraft such as the RB-45, RF-84, RF-100 and RB-57s of their own beginning in late 1953 and early 1954.

The First Far East SAC Replacement …

Weather and very high operational sortie rates of the two RB-45Cs in theater began to take a heavy toll in August through October 1951. Mission accomplish rates plummeted as the Det 2 struggled to keep at least one aircraft available for priority and overflight missions at all times. Equipment failures continued to plague 8014 throughout September and October. First, **8014** developed serious fuel leak in an outboard wing tank, which required that entire portions of the wing be removed for repairs. Then **8013** required a 400 hour inspection and maintenance which grounded it for two full-weeks in September. New maintenance personnel were unfamiliar with the ad hoc, but extremely successful, maintenance procedures and techniques Det 2 had developed to keep their birds airworthy. Until they could be rapidly re-trained, they were effectively net drag rather than a positive resource.

That said, two special overflight missions were carried out in September. On September 21, 1951, now Capt. Hempen made numerous radar and photo runs over Hang-Chou, China. Opposition was negligible and **8014** returned safely to Yokota. A week later, Det 2 Commander Robert Christensen made a photo and radar overflight over Amoy, China in nearly repaired **8013**. Weather was poor throughout the mission, and Christenson had to divert to Formosa for fuel. He finally acquired Amoy on radar and although it was covered by 100% overcast, radar imagery was excellent and the mission was judged a success, even though no visual photos could be obtained. He refueled at Kadena AB, Okinawa on the returned leg, and made a successful recovery at Yokota on instruments.

A long-awaited replacement for **8015**, lost the previous December, arrived in October, 1951. Capt. Floyd Wilhelm's trans-Pacific flight in **8016** was anything but routine. Upon arrival at Travis, the rudder boost failed, requiring immediate replacement. The leg to Hickam was uneventful, but his departure to Midway Island NAF was delayed for five days. When he did get off the ground at Hickam, his bomb bay fuel tanks would not feed and he had to return to Hickam. Parts arrived quickly and he made a successful flight to Midway Island, until upon landing when he struck numerous Albatrosses that constantly covered the runway. Fully inspected and judged airworthy, his next destination was Wake Island. His leg to and stay at Wake Island Airfield were uneventful until departure. While spooling up the engines to take-off power at the end of Wake Island's Airfield, a relatively short runway, the tower rather rudely insisted he leave ASAP as the jet blast was sending chunks of the overrun into Wake Lagoon.

Aloft at last for a planned non-stop leg to Yokota, strong headwinds forced a diversion to Andersen AFB, Guam. Arrival at Guam was again uneventful, but departure the next morning saw **8016** suffer a major engine failure on take-off. Stopping a little more than 500 feet from the end of the runway, **8016** returned to the ramp in need of an engine change. A J47 from Hickam was dispatched and arrived at Andersen 4 days later. It was immediately installed, checked out and a test flight conducted. Capt. Wilhelm departed for Yokota the next morning and arrived successfully on 4 October, although with an inoperable radar, after a marathon 14-day ferry flight from Barksdale AFB to Yokota AB.

It would be another week until the aircraft was on the operations schedule as a result of a major inspection being due, and being used as a Guinee Pig for a defensive armament idea. **8016** rapidly joined the other Yokota RB-45s in carrying nose art, being christened "*Hot to Trot*" **(Photos 11 and 12)**. **8016**'s role as a Guinee Pig arose in no small part due to extreme frustration with the lack of response to the urgent need for airborne self-protection. It involved the acquisition of two AN/M3 .50 caliber machine guns (no doubt through some form of "midnight" requisitioning). The pair was installed into the empty tail cone of **8016**, one straight aft, and one at a 30 degree downward angle. A single flexible belt of ammo was linked to each gun, which were fired by the co-pilot. While performance was lacking, often as few as 4 rounds would fire, it was something and an unpleasant surprise for the Mig-15 that pressed the attack a little too close.

In October 1951, Det 2 completed 13 combat sorties with the three operational RB-45s, but camera parts shortages again kept the Det from having all three aircraft mission-capable at any time throughout the month. Further, on 21 October **8014** suffered a compressor failure on take-off and was grounded for maintenance. One overflight mission was completed during October, a night photographic and radar scope imagery mission. Capt. Robert Hempen and his crew covered the southern portion of Sakhalin Island, in the Soviet Kuriles. Although flown at the vulnerable altitude of 18,000 ft, the crew encountered no opposition.

Photo 11: 8016 *Hot to Trot*, on the left, is shown on the 91st SRS ramp at Yokota AB in early 1951. 8013 *State Side* is in the right. (David Menard Collection)

November and December saw a significant decline in combat sorties, in part due to an enlarged FEAF restriction in flying reconnaissance mission near the Yalu River area and *MiG Alley*. On November 9, a sortie had been boxed in near Haeju, North Korea by Soviet MiGs, the farthest south they had ventured to date. Altitude, between 38,000 and 43,000 feet, saved them from damage, but the interception once again reinforced how vulnerable the RB-45 was against MiGs and just how much the Soviets wanted them. Additional salt was thrown into the wound when FEAF assigned photo-mapping missions over the Japanese home islands in December.

Photo 12: 8016 *Hot to Trot*'s nose art and mission tallies are shown shortly before its loss in June of 1952. (NMUSAF)

The year ended with increased mandatory inspection requirements on each aircraft due after every 50, 100, and 150 flying-hour increments. And a major all-inclusive inspection was now required every 200 flying hours. These new requirements were taken in stride as **8013** and **8014** had not been State-Side for depot level maintenance in over a year. **8014** again developed a serious fuel leak in the port wing, requiring a complete wing removal and replacement outboard of the port engine nacelle for repair. The year ended with Det 2 routinely flying low-threat sorties providing mapping, artillery targeting data, and bridge locations for Allied ground forces in intense combat with PRC and DPRK Army forces.

1952, a New Year Begins Quietly…

A new Det 2 Commander, Capt. Robert Schamber, who had arrived in December, 1951 began preparations for night flash photography tests using the AN/M46 photo flash bomb. The missing parts to begin the effort were special heat shields for installation around the aft bomb bay fuel tank and the bombs. It would be April before live drops could be started. Identification, Friend or Foe (IFF) became another issue during an early January mission after another sortie was aborted when numerous contrails were spotted heading for the RB-45C. They turned out to be friendly Royal Australian Air Force Meteor fighter-bombers, and the new Commander once again raised the issue of tail turrets for the RB-45s. Another interception by several "swept wing aircraft", presumably MiGs, occurred the next day, and FEAF suspended further RB-45 sorties until late January for further investigation into solutions.

February brought another disappointing month with only 16 of 29 scheduled sorties completed, primarily due to weather. **8014** was reconfigured to carry out the AN/M-46 testing during the month with the removal of the forward bomb bay fuel tank and installation of forward bomb racks to carry the flash bombs in its place.. However, the severe lack of the photoflash bombs in theater forced Det 2 to use inert 500 lb. bomb shapes for calibration testing of the radar at night. Two sorties were completed with mixed results including minor damage in the forward bomb bay from the turbulent slip stream. The repairs meant more delays and disruption to the project. All this and other setbacks such as Japanese home island photo-mapping missions resulted in unofficial comments from the 91st SRW that their assets were being ill-used. Further, it might be wiser to bring their RB-45s back to the ZI where they could then be assigned to ongoing temporary assignments at RAF Sculthorpe in the UK. Nothing came of these musings, but the message was clear: "use 'em or lose 'em".

On March 3, 1952 a new Detachment Commander arrived, Capt. Thomas Broughton. Coincidentally, FEAF, perhaps having gotten the message from the 91st SRW, began tasking daytime photo missions near and around *MiG Alley* once again. Broughton complied and 33 missions were flown over the area without Chinese or Soviet opposition. That said, both Broughton and FEAF knew well the vulnerabilities of the RB-45 on these sorties and the Capt. pushed for more nighttime tasking. However, the lack of heat shields, AN/M-46 bombs, and other parts, still hampered Det 2. On 18 March 1952, **8013** served as the high altitude "ferret" on a Top Secret mission to 'light up" (get the operators to turn on and operate) the Soviet radars around the Vladivostok. The radars indeed came to life as planned, and were duly recorded by an RB-50G of the 343rd SRS, 55th SRW on temporary duty with the 91st SRS at Yokota. Similar Electronic Intelligence (ELINT) missions would be flown throughout the Cold War and beyond by pairs of ELINT platforms of ever-increasing performance.

April finally brought the airborne live tests of the AN/M-46 Photoflash Bombs, with mixed results. Three sorties were flown with the first yielding excellent results in all respects. The second and third tests were failures however. Low level haze caused bright reflections which bleached the images on mission two, and the third mission demonstrated severe vibrations at 290 KIAS which made the aircraft instruments and radar scope unreadable and all imagery blurred. Further tests were suspended due to operational tasking and ultimately the use of the AN/M-46 on RB-45Cs was dropped in theater. 21 missions were flown in May, divided between combat photo-recce and photo-mapping sorties over Japan. No known overflights were made during the month. In May, Det 2 also came under the direct control of the Operations Officer of the 91 SRS, a full colonel, to improve the interface with Headquarters, FEAF.

RB-45C **8017** based at WPAFB with the WADC had been working the photoflash problem for well over 18 months since its identification at Yokota AB. WADC had developed an inflight flare dispenser that could be synched with the camera systems on board for night photography. The flare dispensers worked well, allowing over 100 flares to be deployed in the proper sequence.

Photos 13: RB-45C 8017 assigned to the Wright Air Development Center at Wright-Patterson, worked on the issue of flares for illumination to replace the AN/M-46s and the bomb bay turbulence problem. (NMUSAF) Photo 14: The flare pods, two separate fairings running fore and aft and side by side at the aft end of the fuselage under the horizontal stabilizers, dispensed flares synched to the camera system of the RB-45C. The dispenser worked well, however, the flares could not illuminate a target beyond a few hundred feet from the aircraft. (NMUSAF)

However, the flares themselves would not generate enough light to properly expose an image of an object more than a hundred feet or so below the reconnaissance aircraft. While the efforts at Wright-Patt continued until 1954, it was not until a few years into the Southeast Asia War that effective high and low altitude fares and flare dispensers were fielded to all USAF tactical reconnaissance aircraft.

Wear and Tear Catches up with Det 2...

By June 1952, **8013** and **8014** had been in continuous operational use for over 18 months, and **8016** for over 9 months. All were essentially worn out. **8013** had become the cannibalization source for parts to keep **8014** and **8016** airworthy. More importantly, collectively the three aircraft were in non-compliance with 59 TCTO and needed modifications that could only be completed in the ZI. Urgent requests for replacements were repeatedly forwarded to the 91st SRW and HQs SAC. On 6 June the first non-combat loss occurred and was directly attributable to the condition of the airframes and engines.

Seconds after crossing the northern departure end of Yokota's runway, **8016** suffered a catastrophic failure of a compressor disk on number three engine which disintegrated and sent blades through the cowl and into the fuselage. Number 4 engine was damaged and caught fire, threatening the fuel tanks in the starboard outboard wing. The pilot attempted to jettison the canopy to no effect, while the radar navigator was exiting through the nose door. The co-pilot was able to get the canopy jettisoned and achieved a successful ejection and chute deployment. The pilot's seat however only lifted a few feet and remained attached to its rails. He unbuckled and escaped the seat into the crawl space beside the pilot's seats. Fortunately, the forward door and its air deflector had opened as designed and he successfully left the aircraft and manually deployed his chute just before impact. The aircraft impacted less than a mile from the runway and was a total loss, but all three crew members survived with only minor injuries.

The remaining two RB-45Cs were grounded until 11 June and shortly after the Det discovered **8014** had miss-calibrated Exhaust Gas Temperature (EGT) sensors installed which led to power settings that were 17% below actual thrust. Finally, the 18 or months of non-standard maintenance practices required to keep the few airframes available for operational requirements caught up with Det 2. New standard parts received from the ZI didn't fit many of the jury-rigged components on **8013** and **8014**, increasing the Det's non-operational rates. However, by late July the solution was on its way across the North Pacific.

New Jets and a New First...

On 29 July 1952 RB-45C **8042** departed Elmendorf AFB, Anchorage Alaska for a non-stop crossing of the North Pacific to Japan. Commanded by now Major Louis Carrington (the same Carrington who had commanded **8015** on its delivery to Yokota and Det 4149A back in September 1950) with co-pilot Capt. Wallace Vance and Radar Navigator Maj. Frederick Shook, **8042** was air refueled twice enroute by 91st AREFS KB-29Ps. 13 hours and 47 minutes after take-off **8042** touched down at Yokota in very poor weather. The flight was the first multi-engine jet powered non-stop flight across the North Pacific for which the crew earned the 1952 MacKay Trophy. This route, which is some 1,300 miles shorter than the central Pacific track through Hawaii, had had been the course of the pioneering over-water mass deployment of IFR-capable fighters during Operation FOX-PETER ONE earlier that month.

Four days later **8027**, on its second attempt, with Capt. William Kristen in command, co-pilot Lt. Oliver Nasby, and radar-navigator Lt. Joseph Giraudo, landed at Misawa AB from Elmendorf after three IFRs and a 6,700 nm trip against fierce headwinds and icing conditions. **8027** arrived at Yokota and Det 2 on 3 August. At last **8013** was rotated home to Lockbourne AFB in late July after nearly 22 months of continuous operations over North Korea, the PRC and the Soviet Far East. Despite the arrival of two new airframes, maintenance issues kept Det 2's capabilities constrained. Veteran **8014**, kept in the theater as a back-up, had four straight aborts late in the month due to maintenance, and **8042** was grounded for sheet metal work. August brought 18 combat sorties but more maintenance issues for **8014**.

With a rather static ground situation on the Peninsula, August also saw the airfield at K-13, Suwon come into use by Det 2 (**photo 11**). RB-45s would depart Yokota for Suwon, where they would refuel. From there, they would depart north with increased time-over-target in areas of North Korea, Manchuria, and central or southeast China. Then they would have the option to return to Suwon for fuel, or proceed directly home to Yokota. Increased range was the obvious advantage; the only negative being a 4 to 5 hour delay in getting the film back to Yokota AB for processing and distribution after the mission. Maintenance kept **8042** on the ground during September while **8014** and **8027** completed 19 combat sorties.

October brought three overflight missions, one of which, to Vladivostok, was aborted due very effective interceptor response over the Sea of Japan. **8042** came back into service late in October and along with **8027** had the Det 2-developed field mod of two .50 caliber machine guns installed in the tail cone. This mod was first attempted on **8016** (**photo 11**). November brought a new Detachment Commander, Capt. Howard Myers, who flew his first overflight mission on 21 November. Using a 100 inch focal length K-30 camera mounted in the bomb bay, the sortie circumnavigated Sakhalin Island from 25,000 feet. Photography was concentrated on Soviet airfields and they encountered no opposition. The mission recovered at Misawa AB in the far north of Honshu Island, Japan for fuel, and then returned to Yokota. Very poor weather in November held the Det to only 15 combat missions included two Top Secret overflights.

December 1952 was a very good month for the two-aircraft unit, completing 19 combat sorties, including three overflights. One was to the port city of Chingtoa, China in **8027**. The photo and radar scope imagery mission over the port on the southern coast of the Shandong peninsula stopped for fuel at K-13 both before and after the incursion into Chinese airspace. Weather was a major facture for the stop at K-13 on the return leg. 125 knot headwinds slowed the aircraft considerably and forced a 30 to 40 degree crab at times over the target. The month ended with Det 2 having completed 21 combat sorties.

Photo 15: 8027 was photographed on the ramp at K-13, Suwon AB in the Republic of Korea in late 1952. It is carrying its Yokota AB arrival markings and the Det 2 .50 caliber mod in the tail cone. It wasn't much but at least the RB-45 could return a little fire. (NMUSAF)

Photo 16: RB-45C 8042 was photographed in early December 1952 as one of three Det 2 aircraft armed with two fixed .50 M3 machine guns as the first armed RB-45Cs to fly in combat or in any other operations. The weapons were obtained via some form of midnight requisitioning from local RB-29 or RB-50 units at Yokota AB. While the installation was "fixed', there was enough "play' in the installation to allow the weapons to spread their shots in such an erratic pattern that it was enough to keep the MiGs at a much safer distance. The small doors on each side of the tail cone fairing were for casings and shell disposal from the guns. (The Gerald Balzer Collection via Tony Landis from HQ AFMC/HO)

1953 and Additional Resources...

Back in Washington D.C., the New Year saw discussions between FEAF, SAC and the Air Staff about how to increase long range reconnaissance resources in the Far East. FEAF wanted the new Boeing RB-47 as a replacement for the RB-45, while SAC resisted such a decision. SAC needed all the very capable reconnaissance platforms for its world-wide deep penetration sorties and post attack damage assessment requirements. The discussions ended with a commitment by SAC to put two RB-45Cs in the theater as soon as possible and ultimately transfer four RB-45Cs to FEAF permanently. This decision would mean a significant change in the nature of FEAF RB-45 missions within months.

Det 2 completed 19 combat sorties with no Top Secret overflights in January. In addition, the Det continued to grapple with the threat of search lights during high altitude photo missions. Various test sorties in Japan conclusively determined that the safest way to conduct such sorties was with RB-45s that had a black lower surface. The results of these tests would be applied in days

In early February 1953, **8027** became the first of two RB-45Cs at Yokota to be painted over-all gloss black as an anti-search light measure. All national insignias and USAF logos on the wings were painted out along with all squadron markings. The 91st "SRW Square I", the small font U.S. Air Force and tail number, and its Buzz numbers were re-painted in matte red (**photos 12 to 15**). Det 2 flew only 10 combat sorties during the month which included two Top-Secret missions. Both were flown by **8027**. The first on 19 February over airfields in Manchuria; and the second was to Yantai, China on 28 February. No opposition was noted, but the mission was intercepted by two F9F Panthers over the Sea of Japan on the return leg. Being a closely held Top Secret mission, the Navy had no knowledge of an Allied aircraft in the area, and ID'd the RB-45C as an IL-28. The weapons-live interception was only called off when the RB-45 aircraft commander Capt., Myers was able to contact the carrier. While no shots were fired, the Panthers got close enough to rock the RB-45 with their wake and rattle a lot of nerves.

Photo 17: 8027 taxies out at Yokota in February 1953 in this rare image. (Bob Dorr via Steve Ginter)

Photo 18: The newly painted RB-45C 8027 taxies into the 91st SRS Det 2 ramp at Yokota AB after a sortie in February 1952. The gloss black paint was judged the most effective anti-search light measure during tests over ranges in Japan during the previous month. (Mark Natola via Steve Ginter)

Photo 19: 8027 at Yokota, was photographed in February 1952. Note how the wings drooped slightly under the weight of a full fuel load. (Steve Ginter)

Photo 20: RB-45C 8027 is shown at altitude inbound to Yokota after a sortie over North Korea in late February 1953. (Mark Natola via Steve Ginter)

March brought a new CO, Capt. Charles Adams and the arrival of the first two RB-45Cs, **8021** and **8025 (Photo 21)** to be equipped with Bell M-7 twin .50 caliber armed powered turret. The turret installations was designed and engineered by North American, flight-tested on RB-45C **8021** at Inglewood, CA, and at Edwards AFB in B-45C **8002**. The installation was made by the SBAMA, the depot at Norton AFB, CA. A requirement identified nearly two and a half years before, was finally answered and Det. 2's aircraft now had a truly effective means of self-defense. **8021** arrived on 2 March 1953, followed nine days later by **8025**, delayed by weather. The aircraft made good the first part of SAC's commitment to send additional RB-45s to Japan as soon as possible. While mandatory inspections were required upon arrival, both aircraft were soon added to the Operations Schedule and Det 2 completed 21 missions in April.

Photo 21: 8025, one of the first two Yokota AB RB-45Cs to be equipped with self-defense turrets in the tail begins its take-off roll at K-14, Kimpo AB in mid-1953, wearing the colors carried upon arrival. Kimpo came into use as a refueling base for RB-45s just at the end of hostilities. (NMUSAF)

Photo 22: Close up of the Bell M-7 powered turret, without guns installed, on 8021. All M-7 turrets were installed at the San Bernardino Air Materiel Area depot, Norton AFB, CA. (NMUSAF)

Two of the April 1953 missions were leaflet drops over North Korea offering Communist pilots $100,000 US, in gold, to deliver an intact MiG-15 to the ROK. The offer was accepted and delivery completed at K-14 on 21 September 1953 by Lt. No Kum-Sok, DPRK Air Force (**photo 23**). With the arrival of two new RB-45s, finally the workhorse, 8014, went home to Lockbourne AFB late in the month.

Photo 23: Lt No's MiG-15bis 2057 photographed at K-14, Kimpo AB on 21 September 1953. (USAF via the Author)

April saw three Top Secret overflights. Two were flown by Capt. Adams over airfields in Manchuria, and Mukden (now Shenyang), China. The third was flown by Capt F.J. Chiochi and his crew over airfields at Yingkou and An-shan, China. About this time, Det 2 also started carrying Air Force Security Service (AFSS) linguists aboard on overflight missions. Their equipment was located in the forward bomb bay, while they sat in a canvas harness facing aft in front of a tuner in the companion way to the left and below the pilot and co-pilot. In an emergency they would have to exit though the forward crew door. Fortunately that eventuality is not known to have happened. They frequently proved their worth by giving the crew their first notice of impending interceptions. Beyond that though, they said nothing and had virtually no interaction in the air or on the ground with Det 2 crews.

New Unit, new Emphasis...

On 20 April 1953 SAC fulfilled the second of its commitments to provide FEAF with additional recce resources. RB-45Cs **8033** and **8038** (**Photos 24 to 27**) arrived at Yokota but were immediately re-assigned to a new unit there. Although they would be maintained and flown by 91st SRS Det 2 personnel into 1954, these aircraft marked the beginning of a new chapter in FEAF reconnaissance history. The new unit was the 6091st Reconnaissance Flt, reporting directly to and receiving tasking from only HQ FEAF. Shortly after its arrival, **8033** it became the second Yokota RB-45 to receive an all-gloss-black paint job. Unlike **8027** though, **8033** carried only the small font U.S. AIR FORCE and a four digit tail number on the vertical fin and **no** other unit, USAF, or national markings. Such paint changes, while seemingly insignificant, foretold a more covert and clandestine future for the four RB-45s now at Yokota. Both **8033** and **8038** were equipped with Bell M-7 turrets, the third and fourth turret-equipped RB-45Cs to serve in the Far East.

Photo 24: RB-45C 8033 is shown on the ramp at Yokota AB next to 8021. Note the absence of all unit markings. (Mark Natola via David Menard)

Photo 25: 8033 takes on fuel at Yokota before a mission over an undisclosed target on November 1953. Captured from a 35mm USAF film, the image is of poor quality. However, it is a rare image of the overall aircraft and one of the few that can be confirmed as a 6091st Reconnaissance Flight aircraft. (Author's collection)

Photo 26: 8038 was photographed at K-13, Suwon, ROK in mid-1953 carrying the new colors of the 6091st Reconnaissance Flight. The red arrow on the tip tank is a vestige of its 323rd SRS heritage; however the other red markings of that squadron have been replaced by green and yellow. (Norm Taylor via Steve Ginter)

Photo 27: RB-45C 8038 was photographed on its roll out at K-14, Kimpo AB, in the Fall of 1953. The aircraft carried the green and orange-yellow unit markings of the 6091st Reconnaissance Flight. (Courtesy John Starr)

Shortly after the Korean Armistice was signed on 27 July 1953, **8021** and **8033** were outfitted the first ELINT equipment installed on RB-45s. And at this point more and more missions were tasked against FEAF requirements, and an ever-decreasing number against SAC needs. No ELINT specialists were carried, but Russian and no-doubt Chinese linguists from AFSS were constant passengers. **8025** and **8038** remained photo-reconnaissance platforms carrying at least two Fairchild K-30, 100-inch focal-length, cameras. Often the RB-45s flew in concert with 55th SRW RB-50Gs stirring up radar and communications signals for the Superfortresses to collect. Det 2 stood down for a short period after the Armistice until August 1953. They then flew three special reconnaissance flights and one radar calibration sortie. The SAC Det 2 personnel were anxious to get home with combat operation ended, but FEAF prevailed upon SAC to keep the Det in place for at least three more months to train FEAF personnel on the aircraft and the missions.

So until December, Det 2 focused on training FEAF crews in the aircraft and continued flying operational and highly classified missions for FEAF. On 10 September 1953 Major Ed Krum arrived as Det 2's final Commander and began Project Phase Out. While very sensitive and highly classified ELINT, Communications Intelligence (COMINT) and Imagery missions continued into the late fall of 1953, half of the Detachment's flying hours were dedicated to training FEAF crews and maintenance technicians. In September, 19 sorties, including three Top Secret FEAF missions, were flown, despite 13 inches of rainfall at Yokota. In October, 33 sorties were flown, including 18 special missions for FEAF. November brought the last month of DET 2, 91st SRS missions with 46 sorties including 6 highly sensitive FEAF missions, flown. On 29 November 1953, Det 2, 91st SRS was deactivated ending nearly three years of SAC RB-45 operations in the Far East.

On 1 December 1953 the 6091st Reconnaissance Flight was re-designated the 6091st Reconnaissance Squadron, beginning a series of organizational changes, re-equipment drills, and significant growth that would ultimately result in FEAF's intelligence collection and processing capability being on a par with any in the USAF. In February 1954, the 6091st gained a formal parent unit, the 6007th Composite Squadron (R) at Yokota. The 6007th was assigned the four RB-45Cs as well as a number of RB-50Gs. The missions became more sensitive, more far afield (including sorties over restless Java and Indonesia from Clark AFB in the Philippines), and the aircraft more non-descript. Ultimately **8021, 8025** and **8038** lost all national, USAF, and unit markings, carrying only a four-digit tail number, often a false one. **8033**, the all black B-45, lost its small font U.S. Air Force on its vertical fin as well. With these types of changes it was clear that not only was tasking coming from the FEAF and HQ USAF, but also other national-level intelligence agencies. The 6007th became the 6007th Reconnaissance Group (Composite) in August 1956 having gained other squadrons and aircraft such as the Project HAYMAKER RF-86F and SLICK CHICK RF-100As, with Project HEARTTHROB RB-57As soon to follow.

The RB-45's final chapter in the Far East began on 20 March 1956 when 8038 returned to Norton AFB, CA for disposal. **8025** followed to the same fate on 6 December 1956. The last two, **8033** still wearing all black and a false tail number, and **8021** in overall natural metal and a false tail number, and each with only two national insignia on the fuselage for the flight home, arrived at Norton AFB and to their fate on or about 20 January 1957 (**Photo 28**).

Photo 28: RB-45Cs 8021, in front, and 8033 behind, are refueled at Midway Island NAF in mid-January 1957. They are on their final flight to Norton AFB for disposal and still wear their covert colors and markings carried while serving with the 6007th Reconnaissance Group. (NMUSAF)

Photo 29: RB-45C 8033 leads the final flyover of a RB-45C at Yokota AB, Japan in early January, 1957. Off each wing is a pair of Project HAYMAKER RF-86Fs followed by a pair of SLICK CHICK RF-100As with a single HAYMAKER RF-86F. (Doug Gordon via Robert Hopkins)

Colors and Markings of RB-45Cs assigned to Yokota AB, Japan for reconnaissance operations in the Far East....

Overall colors: With the exception of **8027** and **8033**, the RB-45Cs at Yokota were unpainted, flying in worn natural metal.

Anti-glare Panels for 8013, 8014, 8015, 8016, 8027, and 8042: Panels on the nose, engine nacelles and the inboard sides of the tip tanks, and both sides of the tip-tank fins, were largely standard FS 595A color FS 34079, which faded to a very dull FS 595A 34151.

Anti-glare Panels for 8021, 8025, 8033 and 8038: Panels on the nose, engine nacelles and the inboard sides of the tip tanks were re-painted matte black at during their first depot maintenance at the SBAMA prior to departing for Yokota.

The AN/APQ-24 Radome: Initially glossy and very close to today's FS 595A color FS 15526. As it faded and aged, it would pass through the equivalent of FS 595 color 26493, and finally fade to a color near FS 595 36640.

8013, 8014 and 8015 arrived at Yokota on 28 Sep 50 and all three were assigned to Det 4149A, 84th Bombardment Sq., 363rd TRW. None were armed with the M-7 turret. **8013** arrived in theater with a gloss Insignia Red tail band and nose trim outlined in black, from the 84th BS. It carried a gloss Insignia red lightning bolt on the outboard side of each tip tank. **8014** arrived in theater with a gloss Yellow-Orange vertical tail cap and nose trim from the 85th BS, and no markings on the outboard side of either tip tank. **8015** arrival markings are unknown. **8013** and **8014** became SAC assets assigned to the 91st SRW, when Det A was disbanded. The aircraft and personnel were re-assigned to the 91st SRS, Det 2 on 22 Mar 1951. Shortly thereafter they were adorned with full matte-black 91st SRW markings, and color risqué nose art which they carried throughout their tour. When Det 2 was activated, **8013** retained a gloss Insignia Red fin cap and lightning bolt motif on the outboard tip tanks that it had arrived with, and was named "State Side" (**Photos 4 and 7**). **8014** gained shark-mouth and eye markings on the nose and both sides of the tip-tank noses, as well as red & white shark's teeth on the fairing between the forward oblique camera compartment cover and the AN/APQ-24 radome. **8014**'s risqué nose art was painted on the port side of the nose just aft of the crew entry door, and **8014** became "*Split Vertical*" (**Photos 5 and 6**). **8013** and **8014** had all yellow trim markings removed and neither aircraft carried a fuselage band. **8015** was likely named *"Timely Exposure"* (**Photo 3**).

8016 arrived at Yokota on 9 Oct 1951, was assigned to the 91st SRS, Det 2, and was not equipped with the Bell M-7 tail turret. The aircraft arrived in theater with 91st SRW markings and an orange-yellow vertical fin cap, and fuselage band, but no markings on the outboard sides of the tip tanks. It was named "*Hot to Trot*" and acquired the Insignia red & white shark's teeth on the fairing between the forward oblique camera compartment cover and the AN/APQ-24 radome and on the noses of the tip tanks **(Photo 12)**. It was lost in a crash immediately after take-off from Yokota on 6 Jun 1952. "*Hot to Trot*" was photographed on the ramp at Yokota (**Photos 11 and 12**) shortly after arrival in October of 1951. **Photos 4, 5, 7** and especially **41** give good detail of the shark's teeth wrapped around the nose on **8013, 8014, 8015** and **8016**. The covers over the forward oblique camera were painted with a black pupil, surrounded by a bloodshot sclera, outlined in black. The eye was universal on 91st SRW natural-metal RB-45Cs.

8042 arrived at Yokota on 30 July 1952, and was followed three days later by **8027** on 2 Aug 1952. Both were assigned to the 91st SRS, Det 2, however neither was equipped with a Bell M-7 turret. Both were soon fitted with a field-improvised installation of two fixed .50 cal machine guns in the tail cone. Aside from **8016**, they were the only other 91st SRS Det 2 aircraft to carry this mod. **8042** carried the 91st SRW tail identifier, but no fuselage band. Its other arrival markings, if any, are unknown. **8042** returned to the ZI and the 91st SRW on 6 April 1953. Any changes, such as the application of nose art, are unknown. **8027** arrived in standard 91st SRW markings, the Insignia red and white outlined vertical fin and rudder cap, gloss Insignia red fuselage band with a thin yellow zig-zag line through it, and red and white-outlined gloss red trim panel on the fairing between the forward oblique camera compartment cover and the AN/APQ-24 radome. The tip-tanks carried the red arrows of the 323rd Reconnaissance Squadron. **8027** was the first to be painted overall gloss black in early 1953, retaining its 91st SRW tail logo, US Air Force and tail number, and buzz number over-painted in matte insignia red. Note that the gloss black weathered to matte black dull red extremely quickly due to little hangar space, high flying hours, and constant exposure to wind-blown dust in Japan and Korea. All national insignia and USAF markings on the wings were overpainted with black. The aircraft is shown in overall black in **Photos 17 through 20**. It remained in these markings until it returned to the US and the 91st SRW at Lockbourne AFB in April 1953.

8025 arrived at Yokota on 9 Mar 1953, followed by **8021** on 11 March 1953. These were the first RB-45Cs assigned to Det 2 fitted with the Bell M-7 tail turret. Both wore full 91st SRW markings and were assigned to the 91st SRS, Det 2. Both were re-assigned to the 6091st Reconnaissance Flight on 20 April 1953. Upon arrival **8025** carried the Insignia red vertical fin cap and fuselage band of the 323rd SRS. Shark's teeth markings on the nose and tip tanks were added at Yokota. **8021** carried dark blue (approximately FS 595a color 15102) with a white outlined vertical tail cap and nose trim (see **Photo 38**), having been assigned to the 324th SRS at Lockbourne. No fuselage band or outboard tip tank markings were carried; however the outboard sides of the tip tank's forward ventral fins were gloss Insignia red. It is not known if it had shark's teeth markings on the nose and tip tanks added at Yokota. On 11 February 1954, the 6091st and it aircraft were then reassigned to the 6007th Composite Group (R). Sometime between February and August of 1954, it is believed that **8025** was stripped of all its national and unit markings and it likely wore a fictitious four digit tail number. The 6007th Composite Group (R) was redesignated the 6007th Reconnaissance Group (Composite) on 19 February 1956. **8025** returned to the ZI at Norton AFB on 6 December 1956, likely in natural metal overall, a fictitious tail number, and national insignia only on the aft fuselage.

On 20 April 1953, **8021** was reassigned to the 6091st Reconnaissance Flight. The Flight was redesignated as the 6091st Reconnaissance Squadron on 1 Dec 1953. The 6091st Reconnaissance Squadron and its aircraft were re-assigned to the 6007th Composite Group (R) on 11 February 1954. Sometime between February and August of 1954, it is believed that **8021** was stripped of all of all national and unit markings. Now, in only an overall natural metal finish, it likely wore a fictitious four digit tail number only (see **Photo 28**). The 6007th Composite Group (R) was redesignated the 6007th Reconnaissance Group (Composite) on 19 February 1956. A January 1957 photo of **8021** shows it's overall natural metal, carrying a B-45C tail number, 8008 (a B-45C serial), and only aft fuselage national insignia. **8021** (with **8033** in overall black) returned to the ZI at Norton AFB, on or about 20 Jan 1957.

8033 and **8038** arrived at Yokota on 20 April 1953, and were assigned to the 6091st Reconnaissance Flight. Both were equipped with depot-installed Bell M-7 tail turrets. Both wore full 91st SRW markings upon arrival and the red vertical tail cap, fuselage bands and nose trim of the 323rd SRS. Teeth were not added to the aircraft before they were re-assigned to the 6091st Reconnaissance Flight. **8033** was quickly painted overall gloss black with the former 91st SRW markings over-painted. It wore only the small font US AIR FORCE on the upper vertical fin with tail number **8033** in gloss red immediately below (see **Photos 24, 25 and 28**). No other markings or nose art were carried. Note that the gloss black and red weathered to matte black and red extremely quickly due to little hangar space, high flying hours, and constant exposure to wind-blown dust in Japan and Korea. Sometime prior to November 1953, the 323rd Reconnaissance Squadron Insignia red markings on **8038** were over-painted in new trim colors for the 6091st Reconnaissance Flight, matte medium green, approximately FS 595a color 34300 and orange-yellow color 33538. Medium green with orange-yellow pin stripping replaced the red on the nose, fuselage band and vertical fin band. The aircraft was christened "*Baby-San*" in white block letters on both sides of the nose just aft of the forward oblique camera compartment. **8033** and **8038** were re-assigned to the 6091st Reconnaissance Squadron on 1 February 1954. Sometime after 1

February it is believed that **8038** was stripped of all national and unit markings. In only an overall natural metal finish, the aircraft likely wore a fictitious four digit tail number. Both were re-assigned to the 6091st Reconnaissance Sq, 6007th Composite Group (R) on 11 Aug 1954 and were still assigned when the 6007th Composite Group was redesignated a Reconnaissance Group (C) on 19 Feb 1956. **8038** returned to the US at Norton AFB on 20 March 1956 in overall natural metal and fuselage national insignia only. **8033** in black (with **8021** in natural metal) returned to the US at Norton AFB, on or about 20 Jan 1957. A January 1957 photo of **8033** shows it was still in all-black, with the small font US AIR FORCE removed from the vertical fin, and a four digit tail number, likely fictitious (but this has not been confirmed), in matte red. The aircraft was only carrying aft fuselage national insignia as well.

91st Air Refueling Squadron (AREFS) KB-29Ps (photo 31) sent on Temporary Duty Assignments (TDY) to Yokota, carried the 91st SRW "Square I" in black on the vertical tail and a large 91st AREFS emblem on the port side of the forward fuselage. They also carried squadron markings of gloss light blue, the approximate equivalent of FS 595a color 15450) and white diagonal stripes on the vertical fin cap, in a band around the fuselage just aft of the wing, on the fairing surrounding the "flying boom" attachment point on the belly, and on the outboard surfaces of the nose gear well doors.

PART 2: 91ST SRW OPERATIONS IN THE CONUS AND IN THE UK

Training at Barksdale AFB, LA....

Following the immediate departure of **8013**, **8014**, and **8015** for Yokota AB on 25 July 1950, Barksdale was designated the home for all new RB-45Cs and responsible for training new crews and maintenance technicians on the aircraft. The only problem with this plan was that the base didn't have a single Tornado to use for the mission. The first aircraft actually assigned to Barksdale for training duties was **8016** which arrived on 29 August 1950, over a month after the "Det A" birds had departed for the Far East. Nine more aircraft arrived in September, 1950 and three each in October, November, and December 1950. Only two of the first five aircraft sent overseas would ever be returned to the CONUS, **8013** and **8014** to Lockbourne, in November, 1952. Ultimately, all the operational Recce Tornadoes assigned to SAC were delivered to Barksdale by the end of July 1951. These aircraft were delivered to the Wing in a full combat configuration, a new event for the B-45 family. Starting in the fall of 1950, all SAC RB-45C flight crews would be trained at Barksdale using B-45A-1s. These birds were replaced staring in December 1950 and by May 1953 the Wing had nine B-45A-5s to teach basic flying skills and systems for the B-45A-5s and the J47. The RB-45C airframes were used to teach navigation procedures and skills, reconnaissance mission skills and techniques, and in-flight refueling.

Photo 30: RB-45C-1 8031 was delivered to Barksdale on 16 January 1951, the beginning of the second year of the Korean War. It remained at the Louisiana base until 31 March 1952 when the aircraft was transferred to Lockbourne AFB, OH. (Akira Igami collection via Steve Ginter)

Air and maintenance crews trained at Barksdale were responsible for the first turbojet powered long range, high-altitude reconnaissance flights over the Far East and Europe. The missions included sorties over all portions of the Korean Peninsula and far riskier clandestine missions across the Peoples Republic of China, Manchuria, and portions of the eastern Soviet Union including the Kurile Islands, and Kamchatka. At least seven RB-45s and crews from the 91st SRW at Barksdale were reassigned to units based at Yokota AB, Japan for combat sorties in support of SAC and the Far East Air Forces Command between September 1950 and October 1953. Far East Air Forces (FEAF) had a role or controlled almost all combat sorties except Communications (COMINT) and ELINT missions in the Far East during the War. Peripheral Reconnaissance and highly sensitive but critically important overflight sorties were repeatedly conducted over Eastern Europe, the Warsaw Pact and Soviet territory in Europe without a loss.

On 10 February 1951 the 91st Air Refueling Squadron was assigned to the 91st SRW and located at Barksdale AFB to support the air refuelable RB-45C for training and overseas reconnaissance missions in the Far East and Europe. A major difference in the arrival of the RB-45C over the previous versions of the Tornado family was that it had a major mission from the beginning of a difficult and bloody war. And although most of the problems of the B-45 were known, the RB-45C had its own unique ones. Six aircraft were lost to combat or flying accidents in the year it flew from Barksdale with very heavy crew losses. Even so, the exceptional record written by those assigned to the 91st SRW while there, continued and expanded when the unit moved to Lockbourne AFB in September 1951.

Photo 31: This trio of 91st AREFS KB-29Ps was photographed over the CONUS in 1951. (USAF)

PARPRO and Overflight Missions over Europe....

On 10 January 1951, the 91st SRW was alerted for a different type of mission in a far different area of operations than it had trained its crews for. Since 1947 the USAF and US Navy had operated a number of projects to monitor a wide variety of Soviet military operations in areas adjacent to Europe, the Artic and the western Pacific. Many of these "Projects" involved US aircraft equipped for photographic, COMINT and Signals Intelligence (SIGINT) collection equipment flying to distances of 40 miles to as close as three miles of hostile coastlines. In the late 1940s, the Program was ultimately formalized as the Peripheral Aerial Reconnaissance Program (PARPRO) and renamed the Peacetime Aerial Reconnaissance Program in the mid-1970s. SAC initially planned to use Boeing RB-47Bs to replace the propeller driven RB-29s, RB-50s and C-97s, but learned in 1951 that the recce Stratojet would not be available until at least late 1952 or 1953. It was known by SAC that the RB-45C was less than adequate in defending itself against the MiG-15. This was based on its first months of combat experience and its first combat loss to a MiG-15 only 30 days after its first combat mission in Korea. Under many conditions the MiG-15 could match its speed, maximum altitude and maneuverability. Further the most significant shortfall was the lack of a self-defense turret in the tail that was the best means for keeping the MiGs at bay. These short falls notwithstanding, the Tornados were selected for the missions.

PARPRO as an intelligence collection tool was still in its early stages when the RB-45C and the 91st joined the effort. From 1946 until the TDY arrival of the 91st SRW RB-45Cs, RB-17s and RC-47s were the primary USAF photo reconnaissance platforms along the Iron Curtain. Many of these missions came under the category of Bomb Damage Assessment following the War as travel throughout much of Soviet controlled Europe was difficult at best. The goal was to get a better assessment of the severity of the war damage, especially in areas in which the Allies had not significantly operated during the War. With the arrival of the Tornado, its range, altitude capability and speed allowed it to slightly intrude the borders of hostile nations in Soviet-controlled Europe with little chance of interception. The freedom of the Military Liaison Mission teams operating out of Potsdam, Germany to travel throughout Germany confirmed that the Soviets had not de-militarized East Germany but had significantly bolstered its military capabilities and hardware there. The initial goals therefore became to locate areas were the Soviets were building defenses along borders, or the means to rapidly transport mass numbers of troops and armor across terrain or rivers to the West. Next, repair and re-construction of known or suspected facilities that could house, or repair armor or other mobile forces and transportation vehicles were sought. Following these efforts, night-flown radar scope imagery sorties looked for areas that appeared to be significantly built up; areas that would certainly stand out in a group of nations that had been all but destroyed in WW II. It was likely therefore that these areas were military in nature, and would be photographically covered if and when the opportunity arose.

Intelligence collection goals in the northern and eastern Soviet Union focused on any data about the location and operational activities of Soviet Tu-4 atomic-armed bombers, thought to be based in the Siberian Arctic and Kamchatkan Peninsula. Eurasian goals were driven by concerns raised by the Soviets refusing to negotiate over the control of Berlin and their actions during the Berlin Blockade. Knowing what Soviets were storing in the way of a military build-up such as armor, artillery, tactical aircraft and the possible presence of Ilyushin Il-28 atomic capable bombers in the theater were critical to either conventional or Allied atomic war planning. In both theaters, radar scope photography, target verification photography, and target development photography was critical as all atomic strikes were planned using radar to locate target and aim the weapons for an accurate atomic strike. Accurate maps for navigation and mission planning over Soviet held or controlled territory depended on obtaining as much and as accurate radar data as possible.

The first group of sorties flown between 16 January and 11 March 1951 were launched and recovered at RAF Manston, located on the northwestern coast of Norfolk East Anglia. Four SAC RB-45Cs from Barksdale, **8023**, **8026**, **8019**, and **8021**, were assigned to Det 1 at RAF Manston to carry out this first set of PARPRO missions. Four newly modified KB-29Ps from the European-assigned 91th AREFS were also assigned to the Operation. The duration of PARPRO missions were usually 10 hours or more and required one or more IFRs. The recce aircraft's route would, depending on weather and visibility, begin over the northern Adriatic Sea adjacent to Yugoslavia. At the north end of Yugoslavia it would turn almost due East and fly along Yugoslavia's northern border, then north near the Western Hungarian border and then westward across the southern Czechoslovakian border. When it reached the border between East and West Germany it would fly northward again until it reached the Baltic Sea then depart west across Denmark and the Netherlands returning across the North Sea for decent and its approach into RAF Manston.

Once these readily accessible military missions were accomplished, the emphasis shifted to photographing means of transportation. Throughout Western Europe railroad networks, rivers, canals, and roads and autobahns were photographed. The first goal was to build an accurate picture of the condition and connectivity of routes across Europe. This would guide critically important postwar repair projects for military and commercial routes for some time. Ground route maps would also be critically important if war with Soviet and its allied forces were to break out. Such maps and charts would also be immediately needed in case airborne bombing campaigns, either conventional or atomic, had to be conducted. Detailed and accurate maps and charts of all populated centers throughout Western Europe were needed for the same reasons if they had to be freed by force once more. Next, large areas of terrain were mapped using the RB-45C tri-metragon mapping camera systems to begin the huge task of producing new maps for travel and charts for navigation. Most of the mapping of Europe had been accomplished by the German Nazi Luftwaffe before 1939 taking GX imagery to produce the products used by the Wehrmacht and Luftwaffe. While relatively complete in coverage of Europe, the products were significantly out of date.

Photo 32: One the two first RB-45Cs to be seen in the United Kingdom was 8023, photographed at RAF Manston the of its arrival on 16 January 1951. In the background is 8021 which arrived one day later than 8023 in the first Tornado PARPRO Detachment in Europe. (Author's Collection)

Photo 33: This RB-45C was photographed at RAF Manston very shortly after the aircraft's arrival. The crew bags can be seen on the turf just off the hardstand. (Author's Collection)

Photo 34: A very clean near head-on shot of 8023 on its arrival day at RAF Manston in January, 1951. (Author's Collection)

Initial operations were much less than successful. Air Refueling seemed to be the most serious issue. Out of the first eight attempts on PARPRO sorties, only 2 were successful, and **8021** sustained significant damage that kept it on the ground for five months awaiting replacement parts. SAC's concern over the success of the RB-45C continued for the new few months. In March 1951, Det 1 was relocated to RAF Sculthorpe and the unit kept up a strong effort and hard pace to become proficient in its mission. **8026** was lost on 15 May 1951 at Sculthorpe which did little for morale. However, close attention and analysis by the commanders determined that the most critical issue was supply, especially for the J47 engines of the RB-45Cs and the air refueling system of the KB-29Ps. With hard data and specifics about what was needed, these problems began to improve. Two more aircraft were assigned to Det 1 in May 1951, **8027** and **8011**.

By the end of July 1951 real improvement was obtained by the Det and the critical high quality imagery and radar intelligence flowed to SAC and USAFE Headquarters. By 19 Aug 1951 **8023**, **8019**, **8021**, **8027** and **8011** had rotated back to Barksdale AFB. Between the 7th of July and the 10th of August 1951 four new replacements arrived for Det 1 and took up the PARPRO mission in Europe. **8031**, **8037**, and **8038** arrived on 30 March 1951. Three additional RB-45Cs, **8041**, **8040** and **8039**, were sent from Barksdale to boost Det 1 capacity between 12 and 15 August 1951. By the end of 1951, Det 1 was successfully conducting frequent PARPRO missions along the Iron Curtain in central Europe but also in Italy, Spain, France West Germany and Belgium supporting map making and recovery planning efforts across Europe.

The First RB-45C Overflight Deep into Denied Airspace...

8031, **8037**, and **8038** departed Det 1 for Lockbourne AFB between 30 March and 6 April 1952. **8019**, **8034**, **8036**, and **8042** arrived at RAF Sculthorpe between 30 March and 6 April 1952. They were the first elements of what would become Detachment 3, 91st SRW, commanded by Lieutenant Colonel (Lt Col) Marion "Hack" Mixon. But their initial mission was to be far beyond any that had been accomplished since these sorties began in January of 1951. **8041** (once more), **8040** and **8039** departed RAF Sculthorpe between 30 March and 9 April 1952 returning to Lockbourne AFB as a routine rotation.

Both the RAF and USAF now relied almost totally on radar-guided high altitude atomic bombing in their strategic war plans. It was obvious that the first counterblows in a Soviet verses US/RAF war would be by atomic bombers flying from the UK or the US. It was essential that these bombers could first find their way to the targets and secondly strike them accurately. To make these strikes accurate and effective, and to develop peacetime training locations and sorties, cartographic and mosaic products which gave a good representation of what the target would look like on radar were needed. Further the areas surrounding the target areas and approaches might have the scope's radar return blanked out or "shadowed" by nearby high terrain based on the bomber's altitude on the bomb run. The only effective method to fill this requirement was to overfly as many potential targets as possible at an altitude that matched that flown on an actual strike. Throughout the mission, the radar scope would have its returns recorded on film for reference and analysis back at the bombardment bases. The only aircraft with performance approaching what might be needed to safety carry out these missions were the RB-45Cs of Det. 3.

The genesis of the missions that Det. 3 was about to fly, the most sensitive and productive covert sorties of the RB-45's career began in December, 1950 at the beginning the Korean War. At a State Visit to Washington D.C., British Prime Minister Clement Atlee sought an agreement from President Truman that he would not use Atomic Weapons against China in their new invasion of the Korean Peninsula in support of South Korea. Atlee had earlier sold the Soviet's rights to the Rolls-Royce *Nene* turbo-jet engine that ultimately powered the MiG-15 and was developed to power the MiG-17 and other Soviet fighters. He was therefore hardly high on the US list of people to trust. Therefore the President sought and obtained an agreement for US reconnaissance aircraft to be flown from UK bases by RAF crews and used in deep overflights on the Soviet Union. Succeeding National leadership, President Eisenhower in the US, and Prime Minister Churchill in the UK, ultimately agreed to continue the project. Four RAF crews and maintenance personnel were assigned to Barksdale AFB and later Lockbourne AFB initially in November 21, 1951 to learn the RB-45C. These personnel were not "exchange" personnel, but RAF personnel assigned to the 91st SRW. One RAF pilot was replaced in training by another from the UK who had extensive experience in the RB-45C. On 5 April 1952, Squadron Leader John Crampton assumed Command of what then became the RAF Special Duty Flight (SDF), with four RB-

45Cs, **8019, 8034, 8036,** and **8042,** and six KB-29Ps dedicated solely to these missions, and all assigned personnel with the strong support of Lt Col Mixon. None of these RB-45Cs were turret equipped or otherwise armed.

On the 6th or 7th of April 1952, all four RB-45Cs in the SDF were flown to RAF West Raynham, completely stripped of all USAF markings and very nearly all Wing and Squadron markings. They were repainted only with basic RAF national insignia on the top and bottom of the wings, either side of the fuselage and standard flag panels on the vertical fins. The removal of markings was so thorough that it has been impossible to identify specific tail numbers in any operational images or which aircraft flew which routes in the Operation. A day later the aircraft were returned to RAF Sculthorpe and run though all possible maintenance and operational checks. On 21 March 1952 Squadron Leader Crampton had flown a dry run through the Berlin Corridors and the nearby locale of East Germany just after 12:00 a.m. There was no interceptor or anti-aircraft response, and while Soviet and East German ELINT did pick the USAF RB-45 up, nothing significant enough to cancel the mission occurred. On the night of 17 April 1952 Operation **JU JUJITSU** prepared to launch. Three routes had been charted out; first Route Blue across the North and Baltic Seas to the USSR controlled coast north of Poland, then north to the Estonian Soviet Socialist Republic (SSR) city of Tallinn; then east to the eastern boarder of the Estonian SSR, then south through Latvian and Lithuanian SSRs into eastern Kaliningrad, and then into northern Poland. From there the aircraft would turn northwest for Copenhagen to rendezvous with its waiting tanker, and then across the North Sea for a landing at RAF Sculthorpe.

Photo 35: One of the four USAF RB-45A-1s transferred to the RAF's Special Duties Flight at RAF Sculthorpe was photograph right before the first flight over Eastern European and Soviet denied airspace. The aircraft were painted by RAF technicians and since the markings were extremely accurate for those used at the time, it is believed that this is one of the JU JITSU I aircraft. One of the very few unit markings that were not removed prior to the flight were to 84th Bombardment Squadron red arrows on the outboard side of the wing tip drop tanks. (Courtesy the RAF Sculthorpe Heritage Centre via Ian Brown)

The second route was Route Red, again across the North Sea to a point north of the home island of Copenhagen, Sjaelland. Then the aircraft would fly directly to the Polish town of Ostroleka, approximately 60 NM north of Warsaw. From there the aircraft penetrated the Lithuanian and Belarus SSRs between Vilnius and Minsk and continued northwest to the town of Vitsybsk, turning south-southeast to the Russian Soviet Federative Socialist Republic (SFSR) town of Bryansk. From there the aircraft would turn almost due west, proceed across the Desna and Dnepr Rivers, north of the Pripyat River until re-entering Poland continuing to a point approximately 25 NM north of Warsaw. From there the aircraft would return to its awaiting tanker north of the Danish Island of Sjaelland and then directly to its approach and landing at RAF Sculthorpe.

The third route was Route Yellow, the deepest penetration of Soviet territory. This route departed RAF Sculthorpe as the other two, but entered West German territory just after meeting its tanker over the Baltics just northeast of the Netherlands. It then was to fly south-southeast to the East German-Czechoslovak border due west of Prague. It would then fly almost due east just north of Prague to Krakow, Poland, and then to Kiev and Kharkov in the Ukrainian SSR and on to Volgograd in the Russian SFSR. At Volgograd, the aircraft would turn due south heading for the Russian SFSR town of Rostov. Just north of Rostov, the RB-45C would turn due West to re-enter the Ukraine and Czechoslovakia and then back into West Germany to meet its tanker over Wurzburg, Germany then taking a direct northeastern course for RAF Sculthorpe and its approach and landing there.

Once all mission aircraft were launched and all in-flight checks completed, the spare returned to RAF Sculthorpe. Two 91st AREFS KB-29P tankers waited at each outbound air refueling track. One tanker was primary at each track, and one was the back-up. No back-ups were used. After more than 30,000 pounds of JP-1 was off loaded to the recce birds, the RB-45s doused all external lights, rocked their wings to say so-long and began the climb to 36,000-feet over enemy territory. Each sortie conformed to its flight plan and saw no enemy air or ground reactions at all. Allied SIGINT knew that the East German and other Soviet or satellite radars were tracking the RB-45s and had launched many interceptors. However the three Tornados saw no adversaries primarily because none of the Soviet or Satellite aircraft were equipped with airborne radar. Further, neither the Eastern Bloc nor the Russian SFSS had a capable or even minimally effective ground-controlled radar intercept system. With these limitations it was extremely difficult making nighttime interceptions, although they had been accomplished by Soviet, PRC, or North Korean interceptors in the Far East already.

Each sortie lasted between 10 and 11 hours and encountered no mechanical or operational issues, and no air refueling issues. Like the outbound routes, there were three inbound air refueling tracks, again each with one primary and one back-up KB-29P assigned. As before, the back-up tankers were not used. All aircraft recovered successfully at RAF Sculthorpe. The highly sensitive 35mm film images of the radar scope returns were processed at a highly secured and closely controlled facility at RAF West Drayton and shared equally between SAC and RAF Bomber Command.

There would be two more attempts, one successful, duplicating this tremendous covert operation from Sculthorpe. **8019, 8034, 8036,** and **8042** all returned to Barksdale AFB, still wearing RAF markings and flown by their RAF crews on 9 May 1952. This ended the stunning story of 91st SRW Det. 3. The last three RB-45Cs to be assigned to RAF Sculthorpe for standard PARPRO missions were **8041, 8036** and **8029**, arriving on 17 October 1952 as 91st SRW Det. 4. Although it was thought that the RAF SDF had been closed, the re-appearance of Flight Commander Crampton, Lt Col Mixon and eight familiar RAF aircrew members, something similar to the April effort must be about to happen again.

Two months of intense re-qualification flying began in October and by the end of November, the RAF was at a high pitch and ready to duplicate their previous achievements. But suddenly in early December, the SDF was ordered closed and the RAF personnel returned to their former units. Senior leadership at one or both of the two interested countries had changed their minds and JU JITSU II plans were cancelled.

The Det.4 RB-45s had indeed been cover for another clandestine op, but despite this cancellation the aircraft flew some 53 sorties in the nature of what past PARPRO TDYs had accomplished. The three RB-45Cs returned to Lockbourne AFB on 9 December 1952, ending nearly two years of very successful RB-45C contributions and support to Headquarters SAC and Headquarter USAFE PARPRO requirements. There would be a final effort at another incursion, but much would change in the command and organization on the RB-45C world before then,

Barksdale paid the price often paid by organizations that introduce a new aircraft into the inventory. Two were due to flaws in the design of the aircraft's wing and fire prevention methods, and one were due to aircraft operational training and too few hours of practice in basic instrument procedures. The training issues involved the loss of **8024** on take-off at Barksdale on 28 April 1950. The other flying accident occurred on a live instrument approach at Goose Bay AB, Labrador on 1 August 1951 when **8032** hit high terrain that had been given an improper elevation on an approach chart to Goose Bay and never corrected. The two in-flight explosion losses occurred in situations not unlike the loss of the first B-45A, **7001**, in southern CA, when leaking fuel caused an explosion which blew portions of the engine cowls off the aircraft and into the empennage. **8028** was lost near Plymouth, SC on 9 February 1951, and **8030** was lost near Houston, TX on 7 May 1951.

NOTE: A map and layouts of all three sorties and the routes flown on Operation JU JITSU I crafted by Squadron Leader Rex Sanders, Navigator for Flight Commander John Crampton, were completed for the 22-23 February 2001 Early Cold War Overflights, Symposium Proceedings, Volume I: Memoirs. It is used with permission from R. Cargill Hall, retired Historian at the National Reconnaissance Office, for which the Author is extremely grateful.

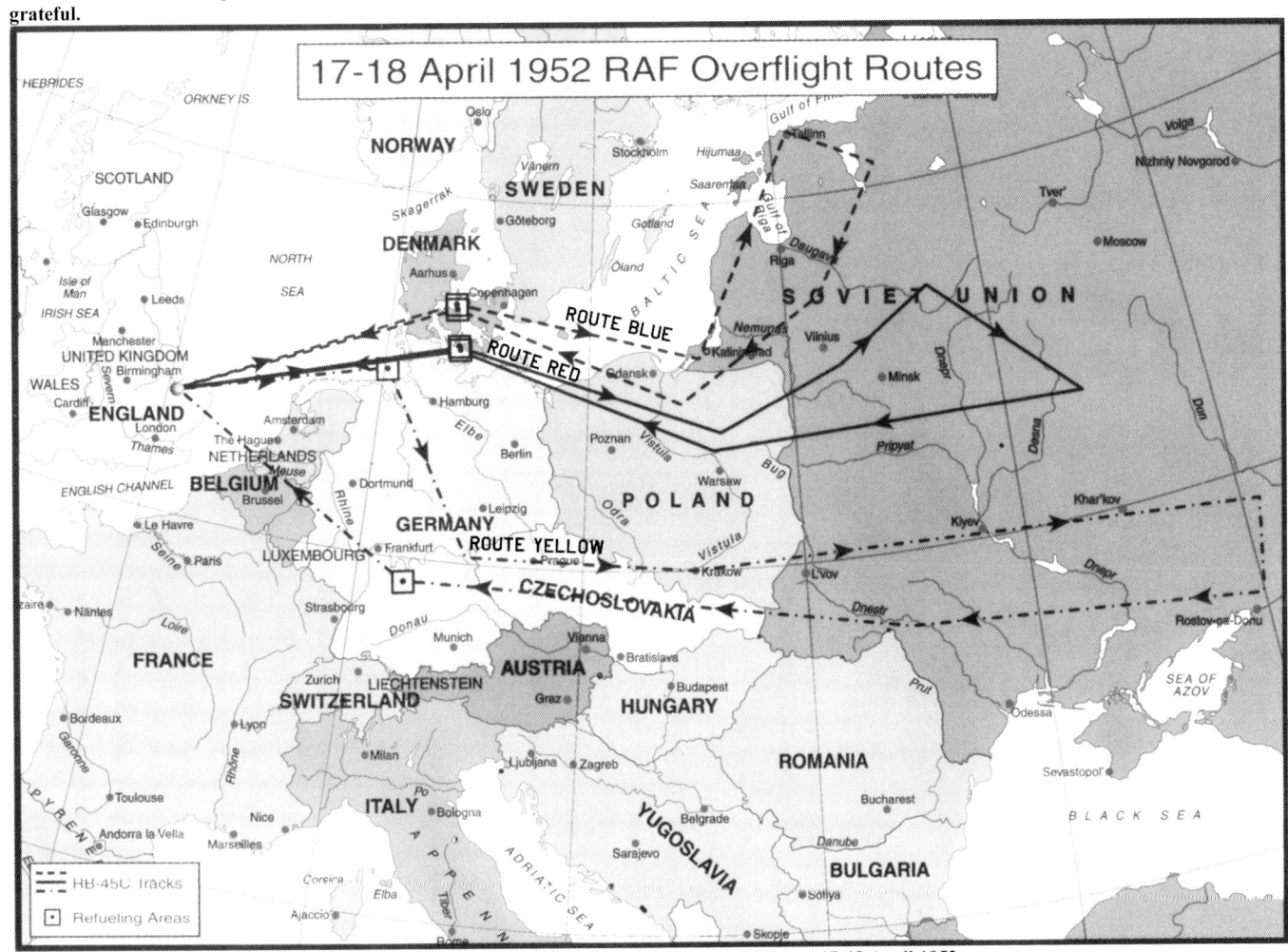

Drawing 1: A map of the Routes flown by three RAF SDF RB-45s on Operation JU JITSU I on 17-18 April 1952.

Barksdale to Lockbourne AFB....

In April 1951 the 91st SRW was alerted that it was changing home bases. In September 1951 it would be relocated to Lockbourne AFB, just south of Columbus, OH. Barksdale had two B-47 Bombardment Wings, the 301st and 376th and more room for additional KC-97 tankers, and maintenance facilities were needed badly. Lockbourne was not nearly as crowded with few large aircraft units located there. The transfer began on 1 September 1951 and was completed 16 April 1953 with 22 RB-45Cs assigned. The KB-29Ps of the 91st AREFS also transferred to Lockbourne in September and shortly began air refueling support to 91st RB-45s deployed to Yokota AB, Japan. In October 1951, Lockbourne assumed immediate responsibility for supporting SAC and FEAF RB-45C operations from Yokota AB Japan. Lockbourne also took oversight of the last two groups of the European PARPRO missions in March and September 1952. Aircrew and maintenance training that had initiated at Barksdale moved to Lockbourne as well, in support of fleet-wide operations in the Far East and Europe. While maintenance remained favorable, airframe and engine issues primarily due to supply support, remained an issue. However Lockbourne only lost one RB-45. **8039** due to an in-flight fuselage fire near Pine Bluff, AR on 24 Dec 52. The 91st spent considerable time and airframe hours supporting multi-service exercises and conducting long range cross-county navigation missions. The Wing also supported numerous civilian airshows with static display and airshow flying demonstrations as frequently as possible.

Photo 36: RB-45C 8033, commanded by Colonel Joseph Preston, the 91st SRW, lands at Lockbourne in mid-September 1951. It was the first 91st RB-45C to arrive. (Steve Ginter) Photo 37 below: By the end of September 1951 all the RB-45Cs from Barksdale had arrived at Lockbourne, and by Sept 1952, all of Barksdale birds in the Far East had rotated home and been replaced by new aircraft from Lockbourne. Visible in this area of Lockbourne are RB-45C 8012, the second RB-45C built, 8025, 8034 and 8027, destined to become the first RB-45C armed in the Far East via a field modification by Det 2. (NMUSAF)

Photo 38: RB-45C 8021 was photographed at Detroit Metro Airport in the summer of 1952. It wears the medium blue and white colors of the 324th SRS. It was the only 324th aircraft to be deployed to Japan during the War. (NMUSAF) Photo 39 below: This image of 8021 was taken at Detroit Metropolitan Airport in 1952 prior to its assignment to Yokota AB. Here, it wore the blue and white squadron markings of the 324th SRS. Note the Douglas D-558-2 Skyrocket in the background. (Balogh via Norm Taylor and Steve Ginter)

Photo 40: RB-45C 8031 taxies out for its demonstration at the MacDill AFB airshow in February 1952. The 91st would take these opportunities to attend airshows across the county in order to conduct long range training missions and demonstrate the Tornado to the American public. In this case 8031 conducted numerous over-land and over-water navigation, air refueling and photo-reconnaissance sorties during the six weeks it was on temporary duty (TDY) in Florida. (Author's Collection)

Air Force Desegregation and the 91st SRW....

In the year following the end of World War II the USAAF began reassigning over 16,000 Black USAAF Officers and Airman from assignments overseas and throughout the United States to a few mid-western airfields. The goal was to discharge as many Black servicemen as possible, but there were quite a number who wished to stay in uniform and pursue careers in the specialties and professions they had mastered during the War. Nearly all who wished to remain in the Army Air Force and continue flying or serving in the aviation branch of the Army were slowly concentrated at an Army Air Field south of Columbus, Ohio. Between 1946 and 1948, while efforts were being made to create segregated units at the airfield, HQ USAAF was less than enthusiastic, however, it had assigned Colonel Benjamin O. Davis, one of its best senior flying officers to the location.

Based on politics but also a personnel desire to improve on a long-standing issue in the Nation, on July 26, 1948 President Harry S. Truman signed executive order 9981 establishing the President's Committee on Equality of Treatment and Opportunity in the Armed Services, committing the government to integrating the segregated military. This order was the first step in desegregating all US Armed Forces and began a series of fundamental changes at the airfield south of Columbus, now known as Lockbourne Air Force Base. Air Force Letter 35-3, Air Force Personnel Policies, dated 11 May 1949 required immediate action to desegregate the U.S. Air Force. Specifically it required testing and written evaluations of all black Air Force personnel which were used to create assignment potential lists. Less than a year later, Secretary of the Air Force Stuart Symington signed master lists of personnel qualified for reassignment to previously all-white units. The requirement was to have 50-50 percent black/white unit assignment ratios within 5 years which was achieved. With the experience gained by, now, Brigadier General Davis at Lockbourne, relating to equality in billeting, base services, health care, mess services, transportation, and Base Exchanges, a standard was established Air Force wide over the next five years.

Photo 41: RB-45C 8016 "Hot to Trot" had its image taken at Yokota in 1952 with its integrated maintenance crew. TSgt Chuck Davidson is the second from the left on the back row. (Keith Davidson) Photo 42: TSgt Davidson is shown in full flight gear at the head of the crew passageway at the forward crew entry door in RB-45C 8016. It is not known whether the exact nature of his airborne work was mission oriented or maintenance. (Keith Davidson)

The 91st SRW being one of the first combat units to be assigned to Lockbourne during the initial desegregation era took full advantage to the existence of the personnel qualification lists and was able to interview and meet many of the top Non-Commissioned Offices at Lockbourne and reassign them to the Wing. The Wing gained a prized outstanding group of enlisted leaders who made the Korean Reconnaissance effort and the PARPRO missions in Europe a success despite great obstacles. One example of the exceptional NCOs the Wing gained was Technical Sergeant (TSgt) Chuck Davidson, father of one of the Author's great contributors to this book, Keith Davidson (**Photos 41** and **42 above**). TSgt Davidson began his service with the 322nd Fighter Group in Italy during World War II as a radio maintenance specialist. After the War he stayed on active duty and was reassigned to Lockbourne in 1946. Until 1950 he remained in radio and navigation system maintenance on the base, until assigned as one of the first maintenance NCO's to enter the airborne maintenance supervisor career field on RB-45C communication and navigation systems. He was among the first jet bomber maintenance specialists in the USAF.

RB-47s Replace the RB-45C in SAC which become TAC assets....

With the end of active combat on the Korean Peninsula, SAC received word that Boeing was close to the delivery of the first production RB-47s. Two field-modified RB-47Bs had been used by the Commend in the start of its RB-47 history with an overflight of eastern Siberia searching for Tu-4 bases between Ambarchlk and Provideniya in 1952. The production RB-47, the RB-47E, would be a state of the art photo reconnaissance platform with a crew of three and greater range, airspeed and ceiling. With these improvements and its better engineering it was a good successor for the RB-45C; but the 240 aircraft would need homes. SAC was down to under 30 RB-45Cs all assigned to Lockbourne AFB, when the RB-47E began to arrive. The 91st SRW was immediately scheduled to be re-equipped with RB-47Es which would force the relocation or disposal of the RB-45Cs. At the urging of senior staff officers of the 363rd Tactical Reconnaissance Wing (TRW) at Shaw AFB and members for the Tactical Air Command Staff at Langley AFB, TAC was granted control of many of the excess RB-45Cs. On 5 June 1953 Hq USAF directed that the 19th Tactical Reconnaissance Squadron be assigned to the 363rd TRW and allocated 12 RB-45Cs at Shaw AFB. All aircraft would have to be processed for an Inspect and Repair as Necessary (IRAN) overhaul at SBAMA and while there fitted with a new Bell M-7 tail turret. All combat-capable would also be modified to carry a pair of the same Fletcher 500-gallon JP-4 tanks when not carrying water/alcohol tanks. At least 17 RB-45C's assigned to the 363rd TRS at Shaw AFB, NC were modified. The newly assigned Squadron Commander, Major John Anderson was an exception to those before him. He was just as capable, if not more so, just far more reserved in his manner and approach to leadership. It shortly became obvious that the unit was in extremely good hands.

The Second RB-45C Overflight Deep into Denied Airspace...

NOTE: A map and layouts of all three sorties and the routes flown on Operation JU JITSU II was crafted by Squadron Leader Rex Sanders, Navigator for Flight Commander John Crampton, were completed for the 22-23 February 2001 Early Cold War Overflights, Symposium Proceedings, Volume I: Memoirs. It is used with permission from R. Cargill Hall, retired Historian at the National Reconnaissance Office, for which the Author is extremely grateful.

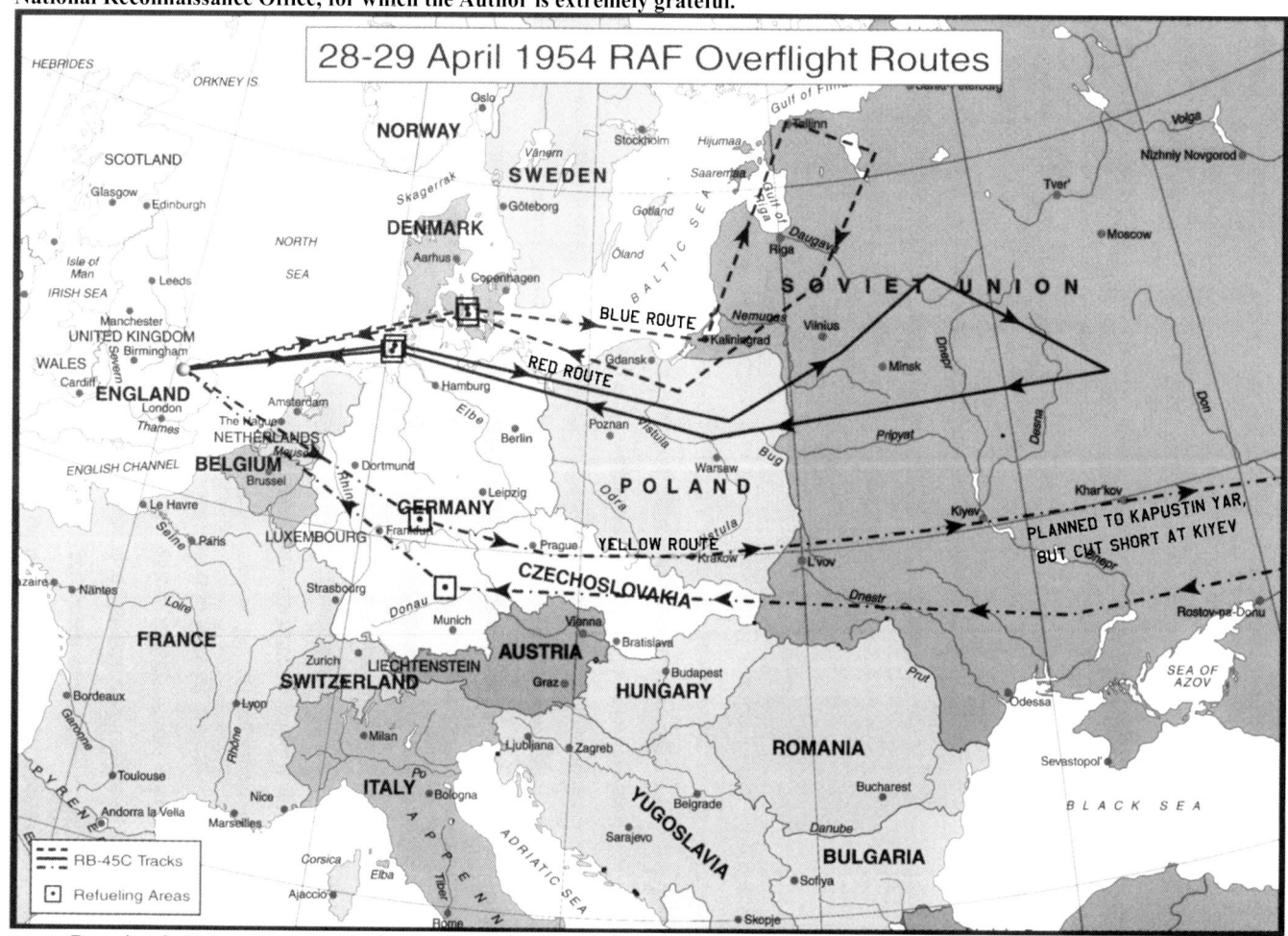

Drawing 2: This map shows the Routes flown by three RAF SDF RB-45s on Operation JU JITSU II on 28-29 April 1954.

Anderson had the advantage of selecting his own people including some who had flown the Tornado for SAC. Since it would be almost a year before all his aircraft had arrived, he had all the time needed to craft a complete flying unit and make it into an outstanding organization. In the autumn of 1953 Major Anderson was alerted that unit would be moved overseas and relocated to RAF Sculthorpe in the spring of 1954, joining the 47th Bomb Wing already there. For the next several months the squadron began the pack up routine of all equipment, spares, tools and other gear they would need overseas. In November 1953 the Squadron received orders to report to RAF Sculthorpe in April of 1954, and the packing continued with a new urgency.

In late winter of 1953, Squadron Leader John Crampton was reassigned to command the RAF SDF for a third attempt at a long distance penetration of East European and Soviet airspace. As before, Operation JU JITSU II was planned to launch and recover at RAF Sculthorpe with air refueling over Denmark and West Germany. The routes were virtually identical to those flown in Operation JU JITSU I. ROUTE BLUE through the Soviet controlled Baltic countries, ROUTE RED across northern Poland and into Russia towards the Moscow area, and ROUTE YELLOW southeast across West Germany and through Czechoslovakia and the Ukraine into the very southern tip of the Russian SFSR. Only this time the YELLOW or southern route was extended deeper into the Russian SFSR to the town and missile test range at Kapustin Yar, just southeast of Volgograd and 175 northwest of the Caspian Sea. HUMINT had reported that the Soviets were testing surface to surface guided missiles there. In 1956, U-2 Imagery Intelligence and earlier SIGINT confirmed that such testing was indeed based there. Lt Col Marion Mixon, now an RB-47E unit commander was once again leading the USAF side of the effort and obtained four newly depot re-furbished RB-45Cs from the 19th TRS at Shaw AFB. The aircraft were **8031**, **8035**, **8037** and **8040**. All were immediately flown to WPAFB in March and early April 1954 for radar and radarscope imagery upgrades, repainting back into RAF markings, and the removal of their newly installed Bell M-7 turrets to give them better high altitude performance.

On 5 April 1954 the four aircraft arrived at RAF Sculthorpe and flight proficiency training began again. Intelligence indicated that airborne interception might be a possibility but at the planned cruising altitude of 39,000 feet, it would be difficult. Further, that altitude would make the impact of flack on the mission a very low concern. The mission launched late on the night of 28 April. All aircraft and the spare launched successfully and the spare was returned to Sculthorpe. All refueling took place on schedule in complete radio silence, although the sound of jet engines over Denmark late at night did occasion a few questions to local police stations. Once again with the aircraft fully topped off, the Recce birds doused their lights, rocked their wings in a "Thanks and So Long" to the tanker crews and began the climb to 39,000 feet. Again all of the missions except the YELLOW Route were scheduled for between 10 and 11 hours. The BLUE and RED routes had no problems, flew a precise course, had no hostile inference, returned and recovered (landed) at RAF Sculthorpe about 10.5 hours later. That said, political channels were called in to land a number of Danish Meteor NF11 night-fighters who had been launched to intercept an unidentified aircraft coming west toward their nation from the Baltics about the time of the end of the BLUE route mission. The YELLOW Route was completely normal until just west of Kiev.

Having seen some "lightning" or "artillery rounds" glow beneath the undercast, Commander Crampton suddenly saw flack bursts at his altitude directly ahead over Kiev. Having been a bomber commander in World War II, he knew exactly what he was seeing and immediately pushed the engines to full throttle and began a tight as possible 180 degree turn to the west. Fortunately none of the flack came within about 10 miles and aircraft and crew were unharmed. He pushed the jet to its airspeed limit and cleared the Ukrainian SSR as fast as possible. His navigator (Rex Sanders) gave him a direct course to their planned air refueling track over the Furstenfeldbruck Air Base in West Germany. Once over the base they rendezvoused with their tankers, but the RB-45C's air refueling receptacle refused to hold the tanker's boom, and they decided to make a precautionary landing at Furstenfeldbruck. They landing without incident, and since Furstenfeldbruck was home to a number of very sensitive USAF reconnaissance units, security was not an issue. Crampton and his crew returned to RAF Sculthorpe a few hours later to meet up with the crews from the BLUE and RED Routes early in the morning of 29 April 1954. Even though Kapustin Yar was not covered, all three routes were judged highly successful. Many years later John Crampton was briefed by MI-6, British Foreign Intelligence, that his portion of the mission had been compromised at RAF Bomber Command Headquarters by Soviet Spy Kim Philby. He then knew that the lightning or ground-to-ground artillery fire all across the Ukraine had been a massive Soviet-forces attempt the shoot his aircraft down. Though closely held, this set of missions became a much honored part of SDF and RAF History.

PART III: RB-45Cs go to Europe in Unit Strength....

The 19th RS Goes to AF Sculthorpe....

On 6 May 1954 the first six 19th TRS RB-45Cs arrived at Sculthorpe adorned in their new strikingly vibrant Insignia Blue and White squadron colors. In the next eight months the unit gained 11 more Tornados including the four from the RAF SDF, some for spares and scrounging no doubt. The 19th's task was to be the primary atomic strike radar and visual imagery unit in Europe. The 49th Air Division (49th AD), which owned all atomic-capable units in Europe (the 20th Fighter-Bomber Wing, the 81st Fighter-Bomber Wing, and the 47th Bombardment Wing) had assigned the 19th the war-time mission to fly "damage assessment" sorties over all targets to be struck by atomic weapons. Those sorties would take radar and perhaps visual images of the drop zones to insure the correct target was struck and that weapons functioned properly.

The mission also included the task of taking pre-strike images of planned atomic targets to provide mission planning materials to the 20th and 81st Fighter-Bomber Wings F-84 crews and the 47th Bombardments Wing B-45A-5s and B-45C-5s crews. The Squadron initially had no facility at Sculthorpe that was capable of handling the processing and production of these kinds of products. But in short, order a full-scale, multi-capable lab that could handle the workload generated by the entire 49th AD for radar and imagery products was developed and achieved an exceptional reputation by the Air Division and all units who used it.

The 19th would remain operational at RAF Sculthorpe until December 1956 and during that time it lost no aircraft and set several operational, maintenance, and flying safety records while exceeding all mission requirements from the 49th AD. It participated in numerous NATO and USAF-conducted atomic mission exercises, teaching atomic weapons tactics and mission planning to NATO allies newly assigned to the world of atomic weapons. But before the Squadron's RB-45Cs folded their wings for the final time, there was one last uninvited intrusion into Soviet-Allied Controlled Airspace from RAF Sculthorpe by the 19th TRS.

Photo 43: RB-45C 8034 takes-off from RAF Sculthorpe in 1954 wearing the Squadron's very eye-catching Insignia Blue and White unit markings and its unit insignia large on the starboard side of the aircraft. (Courtesy Norm Taylor via Dave Menard and Steve Ginter.)

Photo 44: 19th TRS RB-45C 8012 was photographed on the ramp at RAF Sculthorpe during the winter of 1954-1955 being prepared for an evening sortie over the UK. At some point in their service, the Squadron repainted the nose sash in Insignia Red from Insignia Blue. (NMUSAF)

On the evening of 25 March 1955, the 19th launched three RB-45Cs from RAF Sculthorpe. All refueled over West Germany then broke formation and proceeded to separate radar targets in Eastern Europe. The northern-most flight headed to Berlin and the East German locale. The southern-most bird headed to and overflew Zagreb, Yugoslavia, while the middle aircraft commanded by Major Anderson headed to Budapest, Hungary. These aircraft all carried live guns and gunners, and were prepared to be intercepted. Although a number of Soviet interceptors did approach the recce aircraft, no shots were fired and all aircraft recovered safely. Maj. Anderson's aircraft had used most of its fuel reserve out climbing interceptors over Budapest and recovered at Furstenfeldbruck, just west of Munich, West Germany. Once more an intelligence team was waiting ay Furstenfeldbruck and collected all film, charts and notes on the aircraft. The Major and his crew remained overnight and returned to RAF Sculthorpe the following morning.

This was the climax of the RB-45's final three years in Europe. In December 1956, the replacement for the Tornado, the Douglas RB-66B Destroyer, arrived at RAF Sculthorpe and the end was indeed near. By the end of early 1958 all 18 aircraft were disposed of as fire response training airframes at USAF facilities in the UK, France, West Germany, and Turkey. Their ultimate fate was nearly identical to their brothers, the bombers, going on at the same time. Unfortunately none were saved, not even one that had participated in either Operation JU JITSU.

Photo 45: RB-45C 8019 Tornado of the 19th TRS was imaged over the UK in the Spring of 1955, sometime near the Eastern European intrusion sortie in March 1955. (USAF via Author's Collection)

Photo 46: RB-45C 8031 prepares for take-off from RAF Sculthorpe, a cold snow-covered airfield this day. (Courtesy the RAF Sculthorpe Heritage Centre via Ian Brown)

Photo 47: RB-45C 8040, assigned to the 19th TRS, cruises over the quite Norfolk, East Anglia country side of the United Kingdom in late 1956 shortly before the unit converted to Douglas RB-66Bs at Sembach AB, Germany. (NMUSAF)

Photo 48: This Douglas RB-66B of the 19th TRS, assigned to the 66th Tactical Reconnaissance Wing at Sembach AB, Germany in 1957 clearly reflects the clean, sleek lines of the replacement for the RB-45C Tornado. (Author's Collection)

CHAPTER 8 – THE B-45'S PLACE IN USAF HISTORY

Photo 1: The first XB-45 559479 climbs out over Muroc AAF Field in late fall 1947 about eight months after its first flight. The aircraft was now well into its flight test program carrying two full size flight test pitot tubes. (NARA II via Tommy Thomason)

As the late John Fredriksen declared in his 2009 book on the B-45, I also declare my considerable disappointment in not being able to shed more light on all the highly classified and highly important missions these RB-45s flew during their 6-plus years over the Korean Peninsula and across the Far East from their home bases at Yokota AB, Japan and temporary hostel at Clark AB, the Philippines. Due to the extreme sensitivity of their work, many of the records and histories of Det A, Det 2, the 6091st, the 6007th, Det 1 through 4 of the 91st SRW, the RAF SDF and the 19th TRS in Europe remain classified and withheld from historian's eyes. These missions surely brought unknown sacrifice and loss, but certainly great success and achievements as well. And this says nothing of the 47th Bombardment Group, Bombardment Wing, or Tactical Bombardment Wing, the USAF's first multi-engine turbo-jet powered atomic bombardment unit. Much more needs to be said about the lessons the bomber Tornados brought to the service careers of the B-47, B-52 and the B-57. It will be interesting reading when this information is finally released. Perhaps a future author will have the good fortune of full access, and thus the honor to complete this story in a third volume about the North American B-45.

Photo 2: The J35-powered B-45A-1s served primarily as flight test aircraft and as pilot trainers for the 47th Bombardment Group. This B-45A-1, 7015, conducted the initial B-45 extreme cold temperature tests at Ladd Field in Alaska and Whitehorse RCAF Base in Canada. (SDAM via Craig Kaston)

So, with this limitation, this Volume can only be a partial history of a remarkable aircraft. That said, it is still important to correct the significant amount of incorrect data and information that has been published about the Tornado over the years. And in doing so, such efforts may inspire Air Force leadership to increase the budget and manpower resources that are the primary shortfall in the declassification efforts of the Service History Programs today. Further, perhaps this volume, on the design and operational history of the North American B-45, given in a chronological order, will shed a little more light and understanding. I sincerely hope so.

Photo 3: Among the first assignments the B-45 received were as radar targets for Early Warning Radar and Surface to Air Artillery sites across the US and Canada. Later these assignments were expanded to carrying extendable and retractable target tow reels and radar reflector targets to train USAF and RCAF interceptors in high speed and high altitude radar intercepts. This mission lasted into 1958 when this TB-45A,7020, was photographed at Yuma AFB, AZ. (Courtesy B. Hostetter via Davis W. Menard)

The B-45's significance and to a great extent its place in USAF History is first measured by a significant number of firsts for the first operational turbojet bomber and multi-engine turbo-jet reconnaissance platform of the USAF (compiled by Dennis Jenkins and the author):

- The first US turbojet-powered four-engine bomber flown.
- The first US turbojet-powered multi-engine bomber put into full production and entered into operational service.
- The first US turbojet-powered multi-engine bomber to drop bombs at over 500 mph.
- The first US turbojet-powered multi-engine bomber capable of carrying and dropping live atomic weapons.
- The first US turbojet-powered multi-engine bomber to drop a live atomic weapon.
- The first US turbojet-powered multi-engine bomber to provide an atomic strike capability in Europe.
- The first US turbojet-powered multi-engine reconnaissance platform produced and entered into operational service
- The first US turbojet-powered aircraft to be refueled in flight.
- The first US turbojet-powered aircraft to fly non-stop across the northern Pacific Ocean.
- The first US turbojet powered aircraft to fly across the North Atlantic using foreign bases and air refueling.
- The first US turbojet powered multi-engine reconnaissance platform to fly in combat
- The first US turbojet-powered multi-engine reconnaissance aircraft to overfly denied airspace.

But the opposite side must also be stated. The B-45 family had, by percentage, the highest loss rate at 28%, of any operational aircraft in the first decade of the USAF, surpassing the Boeing B-47 by 6%. While it had never been considered unsafe to fly by its crews, fire and wing structural issues were a continuing hazard until the last three years of its service. The fire issues were primarily due to the extremely high operating temperatures of the jet engines that made fuel tank sealants fail faster and caused cooling blankets to crack and allow excessive heating of large portions of the cowls surrounding the engines. The wing issues were not due to weakness per se as the wings met all structural standards placed on the design, but because the original design did not anticipate the much higher and rapid buildup of "G" loading, exacerbated by turbulence, in high speed maneuvers for weapons delivery, and defensive evasion of adversaries flying new high-performance turbojet aircraft.

Experience, a vastly improved spares and supply system, higher quality engines, and better lubricants, sealants, and fuel were all factors in the improvement in the aircraft's sunset years. But the true bottom-lines to almost all of the Tornado's operational problems were the decisions made at Headquarters USAAF and USAF throughout its career. British turbojet-powered aircraft that were fielded during World War II had convinced USAAF and later USAF leadership that turbojets would be the answer to the airspeed, altitude, range and payload goals they sought so desperately to attain. And there was great impatience. The first two or so years of monitoring Soviet forces in East Germany had brought concern about their motives and the strong militarization throughout occupied Europe.

Further, HUMINT from *Wehrmacht* and *Luftwaffe* prisoners of war; prisoners released from Soviet controlled Eastern Europe, and the few former German Scientists released into a type of refugee status, were interviewed at length by Allied Intelligence and Security agents and made clear that the Soviets were actively pursuing development of atomic weapons, radar, turbojet aircraft, surface-to-surface, and surface-to-air anti-aircraft missiles with great energy and urgency. Their open and above-board efforts at offering tremendous amounts of Soviet capital to obtain rights for the production of Western designed and built turbojets, left no doubt about their military goals. This information emphasized just how serious any potential conflict with the Soviets, their Allies and Occupied Nations would be; and that the West was falling behind, if not already seriously behind, in what would become known as the Cold War.

Photo 4: This North American B-45A-5, 7036, was assigned to the 47th Bombardment Group at Barksdale AFB, LA, part of Continental Air Command's Tactical Air Command. It was the USAF's first and only conventionally-armed turbojet-powered bombardment unit. (AF Flight Test Center HO via Steve Ginter)

Photo 5: This J47 powered B-45A-5 BACKBREAKER/SECOND CALL bomber, 8047, was assigned to the 84th Bombardment Squadron, 47th Bombardment Wing (later the 47th Tactical Bombardment Wing) based at RAF Sculthorpe, UK. The wing was the first and only multi-engine USAF turbojet-powered atomic-armed unit in the service for over 5 years. The aircraft was cruising in typical weather conditions for B-45As flying in England. (SDAM via Craig Kaston)

The Air Force needed a first-generation jet bomber to identify and understand the unknowns of jet aviation. Jets were to be the answer to all requirements and the service had to needed to know how they worked as soon as possible. The approach was to build an aircraft that could meet the minimum "known" requirements of altitude, airspeed and payload. It was to be a "conventional" design with very unconventional engines which would push the aircraft into a wide, mostly unknown, set of environments. To accomplish the goals in the minimum amount of time, the Air Staff decided to accept the risks to aircraft and personnel by not fully exploring or questioning what those risks might be from day one. The losses would be high, and the aircraft's operational capabilities would probably be limited, at least initially. But the Air Staff believed that the aircraft would place the service at the highest level of jet powered operational capability before any other adversary or ally.

Photo 6: This image shows the first B-45C-1 8001, later to become the first American turbojet-powered aircraft to drop a live atomic weapon over the Nevada Test Site. The ten B-45Cs were primarily flight test, weapons development and research aircraft for the Air Force. While their design was very close to the later RB-45C, and they could carry wing tip drop tanks, but were not fitted with an in-flight air refueling receptacle. (NARA II via Tommy Thomason)

These decisions resulted in various shortfalls because its development and design was rushed. "Get it on the ramp first and we'll fix it later" was the common call from higher headquarters. Engines, supply, training, and components had serious faults not found before active service. Research into structures that could withstand and resist failure during the rapid onset of very high G forces and high speed flight control systems were incomplete and hurried. The limited structure, although it met all contract specifications, did not give the B-45 enough strength to conduct air to ground close air support bombing, or conduct defensive maneuvers in a high threat environment while carrying weapons. This took away two important capabilities in the bomber. There were also serious faults, unknowns and shortfalls in power plant maintenance, ground support equipment, life support, communications, navigation, and many other important aircraft systems.

These shortfalls nearly brought an early and ignominious end to the Tornado's career. A scientific development, the creation of a new family of atomic weapons, Operation SANDSTONE, resulted in the aircraft fulfilling its most important role in Europe as the first atomic-armed multi-engine bomber to support NATO. Then the unanticipated start of the Korea War forced the conversion of a number of B-45C airframes into a new model, the RB-45C which became a highly successful, air-refuelable, tactical and strategic reconnaissance airframe for nearly seven-plus years in both the Far East and Europe.

It was not the Air Staff that brought these new developments about, but a number of highly motivated and dedicated operators who saw the mission potentials, the air crews who grew to know and use the airframe and determine what would be needed to make this machine perform these duties. The people who made this aircraft successful were the laboratory, manufacturing company, and depot engineers who developed the means to make the B-45s capable of performing these missions. And perhaps most importantly it was the maintainers who developed the maintenance requirements, the tasking and supply needs to make the aircraft continually operational and mission capable far beyond anything conceived of in the late 1940s for nearly eight hard years. This becomes the B-45's key place in USAF history: It was the prime example of how an aircraft with multiple shortcomings in design and operations became able to accept and carry out a number of original missions for the Air Force and become an example to teach others on how it could be done. It was also an example how important flying, maintenance, depot, manufacturing, and aircraft engineering crews, military, civilian and corporate, were to making an aircraft fully operationally capable despite so many setbacks and the lack of full or even minimal support in the ranks of higher levels of command. It served its personnel as they served it.

Photo 7: RB-45C, 8027, of Det 2, 91st SRS, takes-off from Yokota AB, Japan in the spring of 1953 on another high altitude night overflight of China or Manchuria toward the end of hostilities on the Korean Peninsula. The heavy smoke was common, even when not using water/methanol injection to improve take-off performance. (Mark Natola via Steve Ginter)

Photo 8: RB-45C, 8027, sans it's all black paint scheme from Korea, was imaged over the UK in late 1954 when newly assigned to the 19th TRS based at RAF Sculthorpe. There, the 19th was responsible for photographing atomic strike missions for the 49th Air Division and the 20th and 81st Fighter-Bomber Wings. (Steve Ginter Collection)

CHAPTER 9 – DATA AND TABLES

1. Mission, Design and Series (MDS):

a. Mission refers to the military operational mission of the aircraft, with a single or two letters assigned by Headquarters USAAF to September 1947. Beginning in early 1948, Headquarters USAF used one letter only to define the mission. In the case of the B-45, the single letter B was used by both services to indicate that the mission of this aircraft was bombardment.

b. The B-45 had two prefixes to the mission designation letters during production, also assigned by HQ USAAF and HQ USAF:

    i. First was a "T" for trainer, in this case Target Tow Trainers for gun or unguided rocket training by interceptors, i.e. TB-45A.

    ii. The second prefix was "R" for strategic or tactical aerial reconnaissance, i.e. RB-45C.

c. The number following the initial letter in the MDS stood for the sequential number of an aircraft being designed for that mission. The USAF retained the sequential number assigned by USAAF. The B-45 was the 45th bombardment design selected for development by the USAAF.

d. The suffix letters following the design number represented minor design or mission changes made to the basic aircraft during its service life. Called model changes, they were identified by the assignment of letter designations in alphabetical order. The B-45 design had three series or model changes: The A model for the production bomber versions powered by J35s and J47 engines; the B model which was to have been a guided rocket interceptor, was not produced; and the C model were reconnaissance conversions made on the production line at the Long Beach plant.

2. Block Numbers: Block numbers, designated in 5-digit groups (-1, -5, -10, and -15 etc.) were used to identify equipment or structural changes made to aircraft during production blocks by the manufacturer. Field modifications of significance used numbers between the 1 and 5 numbers; for example, some field modified TB-45A-1s were Block 2 airframes as TB-45A-2. The (–NA) suffix indicated that the aircraft was built at either North American's plant at Inglewood or Long Beach, CA.

TABLE 1: DIMENSIONS:

	B-45A-1 and -5, and TB-45A-1 and -5	BACKBREAKER and SECOND CALL B-45A-5 w/ Emerson A-6	B-45C w/ Tip Tanks	RB-45C w/ Tip Tanks	RB-45C w/ Bell M-7 w/ Tip Tanks	B-45C-5 w/ Bell M-7 w/ Tip Tanks	B-45C-5 w/ Emerson A-6 w/ Tip Tanks
Length	75.3 ft.	75.6 ft.	75.3 ft.	75.9 ft.	75.5 ft.	75 ft.	75.6 ft.
Wingspan	89 ft.	89 ft.	95.9 ft.	95.9 ft.	95.9 ft.	95.9 ft.	95.9 ft.
Height	25.2 ft.	25.2 ft.	25.2 ft.	25.2 ft.	25.2 ft.	25.2 ft.	25.2 ft.
Tread	22.4 ft.	22.4 ft.	22.4 ft.	22.4 ft.	22.4 ft.	22.4 ft.	22.4 ft.

TABLE 2: FUEL CAPACITIES:

	Tip Tank	4 Lt. Wing Tanks	Fletcher Tank	Fixed Fus. Tank	Fwd. Upper BB Tank	Fwd. Lower BB Tank	Aft Upper BB Tank	Aft Lower BB Tank	Fletcher Tank	4 Rt. Wing Tanks	Tip Tank	Total Load (Gal)
B-45A-1		1511		344	318	872	318	872		1511		5752
B-45A-5		1511		344	318	872	318	872		1511		5752
B-45A-5 BB *		1511	500	344	318	**	318	**	500	1511		4702
B-45A-5 SC *		1511	500	344	318	**	318	***	500	1511		4702
B-45C-5	1125	1511		344	318	872	318	872		1511	1125	7996
B-45C	1125	1511		344	318	872	318	872		1511	1125	7996
RB-45C	1125	1511	500*****	355	318****	872****	318****	872****	500*****	1511	1125	7633

*BB refers to BACKBREAKER and SC refers to SECOND CALL aircraft.
**872 gallon tanks could only be carried at these locations if no Atomic Weapons were carried in BACKBREAKER B-45A-5s.
***An 872 gallon tank could only vary in this location if no Mk 7 weapon was carried in SECOND CALL B-45A-5s.
****Tanks could only be carried in these RB-45C locations if no or only a limited number of photo-flash weapons were carried.
***** 500-gallon under nacelle tanks could be carried if converted by SECOND CALL Teams from SBAMA. Total max fuel load became 7633 gallons.

Fuel amounts are listed in gallons. All fuel tanks carry a certain amount of fuel which cannot be transferred to the engines due to tank structure or the location of fuel intakes or pumps. Fuel which can be transferred is called usable fuel, and it is the amount of usable fuel that is listed for each tank and used for flight planning. The conversion factor to calculate the weight of the fuel load is 6.84 pounds to the gallon of JP-4.

TABLE 3: WEIGHTS AND PERFORMANCE DATA:

		XB-45	B-45A-1	B-45A-5	B-45A-5 BACKBREAKER	B-45A-5 SECOND CALL	B-45C	RB-45C
Weights (lbs.)	Empty		45,206		47,022	47,022	48,969	49,984
	Max T/O				92,745	92,745	112,965	110,721
	Normal T/O	82,600	85,451	85,895	67,820	92,403	103,829	110,279
	Max Landing				90,428	90,428	112,965	110,721
	Max Payload	8,352	22,000		3,500	3,500		5,161
Performance (At combat wt. in KIAS)	Cruise Speed	470			401	401	436	436
	Combat Speed	501	442	540	435	415	442	442
	Max Speed	536 @ 16,000	504	539 @ S.L.	492 @ S.L.	492 @ S.L.	498 @ S.L.	495 @ 4000 ft
	Max Combat Ceiling (ft.)	38,600	33,900	38,500	39,000	32,500	31,500	33,500
	Max Service Ceiling (ft.)	43,000	42,100	41,600	41,100	38,000	37,550	37,800
Range (Nautical miles)	Ferry Range (unrefueled)	2,900			1804	1807	2108	2164
	Basic Combat Radius				764	764	783	916
	Max Combat Radius		742		795	795	876	1107
	Max Combat Range (unrefueled)	3129	3150		1807	1807	1415	2250

TABLE 4: PRODUCTION AND UNIT ASSIGNMENTS, MODEL, BLOCK NUMBERS, AND NOTES:

Block Number & NAA Model Number	Tail Numbers	Notes
B-45A-1-NA, NA-147	7001 through 7022.	- Total Production: 22 airframes produced at the Inglewood Plant - Engines 1 & 3 two GE J35-9 engines, 4,000 lbs. thrust, dry - Engines 2 & 4 two GE J35-11 engines, 4,000 lbs. thrust, dry - No combat capable radar or navigation systems - Fitted to carry two Aerojet XLR13-AJ-5 ATO pods; not used operationally due to hazards of handling, firing and disposing of the ATO rockets - Fitted with vertical bomb bay structures, but no bomb racks or shackles - Fitted with NAA designed, built and installed ejection seats.
TB-45A-1-NA and TB-45A-2, NA-147	7003, 7004, 7006, 7007, and 7009 through 7022.	- Converted to Target Tow Aircraft by SBAMA maintenance teams at Biggs, George, Castle, Yuma, Norton and EAFB. - 18 airframes converted. - Available records do not allow the number of reels carried by specific aircraft to be determined. - Engines 1 & 3 two GE J35-9 engines, 4,000 lbs. thrust, dry - Engines 2 & 4 two GE J35-11 engines, 4,000 lbs. thrust, dry

B-45A-5-NA, NA-147	7023 through 7096.	- Total Production: 75 airframes - 13 (7023 through 7035) produced at the Inglewood Plant - 62 (7036 through 7096) produced at the Long Beach Plant - Engine 1 and 3 two GE J47-7, or two GE J47-13, 5,220 lbs. thrust, dry - Engine 2 and 4 two GE J47-9, or two GE J47-15, 5,220 lbs. thrust, dry - 53 B-45A-5-NA were subject to later conversion to BACKBREAKER or SECOND CALL Atomic modifications. - Most not fitted with AN/APN-84 SHORAN or AN/APQ-24 radar - Not equipped with inflight aerial refueling - Not equipped with water/alcohol injection - Not equipped to carry wing tip tanks - Could carry one installed fuselage tank, and two removable and two jettisonable fuel tanks in the bomb bay - Fitted to carry two Aerojet XLR13-AJ-5 ATO pods; not used operationally due to hazards of handling, firing and disposing of the ATO rockets - Fitted with NAA designed, built and installed ejection seats
B-45A-1-NA, 47th Bomb Gp, 25/3/49 to 19/10/49, NA-147 B-45A-5-NA NA-147	7004, 7006, 7007, 7008, 7009 7013, 7018, 7019, 7020 and 7022. 7027	- These ten B-45A-1s and the single B-45A-5 were assigned to the 47th BG, 84th and 85th BS at BAFB between 25 March 1949 and 19 October 1949. This Group was the first turbojet–powered BG in the USAF. It was dis-established on 18 October 1949 and its aircraft and personnel re-located to LAFB (VA) and assigned to the 363rd TRG, 84th and 85th BS on 19 October 1949.
B-45A-5-NA, NA-147	7097	- 1 static test airframe
TB-45A-5-NA, NA-147	7023, 7024, 7038 and 7063.	- 4 airframes converted in the field. - Modified to carry a single or two metal cable reel assemblies in the Bomb Bay powered by engine driven hydraulic pumps. - Available records do not allow TB-45A-5 serial numbers to be determined. Photographic evidence is the only way to make a determination at this time, and is a very limited source of information. - Engine 1 and 3 two GE J47-7, or two GE J47-13, 5,220 lbs. thrust, dry - Engine 2 and 4 two GE J47-9, or two GE J47-15, 5,220 lbs. thrust, dry
BACKBREAKER B-45A-5-NA, NA-147	7027, 7029, 7030, 7035, 7050, 7054, 7058, 7059, 7061, and 7065 through 7095.	- 4 lost in service (4770, 4775, 4785, and 4788) after conversion to BACKBREAKER aircraft and did not survive long enough to become SECOND CALL capable aircraft. All other BACKBREAKER aircraft converted to SECOND CALL configuration and served to the retirement of the B-45A-5.. - BACKBREAKER aircraft fitted to carry one Mk 5 Atomic weapon or a single Mk 7 in the forward and center areas of the bomb bay. The third rack for a Mk 8 was to be installed in the aft bomb bay, but this planned arrangement lost the aircraft the ability to carry either of the two 872 gallon jettisonable tanks. BACKBREAKER aircraft, therefore, could carry only the permanently installed fuselage fuel tank and the forward and aft upper installable bomb bay tanks. This meant loss of a total 1,744 gallons of usable fuel. - BACKBREAKER aircraft carried two 500 gallon droppable fuel tanks, one under each nacelle, for a gain of 1,000 gallons of fuel. Engine 1 and 3 two GE J47-7, or two GE J47-13, 5,220 lbs. thrust, dry. - Engine 2 and 4 two GE J47-9, or two GE J47-15, 5,220 lbs., thrust dry. - Not equipped with inflight aerial refueling

BACKBREAKER B-45A-5-NA, NA-147 (continued)	7027, 7029, 7030, 7035, 7050, 7054, 7058, 7059, 7061, and 7065 through 7095. (continued)	- Not equipped with water/alcohol injection - Equipped with AN/APN-84 SHORAN and AN/APQ-24 bomb/nav Radar. - Aircraft fitted with newly installed Emerson A-6 turrets. With twin M.3 .50 caliber turrets, and 300 rounds for each weapon.
SECOND CALL B-45A-5-NA, NA-147	7039, 7042, 7043, 7045, 7046, 7047, 7048, 7051, 7053, 7055, 7056, 7057, and 7062. BACKBREAKER Aircraft brought up to full SECOND CALL standards: 7027, 7029, 7030, 7035, 7039, 7050, 7054, 7058, 7059, 7061, 7065 through 7069, 7071 through 7074, 7076 through 7084, 7086, 7087, 7089, and 7090 through 7095.	- 13 totally unmodified airframes converted to atomic capability; while 37 BACKBREAKER airframes were brought up to SECOND CALL standards for a total of 49 SECOND CALL capable aircraft. - Atomic capable aircraft conversions. - Second USAF turbojet bomber atomic conversion project carried out by the by SBAMA, NAA, BAC, the USAF depot at RAF Burtonwood, and maintenance teams at RAF Sculthorpe. - Three lost in service after conversion by SECOND CALL (7039, 7045, and 7046), and no BACKBREAKER aircraft converted by SECOND CALL were lost. - SECOND CALL modifications allowed carriage of a single Mk 5, Mk 7 or Mk 8 Atomic Weapon on specific atomic-capable racks designed for each weapon in the forward bomb bay. - SECOND CALL aircraft carried the permanently installed fuselage fuel tank just aft of the cockpit, a 318 gallon tank at the top of the forward bomb bay, and the aft installable and aft jettisonable fuel tanks in the aft bomb bay carrying 1,174 gallons of usable fuel. - SECOND CALL aircraft carried two 500 gallon droppable fuel tanks, one under each nacelle, for a gain of 1,000 gallons of fuel. - Engine 1 and 3 two GE J47-7, or two GE J47-13, 5,220 lbs., thrust, dry - Engine 2 and 4 two GE J47-9, or two GE J47-15, 5,220 lbs., thrust, dry - Not equipped with in-flight-aerial refueling - Not equipped with water/alcohol injection - Fitted with AN/APN-84 SHORAN and AN/APQ-24 bomb/nav radar - All aircraft fitted with newly installed Emerson A-6 turrets, with twin M.3 .50 caliber turrets, and 300 rounds for each weapon.
B-45C-1, NA-153	8001 through 8010.	- Total Production: 10 airframes produced at the NAA Long Beach - Plant - Flight test and research and development aircraft - Fitted with two 1125 gallon jettisonable tip tanks and single point ground refueling - Could carry one installed fuselage tank, and two removable and two jettisonable fuel tanks in the bomb bay - Engine 1 and 3 two GE J47-7, or two GE J47-13, 5,220 lbs., thrust, dry - Engine 2 and 4 two GE J47-9, or two GE J47-15, 5,220 lbs., thrust, dry, - AN/APQ-24 bomb/nav radar - Four modified for atomic capability: 8001, 8004, 8005 and 8008. - First US turbojet to carry and drop an armed atomic weapon - Not equipped with inflight aerial refueling - Not equipped with water/alcohol injection
RB-45C-1-NA	8011 through 8043.	- 33 airframes produced at the NAA Long Beach Plant - Engines 1 and 3 two GE J47-7 or two GE J47-13 - Engines 2 and 4 two GE J47-9, or two GE J47-15 - Fitted with two 1125 gallon jettisonable tip tanks - Fitted with single point ground refueling - Fitted with In-flight-Aerial Refueling

RB-45C-1-NA (continued)	8011 through 8043. (continued)	- Fitted with AN/APQ-24 photo-nav radar - Carried the permanently installed fuselage tank behind the cockpit; the fwd. and aft installable upper bomb bay tanks and fwd. and aft jettisonable bomb bay tanks. - Fitted for two Picatinney Arsenal T-34 4,000 lb. thrust ATOs, one under each nacelle; not used operationally due to hazards of handling, firing and disposing of the ATO rockets; and two 214-gallon water/alcohol tanks for engine injection, one under each nacelle. - No turrets installed on production line.

RB-45C-1-NA, 91st SRW, assigned TDY or PCS to Yokota AB, Japan	8013 28/9/50 to 24/11/52
	8014 28/9/50 to 24/11/52
	8015 28/9/50 to 1/11/50
	8016 9/10/51 to 6/6/52
	8042 30/7/52 to 6/4/53
	8027 2/8/52 to 6/4/53
	8025 9/3/52 to 6/12/56
	8021 11/3/53 to 20/1/57
	8033 20/4/53 to 20/1/57
	8038 20/4/53 to 20/3/56
		> 8015 shot down by MiG-15
		> 8016 totally destroyed in take-off crash at Yokota; crew survived
		> 8042, 8027, and 8016 field mod with twin .50 calibers
		> 8025, 8021, 8033 and 8038 fitted with Bell M-7 turrets by SBAMA
		
		Units flying RB-45Cs at Yokota:
		  1. Det A, 4149A, 84th BS, 363rd TRW, TAC. 28/9/50 to 22/3/51.
		  2. Det 2, 91st SRS, 91st SRW, SAC. 23/3/51 to 30/11/53.
		  3. 6091st RF, FEAF. 20/4/53 to 30/11/53.
		  4. 6091st RS, FEAF. 1/12/53 to 1/2/54.
		  5. 6091st RS, 6007th Composite Sq. (R), 11/8/54 to 19/2/56.
		  6. 6091st RS, 6007th RG, 19/2/56 to 20/1/57 (with RB-45Cs).

RB-45C-1 91st SRW NA-153	8011, 8012, 8013, 8014, 8015, 8016, 8019, 8020, 8021, 8022, 8023, 8024, 8025, 8026, 8027, 8028, 8029, 8030, 8031, 8032, 8033, 8034, 8035, 8036, 8037, 8038, 8039, 8040, 8041, 8042 and 8043.	- Wing based at BAFB, from 25 July 1950 to 30 September 1951. - Wing based at LAFB (OH), from 1 October 1951 to 5 June 1953. - 28 RB-45Cs delivered to the 91st SRW at BAFB, and 3 delivered to YAB while TDY, and assigned to the 91st SRW at BAFB. - 15 B-45A-5s were also assigned to the 91st SRW between 2 December 1950 and 22 December 1952 for pilot training and proficiency. - 8021, 8025, 8033 and 8038 fitted with Bell M-7 turrets by SBAMA - Fitted to carry two 500-gallon drop tanks, one under each nacelle.

RB-45C-1-NA 91st SRW, PARPRO & RAF SDF NA-153	8019, 8021, 8023, 8026
	
	8011, 8027
	
	8031, 8037, 8038, 8039, 8040, 8041
	
	8019, 8034, 8036, 8042
	8029, 8036, 8041
	
	8031, 8035, 8037, 8040
		>PARPRO, RAF Manston, then RAF Sculthorpe 16/1/51 through 19/8/51
		   >8026 damaged in landing accident 15/5/51, SOC; 8021 damaged
		> PARPRO, added at RAF Sculthorpe, 20 May 51 to 19/8/51
		
		> PARPRO, RAF Sculthorpe, 7/5/51 to 9/4/52
		
		> RAF SFD JU JITSU I, 30/3/52 to 9/5/52
		>PARPRO, RAF Sculthorpe, 17/10/52 to 9/12/52
		  > Cover for planned JU JITSU II, PARPRO flown as well
		>RAF SDF, JU JITSU II, 5/4/54 to 10/5/54. Four aircraft transferred to 19th TRS on 10 May 54

B-45C-5-NA SECOND CALL, NA-153	8013, 8014, 8023 and 8035	- Due to shortage of convertible B-45A-5s, four RB-45Cs converted to full B-45A-5 SECOND CALL standards. - Engine 1 and 3 two GE J47-7, or two GE J47-13, 5,220 lbs., thrust, dry, 6,000 lbs. wet - Engine 2 and 4 two GE J47-9, or two GE J47-15, 5,220 lbs., thrust, dry, 6,000 lbs., wet - 8013, 8014, and 8023 fitted with Emerson A-6 turrets with twin .50 cal. guns and 300 rounds per weapon. - 8035 fitted with a non-radar equipped Bell M-7 turret with twin .50 caliber guns with 300 rounds per weapon. - Fitted with AN/APN-84 SHORAN and AN/APQ-24 photo/nav radar

B-45C-5-NA SECOND CALL, NA-153 (continued)	8013, 8014, 8023 and 8035 (continued)	- Interior fitted exactly like B-45A-5-NAs. - Aircraft retained tip tanks, In-Flight-Refueling and Single-Point ground refueling. - Bomb bay fuel tank arrangement was as-fitted to B-45A-5s. - Fitted with two 500-gallon Fletcher tanks, one under each nacelle.
<u>RB-45C-1-NA for 19th TRS, NA-153</u>	<u>8011, 8012, 8013, 8014, 8019, 8022, 8023, 8027, 8029, 8031, 8034, 8035, 8036, 8037, 8040, 8041 and 8043.</u>	- 17 airframes relocated to RAF Sculthorpe in May 1954, later assigned to the 47th BW, and assigned atomic target reconnaissance and strike assessment for atomic-capable units assigned to USAFE. - Fitted with AN/APN-84 SHORAN and AN/APQ-24 photo/nav radar - Fitted with two 500-gallon Fletcher tanks under each nacelle, or 214-gallon water/alcohol tanks for engine injection, one under each nacelle. - All aircraft fitted with non-radar equipped Bell M-7 turret with twin .50 caliber guns with 300 rounds per weapon.

TABLE 5: RESEARCH AND DEVELOPMENT MODEL BLOCK NUMBERS AND NOTES:
The following Prefixes were used for aircraft converted to Research and Development roles to 1962:
    D - Aircraft fitted as Drone Director
    E - Aircraft bailed to non-US government organizations
    J - Aircraft fitted for temporary special tests from 1956
    N - Aircraft permanently modified for special tests from 1956
    Q – Radio or Self-Controlled Drone

EB-45A-1	4702 on 29 Dec 49 4703 on 2 Jun 50 to 4 Mar 51 4705 on 2 Jun 49 to 4 Apr 50 4710 on 25 Jul 49 to 11 Apr 56 4711 on 12 Jun 50 to 25 Apr 54 4712 on 11 Feb 49 to 11 Apr 56 4714 on 17 Jun 49 to 13 Dec 54 4716 on 10 Sep 52 to 17 Dec 56 4721 on 2 Feb 49	> Bailed to NAA, Inglewood, CA; Flt Test data recording > 2750th ABW; Flt Test and data recording at WPAFB. > 2750th ABG; Flt Test and data recording at WPAFB > 2750th ABG; Flight Test and data recording at WPAFB > 2750th ABG; Flight Test and data recording at WPAFB > 2750th ABG; Flight Test and data recording at WPAFB > 2759th ABG; Flight Test and data recording at Muroc AFB > 6531st FTS; Flight Test and data recording at GAFB > NACA; Flight Test and data recording at LAFB (VA)
EDB-45A-1	4702 on 1 May 50 to 30 Nov 56 4705 on 29 May 50	> 2750th ABW, WPAFB; SNARK Program > 2750th WPAFB
JDB-45A-1	4702 on 30 Nov 1956	> Bailed to NAC, PAFB; SNARK Program
JQB-45A-1	4702 on 3 Dec 1956	> Bailed NAC, PAFB; SNARK Program
EB-45A-5	4723 on 22 Jun 50 to 11 Apr 56 4724 on 18 Oct 49 to 23 Jun 55 4736 on 26 Dec 50 to 7 Dec 56 4763 on 5 Dec 50 to 22 May 56 4796 on 26 Feb 51	> Bailed to NAA, Inglewood; Flt test and data recording > 2750th ABW; Flight Test and data recording at WPAFB > 2750th ABW; Flight Test and data recording at WPAFB > 4925th SWG; Atomic Weapon's flt test at KAFB > 2750th ABW; Flt Test and data recording at WPAFB
JB-45A-5	4723 on 12 Apr 56 to 5 Oct 56 4736 on 8 Dec 56 4725 on 30 Nov 56 4738 on 30 Nov 56	> 17th TTS; equipment development at Yuma AFB > 6515th Maint. Gp; at EAFB Flight Test > OCAMA; based on BAC work, 16 Apr 54 to 13 May 55 > Bailed to NAC, Hawthorne, CA; airborne radar
EB-45C-1	8001 on 9 Dec 50 to 14 Jan 56 8002 on 18 Oct 49 to 4 Dec 56 8004 on 6 Oct 51 to 14 Jan 56 8005 on 10 Jun 1 to 14 Jun 52 8008 on 5 Oct 50 to 8 Nov 50 8008 14 Aug 51 to 7 Dec 56 8009 on 29 Jun 53 to 3 Dec 56	> 4925th SWG; atomic weapons testing at KAFB and NTS > Bailed to Allison Div, GM; Indianapolis AP, engine flt. testing > 4925th SWG, KAFB; atomic weapons testing at NTS > 2750th ABW; Flight Test and data recording at WPAFB > 4925th SWG, KAFB; atomic weapons testing at NTS > Bailed to GE, Schenectady, NY; engine flt. testing for GE

EB-45C-1 (continued)	8010 on 1 May 50 to 29 Nov 56	> 2750th ABW; Flight Test and data recording at WPAFB > Bailed to P&W, Hartford, CN, engine testing
JB-45C-1	8001 on 15 Jan 56 8002 on 5 Dec 56 8004 on 15 Jan 56 to 9 Jan 58 8008 on 8 Dec 56 8009 on 4 Dec 56 8010 on 30 Nov 56 to 9 Apr 57	> SBAMA; atomic weapons testing at KAFB and NTS > Bailed to Allison Div, GM; Indianapolis AP, engine flt testing > SBAMA; atomic weapons testing at KAFB and NTS; GAFB; Flt testing for Rome Air Dev. Center > Bailed to GE, Schenectady, NY; engine flt testing > Bailed to Pratt & Whitney, Hartford, CT; engine flt testing. > Bailed to NAC, Los Angeles IAP
B-45C-1	8004 on 10 Jan 58	> Griffiss AFB; Flt testing for Rome Air Dev Center
EDB-45C-1	8005 on 15 Jun 52 8007 on 15 Jun 52 to 29 Nov 56	> Bailed to NAC, Hawthorne, CA; SNARK Program > Bailed to NAC, Hawthorne, CA; SNARK Program
JDB-45C-1	8007 on 30 Nov 56	> Bailed to NAC, Los Angeles Int'l; SNARK Program
NB-45C-1	8010 on 10 Apr 57	> Bailed to P&W, Hartford, CT; engine Flt. testing
ERB-45C-1-NA	8017 on 13 Oct 50 to 29 Nov 56 8023 on 14 Jan 1952 to 21 Aug 54	> 2750th ABW; Flight Test and data recording at WPAFB > 2750th ABW; Flight Test and data recording at WPAFB
JRB-45C-1-NA	8017 on 30 Nov 56 to 9 Apr 57 8029 on 12 Jan 56 8041 on 4 Dec 56 8042 on 30 Nov 55 to 18 Jan 56	> Bailed to P&W, Hartford, CT; engine Flt. Testing > Bailed to NAA, Inglewood, CA; Flt test and data recording > Bailed to NAA, Inglewood, CA; Flt test and data recording > Bailed to NAC, Los Angeles IAP; SNARK Program
NRB-45C-1-NA	8017 on 10 Apr 57	> Bailed to P&W, Hartford, CT; engine flt. Testing.

TABLE 6: AIRCRAFT IN LONG-DURATION STORAGE AT THE NAA LONG BEACH PLANT: (These 32 aircraft were bailed to North American Aviation for outdoor storage over two USAF Contracts. The purpose of the storage contacts was to preserve the aircraft in an airworthy condition while the Air Staff made formal decision on the ultimate future of these B-45A-5s. Once removed from storage, the aircraft were cycled through the NAA plant at Long Beach for installation of GFE which had not been available when they were on the initial production line. They were then flown at Langley AFB by the 47th BW and then cycled through the SBAMA for BACKBREAKER modifications.)

B-45A-5-NA	7064 through 7095	32 airframes cocooned and stored outdoors at the Long Beach Plant for individually varying lengths via two contacts between 15 February 1950 and 23 August 1951.

TABLE 7: ACCIDENTS AND LOSSES: (Note: Dates may not reflect the exact date of the accident.)

Date	Aircraft Type Tail Number	Location of Accident	Survivors	Notes and Cause
20 Sep 48	B-45A-1 7001 totally destroyed	Alpaugh, CA	None	Wing tanks above port nacelle leaking; explosion in port engine blankets; lower nacelle doors blew off tearing off port stabilizer and starboard stab broke; impacted out of control.
14 Jun 49	B-45A-5 7033 totally destroyed		None	Aircraft executed a planned approach go-around; on the downwind leg the pilot called for an immediate bail out, and the aircraft banked right and entered a 20 degree nose down decent and impacted the ground; control issues were traced to a flap torque tube failure.
13 Jul 49	XB-45 559479 damaged beyond repair	WPAFB	Two (all)	Could not retract the main landing gear and the right hand wing flaps malfunctioned; aircraft belly landed.

Date	Aircraft	Location	Fatalities	Description
12 Aug 49	B-45A-5 7037 damaged beyond repair	BAFB	All	Right main gear collapsed on touch down and pilot attempted a go-around; engine fire began: and aircraft was landed and severely damaged on remaining runway contact.
31 Aug 49	B-45A-5 7044 totally destroyed	BAFB	None	Right nacelle fire reported by witnesses immediately after takeoff; aircraft went into a glide and impacted the ground.
10 Sep 49	B-45C-1 8003 damaged beyond repair	Long Beach AP, CA	Unk	Aircraft suffered an engine over speed caused by an emergency fuel regulator failure; resulting in engine disintegration, loss of airspeed and crash on a public golf course adjacent to the NAA Long Beach Plant.
4 Dec 50	RB-45C-1 7015 Combat loss	North Korea	None	Shot down by a MiG-15 of the 523rd Fighter Aviation Regiment near Andong, Korea.
3 Feb 50	B-45A 7040 damaged beyond repair	LAFB (VA)	Unk	Aircraft lost engine power during approach due to likely mismanagement of fuel system controls; aircraft broke up during skid along the runway and berm.
24 Feb 50	B-45C 8006 totally destroyed	Vandalia, OH	One	During a landing missed approach and go-around; power was added too fast causing compressor stalls; right nacelle caught fire immediately and aircraft bounced on the runway loosing right nacelle and right wing broke into three parts. Wing was previously damaged by corrosion; fuel regulators and fuel bypass valve was not repaired from a recent delivery.
28 Apr 50	RB-45C-1 8024 damaged beyond repair	BAFB	Unk	Damaged beyond repair by a severe landing accident.
28 May 50	B-45A-5 7032 totally destroyed	LAFB (VA)	Unk	Aircraft crashed near Langley AFB; no details obtained.
7 Jul 50	B-45A-5 7034 totally destroyed	New Albany, IN	Two	Elevator control was frozen after a routine release of the auto-pilot; aircraft climbed, stalled, spun, recovered and climbed again; bail out ordered; Auto-pilot had several reported issues in previous flights.
9 Feb 51	RB-45C-1 8028 Totally destroyed	Plymouth, NC	None	Destroyed by an in-flight explosion; cause unknown.
7 May 51	RB-45C-1 8030 totally destroyed	Houston, TX	One	Bomb bay tank release arm failed causing bomb bay tanks to fall and hit bomb doors; caused buffet and disintegration, then loss of right hand outer wing panel and two major explosions.
15 May 51	RB-45C 8026 damaged beyond repair	RAF Sculthorpe	All	Landing accident; unable to stop on runway due to slick conditions.
24 July 51	B-45A-5 7041 damaged beyond repair	BAFB	None	Aircraft lost two engines in an emergency go-around; stalled and crashed and burned uncontrolled.
12 Dec 51	B-45A-5 7064 totally destroyed	LAFB (VA)	None	Shut down numbers three and four engines; ditched into the Chesapeake Bay just east of Langley; likely ran out of fuel.
21 Mar 52	B-45A-5 7075 totally destroyed	Paducah, KY	None	Aircraft impacted the ground from high altitude at 45 degree right roll; canopy leak covered by masking tape; oxygen masks not used at 40,000 foot cruising altitude; aileron boost hydraulic leak not repaired; and engine issue not repaired.

Date	Aircraft	Location	Fatalities	Description
26 Mar 52	RB-45C-1 8020 damaged beyond repair	Goose Bay AB, Labrador	Unk	Pilot under shot an ILS approach and hit a tall snow bank of plowed snow just off the approach end of the runway; the crew survived but the aircraft was damaged beyond repair.
29 Mar 52	B-45A-5 7085 totally destroyed	East of LAFB (VA), Chesapeake Bay	None	Aircraft made an abrupt high-G turn to avoid an airborne collision and both wings failed; the aircraft immediately crashed into the Chesapeake Bay three minutes after take-off from Langley.
17 Apr 52	B-45A-1 7005 totally destroyed	Holloman AFB	One	On emergency landing approach with right wing on fire, advanced throttles to fast with likely compressor stalls for a go around and the aircraft crashed and burned.
20 Apr 52	EDB-45A-1 7005 totally destroyed	Holloman AFB, NM, Bailed Northrop Aircraft	One	Crashed on approach to Holloman AFB, due to fuel starvation.
26 May 52	B-45A-5 7088 totally destroyed	Carrsville, VA	None	Possibility of imminent collision with another aircraft caused a rapid pull up; right hand wing immediately failed and exploded.
6 Jun 52	RB-45C-1 8016 totally destroyed	YAB	All	Unk engine exploded immediately after take-off and spread fire across the aircraft; aircraft crashed within approximately three miles of Yokota and crew was able to bail out successfully.
4 Aug 52	RB-45C-1 8032 totally destroyed	Between Presque Isle and Goose Bay, Canada	None	Aircraft received a fire warning light on # 3 engine and crashed on approach to Goose Bay AB during the emergency landing. The aircraft hit high terrain that had been given an improper elevation on the approach chart and the approach was never corrected to make amendments for the high terrain error.
15 Aug 52	B-45A-1 7021 totally destroyed	LAFB, (VA) NACA	One	9-G pull up; poss. elevator boost system failure suspected; both wings broke off, then general airframe breakup.
6 Oct 52	B-45A-5 7028 totally destroyed	LAFB (VA)	None	Severe fire in port nacelle followed by general in-flight explosion; fires possible from fuel leaks into engine blankets.
20 Oct 52	B-45A-5 7052 totally destroyed	LAFB (OH)	None	Failed to rotate by the end of the runway at Lockbourne and crashed into a large pile of concrete block debris.
14 Dec 52	B-45A-5 7060 totally destroyed	LAFB (OH)	Two	Snap roll to the right; fire in cockpit and right wing; possible nacelle fires and starboard stab failure following severe landing.
15 Dec 52	B-45A-5 7070 totally destroyed	Munich, Germany	Two	High-G incident in which port horizontal stab damaged, and then both wings failed and departed the aircraft. Similar to incidents with 7085, 7075, and 7038.
24 Dec 52	RB-45C-1 8039 totally destroyed	Pine Bluff, AR	All	Aircraft suffered an explosion and fire in the number four engine during engine run up which damaged the aircraft beyond repair.
14 Jan 54	B-45C-1 8005 totally destroyed	Los Angeles Intl AP, CA	Unk	Bailed to Northrop Aircraft; destroyed near Los Angeles Intl AP, CA; no other details obtained.
18 Feb 54	B-45A-5 7060 totally destroyed	LAFB (OH)	Unk	Destroyed in a crash at LAFB (OH); no other details obtained.
9 Mar 54	B-45A-5 7039 totally destroyed	LAFB (VA)	Unk	No details obtained.
15 Sep 56	B-45A-5 7026 totally destroyed	Eglin AFB	Unk	No details obtained.

18 Apr 57	B-45A-5 7059 totally destroyed	RAF Sculthorpe	Unk	Destroyed in a crash at RAF Sculthorpe; no other details obtained.	
20 Jun 57	B-45A-5 7071 totally destroyed	RAF Sculthorpe	Unk	Crashed at RAF Sculthorpe; no other details available.	
20 Sep 57	B-45A-5 7083 totally destroyed	RAF Sculthorpe	Unk	Crashed at RAF Sculthorpe; no other details available.	
7 Nov 57	TB-45A 7016 totally destroyed	Vincent AFB	Unk	Crashed near Vincent AFB, AZ.	
20 Jun 58	B-45A-5 7046 totally destroyed	RAF Alconbury	Unk	Destroyed by a crash at RAF Alconbury; no other details obtained.	

TABLE 8 <u>ARMAMENT MODIFICATIONS TO B-45A-5s, B-45C-5s AND RB-45Cs:</u>

Bell M-7	Turret, manual azimuth and gun-sight ranging.	600 rounds; two .50 caliber, M3 guns	RB-45C 8021, 8025, 8033 and 8038; 91 SRS, Det 2, and 6019th RFlt., 6019th RSq., 6007th RSq, and 6007th RG, YAB.	NAA designed, engineered, ground-tested the installation at Inglewood CA, and flight-tested the turret installation at Inglewood in RB-45C 8021 and at EAFB in B-45C 8002. All installations were conducted by the SBAMA, NAFB.
Bell M-7	Turret, manual azimuth and gun-sight ranging.	600 rounds; two .50 caliber, M3 guns	All RB-45Cs assigned to the 19th SRS and re-located to RAF Sculthorpe, UK in 1954. 17 airframe serial numbers listed in Table 2.	NAA designed, engineered, ground-tested the installation at Inglewood CA, and flight-tested the turret installation at Inglewood in RB-45C 8021 and at EAFB in B-45C 8002. All installations were conducted by the SBAMA, NAFB.
Bell M-7	Turret, manual azimuth and gun-sight ranging.	600 rounds; two .50 caliber, M3 guns	One B-45C-5, 8035, converted under SECOND CALL by SBAMA prior to relocation to RAF Sculthorpe in April, 1954. Assigned to the 47th BW.	NAA designed, engineered, ground-tested the installation at Inglewood CA, and flight-tested the turret installation at Inglewood in RB-45C 8021 and at EAFB in B-45C 8002. All installations were conducted by the SBAMA, NAFB.
Emerson A-6	Turret, gun sight azimuth and radar ranging.	600 rounds; two .50 caliber, M3 guns	All B-45A-5s assigned to the 47th BW at LAFB (VA) during to Operation BACKBREAKER/SECOND CALL prior to relocation to RAF Sculthorpe, UK in 1952 through 1954. Serial numbers for 40 BACKBREAKER and 13 SECOND CALL aircraft modified listed in Table 2. At LAFB (VA) SBAMA modified three 47th BW B-45C-5s with Emerson A-6 turrets. B-45C-5 serials listed in Table 2.	NAA designed, engineered, and ground-tested the installation at Inglewood CA, and flight-tested the turret installation at Inglewood in B-45A-5 7091 All installations were conducted by the SBAMA, NAFB.
M.3 .50 Cal	Two fixed M3 guns, with 300 rounds each, field mod with assistance from armed units at Yokota AB, Japan.	Two fixed, .50 caliber 300 round, M3 weapons fixed in the tail cone of 8027, 8016 and 8042 for protection against MiG-15 interceptors.	RB-45C field modified by maintenance teams of the 91st SRS, Det 2, at YAB.	Field mod completed without authorization two years after first requests for defensive armament sent forward to FEAF and SAC; and six months before the first M-7 turret equipped aircraft arrived at Yokota, AB.

CHAPTER 10 – ABBREVIATIONS and ACRONYMS

AAF	Army Air Field	PRC	People's Republic of China
AEC	Atomic Energy Commission	PTG or PTW	Proof Test Group or Wing
AFB	Air Force Base	P&W	Pratt & Whitney
AFSWC	Air Force Special Weapons Center or Command	RADAR	Radio Detection and Ranging
AMC	Air Materiel Command	RAF	Royal Air Force
APU	Auxiliary Power Unit	RFP	Request for Proposal
ARDC	Air Research and Development Command	R&D	Research and Development
ATO	Assisted Takeoff	ROK	Republic of Korea
BG	Bomb (Bombardment) Group	SAC	Strategic Air Command
BS	Bomb (Bombardment) Squadron	SBAMA	San Bernardino Air Materiel Area
BW	Bomb (Bombardment) Wing	SDAM	San Diego Air and Space Museum
CAC	Continental Air Command	SHORAN	Short Range Navigation
CG	Center of Gravity	SIGINT	Signals Intelligence
COMINT	Communications Intelligence	SRS, SRG or SRW	Strategic Reconnaissance Squadron, Group or Wing
DIV	Division	TAC	Tactical Air Command
DPRK	Democratic People's Republic of Korea	TCTO	Time Compliance Tech Order
ECM	Electronic Counter-Measures	TDY	Temporary Duty Travel or Assignment
ELINT	Electronic Intelligence	TRS, TRS, or TRW	Tactical Reconnaissance Squadron, Group, or Wing
FEAF	Far East Air Forces	TTS	Target Tow Squadron
FLT	Flight	UK	United Kingdom
GFE	Government Furnished Equipment	UNK	Unknown
GE	General Electric	USAF	United States Air Force
GM	General Motors	USAFE	United Sates Air Forces Europe
HQ	Headquarters	USAAF	United States Army Air Force
HUMINT	Human Intelligence		
IAP	International Airport		
IFI/IFE	In-Flight Insertion//In-Flight Extraction	US AIR FORCE BASES	
ILS	Instrument Landing System	BAFB	Barksdale AFB, LA
NAA	North American Aviation	EAFB	Edwards AFB, CA
NTS	Nevada Test Site	KAFB	Kirtland AFB, NM
NACA	National Advisory Committee for Aeronautics	LAFB (VA)	Langley AFB, VA
NARA II	National Archives and Records Administration, College Park, MD	LAFB (OH)	Lockbourne AFB, OH
NATO	North Atlantic Treaty Organization	PAFB	Patrick AFB, FL
NMUSAF	National Museum of the United States Air Force	WPAFB	Wright-Patterson AFB, OH
NTS	Nevada Test Site		

CHAPTER 11 – SOURCES AND REFERENCES

- Alling, Frederick A., **History of Modification of USAF Aircraft for Atomic Weapons Weapon Delivery, 1948-1954**, Historical Division, Office of Information Services, Air Materiel Command, February 1955, Wright-Patterson AFB, OH
- Anderson, Fred, **Northrop**, *An Aeronautical History*, Northrop Corporation, Los Angeles, CA, 1976,
- Andrade, John M., **U.S. Military Aircraft Designations and Serials, 1909 to 1979**, Midland Counties Publications, Earl Shilton, Leicester, UK, 1997, ISBN 0 904597 22 9
- Arvin, G.H., **Summary of B-45 Accidents as Compiled From NAA Field Service Reports,** *AAF Models B-45A, B-45C and RB-45C*, Report NA-53-10, North American Aviation, Municipal Airport, Los Angeles, CA, January 5, 1953
- Blanchard, Peter, Chinnery, Philip, and Swann, Martyn, **MASDC**, *Military Aircraft Storage & Disposition Center, Davis-Monthan A.F.B., Arizona*, Aviation Press Ltd, London, 1983, ISBN 0 907898 02 8
- Bowen, Lee and Little, Robert D., **The History of Air Force Participation in the Atomic Energy Program, 1943-1953**, Five Volumes, Washington, Office of Air Force History, 1959
    Vol. III: Little, Robert D., **Building an Atomic Air Force, 1949-1953, Parts 1 and 2**
    Vol IV: Bowen, Lee, **The Development of Weapons, Parts 1 and 2**
    Vol V: Alling, Frederick A., **Atomic Weapon Delivery Systems, Parts 1 and 2**
- Bowers, Peter M., **Boeing Aircraft Since 1916,** Naval Institute Press, Annapolis, MD, 1989, ISBN 0-87021-037-8
- Bullington, Richard W. and Mehler, Carl, **National Geographic Atlas of the World, Eight Edition**, The National Geographic Society, 2005, 0-7922-7543-8
- Burnham, Frank, and Rogers, Ted, **I-A XP-59A**, *"We piloted the first jet airplane coast-to-coast"*, General Electric Aircraft Engine Business Group, Cincinnati OH, 1962
- Campbell, Richard H., **The Silverplate Bombers**, *A History and Registry of the Enola Gay and Other B-29s Configured to Carry Atomic Bombs*, McFarland & Company, Inc., Jefferson, NC, 2008, ISBN 0-7864-2139-8
- Carpenter, David M., **P-59 Airacomet First Jet**, **Flame Powered**, *The Story of America's First Super Secret Jet, October 2, 1942*, Jet Pioneers of America, 1992, ISBN 0-9633387-1-4
- Carpenter, David M., **Flame Powered, The Bell XP-59A Airacomet and The General Electric I-A Engine**, *The Story of America's First Jet Powered Aircraft*, Jet Pioneers of America, 1992, ISBN 0-9633387-0-6
- Chilton, E. H., **INVESTIGATION OF TAILPIPE THERMAL INERTIA PROPERTIES CONDUCTED ON THE ENGINE TEST STAND**, *Report NA-46-963*, North American Aviation, Inglewood, CA, October 1, 1947
- Maj. Ret. Elliot, John M., **The Official Monogram US Navy & Marine Corps Aircraft Color Guide**, *Vol. 3, 1950-1959*, Monogram Aviation Publications, Sturbridge, MA, 1991, ISBN 0-914144-33-2
- Fletcher, Harry R., **Reference Series, Air Force Bases, Volume II, Air Force Bases Outside the United States of America**, Center for Air Force History, United States Air Force, Washington D.C., 1993, ISBN 88-600231
- Dr. Francillon, R. J., **Japanese Aircraft of the Pacific War**, Funk & Wagnalls, New York, 1970
- Dr. Francillon, R.J., **McDonnell Douglas Aircraft Since 1920**, Putnam, London, 1979, ISBN 0-370-00050-1
- Dr. Francillon, Rene, and Roth, Mick, **Douglas B-66 Destroyer**, *Aerofax Minigraph* 19, Aerofax, Inc., Arlington, TX, 1988, ISBN 0-942548-38-8
Fredriksen, John C., **The B-45 Tornado, An Operational History of the First American Jet Bomber**, McFarland & Company, Inc., Jefferson, NC, 2009, ISBN 978-0-7864-4278-2
- Friedlan, D., **Insulation Blankets For FY-93A, F-86, and B-45 Airplanes – Investigation, Repo**rt, *NA-50-1332*, North American Aviation, Municipal Airport, Los Angeles, CA, 12-12-50
- Futrell, Robert Frank, **The United States Air Force in Korea 1950-1953**, USAF Historical Division, Research Studies Institute, Air University, United States Air Force, Duell, Sloan, and Pearce, New York, 1961
- General Electric Engines, Editorial and Creative Directors: Brown, Wayman E., Falk R. Eric, Garner, Howard J., Parrott, Wanda J., Roush, William B., and Weber, Dwight E., **Eight Decades of Progress**, *A Heritage of Aircraft Turbine Technology*, The Hennegan Company, Cincinnati, 1990
- Green, William, and Cross, Roy, **The Jet Aircraft of the World**, Hanover House, Garden City, NY, 1955
- Green, William, **The Aircraft of the World**, Doubleday and Company, Garden City, New York, 1965,
- Hall, R. Cargill, and Laurie, Clayton D., editors, **Early Cold War Overflights**, *Symposium Proceedings, Volume I: Memoirs*, Office of the Historian, National Reconnaissance Office, Government Printing Office, Superintendent of Documents, Washington, D.C., 2003, ISBN 0-9724322-0-5
- Hall, R. Cargill, and Laurie, Clayton D., editors, **Early Cold War Overflights**, *Symposium Proceedings, Volume II: Appendixes*, Office of the Historian, National Reconnaissance Office, Government Printing Office, Superintendent of Documents, Washington, D.C., 2003, ISBN 0-9724322-0-5
- Hansen, C. J., **THE DEVELOPMENT OF THE B-45 PROGRAM**, *A Lecture to be Delivered to the Air Command and Staff School, Air University, Maxwell, AFB, AL, Report NA-48-911 (partial copy)*, North American Aviation, Municipal Airport, Los Angeles, CA, 16 August 1948
- Hansen, Chuck, **U.S. Nuclear Weapons**, *The Secret History*, Aerofax Publishers, , Arlington, TX, 1988, 0-517-56740-7
- Hardison, John D., **The Megaton Blasters**, *Story of the 4925th Test Group (Atomic)*, Boomerang Publishers, Arvada, CO, 1990, ISBN 9780960590049
- Hays, Geoffrey, **Boeing B-50**, *Air Force Legends Number 215*, Steve Ginter, Simi Valley, CA, 2012, ISBN 10 9846114-9-5

- Hopkins, Robert S., III, **Spyflights and Overflights,** *US Strategic Aerial Reconnaissance, Volume 1, 1945-1960*, Hikoki Publications, Manchester, UK, 2016, ISBN, 9 781902 109503
- Hopkins, Robert S., III, **Strategic Air Command in the UK**, *SAC Operations 1946-1992,* Hikoki Publications, Manchester, UK, 2019, ISBN 9 781902 109565
- Holton, E.C., and Pomeroy, G.E., **A-6 REMOTE CONTROL TURRET SYSTEM GROUND AND PRELIMINARY FLIGHT TESTS**, *Tail Cabin Test Stand and Airframe AF47-091, Report NA-52-178*, North American Aviation, International Airport, Los Angeles, CA, 2-27-52
- Holton, E.C., and Horvat, W., **M-7 BELL TAIL TURRET SYSTEM GROUND AND FLIGHT TESTS** on *Air Force Airplane RB-45C NO. 48-021, Report NA-52-905*, North American Aviation, International Airport, Los Angeles, CA, 8-26-52
- Jackson, Robert, **F-86 Sabre**, *The Operational Record*, Airlife Publishing, Shrewsbury, UK, 1994, ISBN 1 85310 402 7
- Jones, Alwyn T., **Boeing's B-47 Stratojet**, Specialty Press, North Branch, MN, 2005, ISBN 1-58007-071-X
- Kinder, Herman, and Hedgeman, Werner, **The Anchor Atlas of World History, Volume II**, From the French Revolution to the American Bicentennial, Anchor Press/Doubleday, Garden City, New York, 1978, ISBN 0-385-13355-3
- Lines, Richard, and Hellström, Leif, **Frog Model Aircraft, 1932-1976**, *The Complete History of the Flying Aircraft & the Plastic Kits,* New Cavendish Books, London, 1989, ISBN 0 904568 63 6
- Mann, Robert A., **Aircraft Record Cards of the United States Air Force**, *How to Read the Cards*, McFarland & Company, Inc., Jefferson, NC, 2008, ISBN 978-0-7864-3782-5
- Maurer, Maurer, **Combat Squadrons of the Air Force, World War II**, USAF Historical Division, Air University, Department of the Air Force, 1969
- Maurer, Maurer, editor, **Air Force Combat Units of World War II**, Office of Air Force History, Washington, D.C., 1983, ISBN 0-912799-02-1
- Mueller, Robert editor, **Reference Series, Air Force Bases, Volume I, Active Air Force Bases Within the United States of America on 17 September 1982,** United States Air Force Office of Air Force History, Washington, D.C., 1989, ISBN 0-912799-53-6
- No, Kum-Sok, with Osterholm, Roger, **A MiG-15 to Freedom,** McFarland & Company, Inc., Jefferson, NC, 1996, ISBN 0-7866-0210-5
- North American Aviation, **AN 01-60GFA-1, Pilot's Handbook**, *for Army Model B-45A Airplane,* NAA, Inglewood, CA, 15 Feb 1948
- Office of Legacy Management, **DOE History Timeline, Timeline of Events: 1938-2016,** U.S. Department of Energy, Washington, D.C. 2017
- Ravenstein, Charles A., **Reference Series, Air Force Combat Wings**, *Lineage and Honors Histories*, *1947-1977*, Office of Air Force History, United States Air Force, Washington, D.C., 1984, ISBN 0-912799-12-9
- Ricci, Mark, Zmijewsky, Boris, and Zmijewsky, Steve, **The Films of John Wayne**, The Citadel Press, New York, 1970, ISBN 0-8065-0222-3
- Rice, R.H., **FLIGHT TEST PROGRAM FOR A JET PROPELLED HIGH PERFORMANCE MEDIUM BOMBARDMENT** AIRCRAFT, (A.A.F. MODEL XB-45), Report NA-8641, North American Aviation, Inglewood, CA, July 24, 1945
- Rodrigues, Rick, **Aircraft Markings of the Strategic Air Command, 1946-1953**, McFarland & Company, Inc., Jefferson, NC, 2006, ISBN 978-0-7864-2496-6
- Simmons, Graham M., **OPERATION LUSTY**, *The Race For Hitler's Secret Technology,* Pen and Sword Books, Ltd, Barnsley, UK, 2016, ISBN 9781473847378
- Size Knaack, Marcelle, **Post World War II Bombers**, Office of Air Force History, US Government Printing Office, Washington, D.C., 1988, ISBN 0-912799-59-5
- Smith, J. Richard, and Creek, Eddie J., **ARADO 234 BLITZ, Vol. 3, 1950-1959,** *Monogram Monarch 1,* Monogram Aviation Publications, Sturbridge, MA, 1992, ISBN 0-914144-51-0
- Staerck, Christopher, and Sinnott, Paul, **Luftwaffe**, *The Allied Intelligence Files*, Brassey's, Inc., Washington, D.C., 2002, ISBN 1-57488-387-9
- Swanborough, Gordon, and Bowers, Peter M., **United States Military Aircraft Since 1909**, Smithsonian Institution Press, Washington, D.C. 1989, ISBN 0-87474-880-1
- T-2, Historical Section, Historical Davison, **SUMMARY OF CASE HISTORY OF TURBOJET ENGINE J35 (TG-180) SERIES**, USAAF, Wright Field, 1 July 1946
- USAF Air Materiel Command/HO, Boetteche, Lt. Col., B.K., **B-45 Historical Summary**, 1 January 1951 to 30 June 1951, Directorate, Procurement and Industrial Planning, HQ, AF AMC, released 15 August 1957
- USAF Air Materiel Command, **AN 01-60GFA-1, Handbook Flight Operating Instructions (multiple)**, *B-45A and C series Aircraft*, and subsequent Revisions
- USAF Air Materiel Command, **AN 01-60GFA-2, Handbook, Maintenance Instructions (multiple)**, *RB-45C Series Aircraft*, and subsequent Revisions
- USAF Air Materiel Command, **AN 01-60GFB-1 Handbook Flight Operating Instructions**, *RB-45C Series Aircraft*, 25 May 1950 and subsequent Revisions
- USAF Air Materiel Command, **AN 01-60GFB-3 Structural Repair Manual**, *B-45C and RB-45C Series Aircraft*, and subsequent Revisions
- USAF Air Materiel Command, **AN 01-60GFA-4 Illustrated Parts Breakdown (multiple)**, *B-45A, C and RB-45C Series Air*craft and subsequent Revisions

- USAF Air Materiel Command, **Characteristics Summary, Medium Bomber B-45A**, Wright-Patterson AFB, OH, 6 June 1949
- USAF Air Materiel Command, **Characteristics Summary, Bomber B-45A BACKBREAKER,** Wright-Patterson, AFB, OH, 25 January 1957
- USAF Air Materiel Command, **Characteristics Summary, Bomber B-45C,** Wright-Patterson AFB, OH, 11 July 1952
- USAF Air Materiel Command, **Characteristics Summary, Bomber B-45A-5, SECOND CALL**, Wright-Patterson AFB, OH, 25 January 1957
- USAF Air Materiel Command, **Characteristics Summary, Bomber Reconnaissance RB-45C,** Wright-Patterson AFB, OH, 25 January 1957
- USAF Air Materiel Command, **Standard Aircraft Characteristics B-45C Tornado**, Wright-Patterson AFB, OH, 9 July 1951
- USAF Air Materiel Command, **Standard Aircraft Characteristics B-45A (BACKBREAKER) Tornado,** Wright-Patterson AFB, OH, 11 July 1952
- USAF Air Materiel Command, **Standard Aircraft Characteristics RB-45C Tornado**, Wright-Patterson AFB, OH, 18 February 1954
- USAF Air Materiel Command, **Standard Aircraft Characteristics B-45A SECOND CALL Tornado**, Wright-Patterson AFB, OH, 25 January 1957
- USAF Air Materiel Command, **T.O. 1B-45A-1, Flight Handbook**, USAF Series B-45A & C Aircraft, 25 Nov 1950 and subsequent Revisions
- USAF Air Materiel Command, **T.O. 1B-45A-1, Flight Handbook**, *USAF Series B-45A & C Aircraft,* 29 April 1957 and subsequent Revisions
- USAF Air Materiel Command, **T.O. 1B-45A(T)-1, Partial Flight Handbook**, *USAF Series TB-45A Aircraft*, 6 April 1956 and subsequent Revisions
- USAF Air Materiel Command, **T.O. 1B-45A-2, Handbook**, **Maintenance Instructions**, *USAF Series B-45A and C Aircraft*, and subsequent Revisions
- USAF Air Materiel Command, **T.O. 1B-45A-4, Illustrated Parts Breakdown (multiple)**, *USAF Series B-45A, C and RB-45C Aircraft* and subsequent Revisions
- USAF Continental Air Command, 47th Bombardment Group/Wing, **HISTORICAL DATA, JULY 1949**, Barksdale AFB, LA, July, 1949
- USAF Continental Air Command, 47th Bombardment Group/Wing, **HISTORICAL DATA, AUGUST, 1949**, Barksdale AFB, LA, August, 1949
- USAF United States Air Forces Europe, 47th Bombardment Wing (L), **HISTORY OF THE 47th BOMBARDMENT WING, LIGHT**, 1 January – 30 June 1954, RAF Sculthorpe, 29 July 19544
- USAF United States Air Forces Europe, 49th Air Division, **OPERATIONS ANALYSIS NO. 2, RADAR BOMB SCORING ACCURACY OF MSQ-1 CONTROLLED B-45 AIRCRAFT**, RAF Sculthorpe, 3 August 1954,
- USAF United States Air Forces Europe, 49th Air Division, **49th AIR DIVISION HISTORY, SUMMARY, 1 JAN 56 -1 JUL 56,** RAF Sculthorpe, UK, July 1956
- USAF United States Air Forces Europe, 49th Air Division, Directorate of Operations, **LETTER, DECEMBER 1954, TO 49TH AIR DIVISION, COMMANDER, SUBJECT: Problem Areas,** 49th Air Division/DO, RAF Sculthorpe, UK, December 1954
- USAF United States Air Forces Europe, 19th Tactical Reconnaissance Squadron, **HISTORY OF 19TH TACTICAL RECONNASSANCE SQAUDRON NIGHT PHOTO JET, 20 JULY 1953 –11 MAY 1954**, 47th Bombardment Wing, RAF Sculthorpe, UK, 20 May 1954
- USAAF Air Materiel Command, and USAF Air Materiel Command, **USAAF and USAF XB-45 Individual Aircraft Record Cards**, *3 Sets*, USAAF and USAF AMC, Wright-Patterson AFB, OH, 1947 to 1951
- USAF Air Materiel Command, **USAF B-45A-1 Individual Aircraft Record Cards**, *22 Sets*, USAF AMC, Wright-Patterson AFB, OH, 1947 to 1959
- USAF Air Materiel Command, **USAF B-45A-5 Individual Aircraft Record Cards**, *76 Sets*, USAF AMC, Wright-Patterson AFB, OH, 1947 to 1959
- USAF Air Materiel Command, **USAF B-45C-1 Individual Aircraft Record Cards**, *10 Sets*, USAF AMC, Wright-Patterson AFB, OH, 1949 to 1962
- USAF Air Materiel Command and Air Force Logistic Command, **USAF RB-45C-1 Individual Aircraft Record Cards**, *33 Sets*, USAF AMC, Wright-Patterson AFB, OH, 1950 to 1962
- U.S. General Services Administration, **Fan Deck, Federal Standard 595C Colors**, U.S. Government Printing Office, Washington, D.C., 2008
- Vorce, Ruth E., **CASE HISTORY OF THE J-47 TURBOJET ENGINE**, Historical Office, Executive Secretariat, Air Materiel Command, Wright-Patterson AFB, OH, July, 1950
- Wack, Fred John, **The Secret Explorers**, *Saga of the 46th/72nd Reconnaissance Squadrons*, Seegers Printing, Turlock, CA, 1992
- White, Ken, **World in Peril**, *The Origin, Missions, & Scientific Findings of the 46th/72nd Reconnaissance Squadron*, K.W. White & Associates, Elkhart, IN, 1992, ISBN 0-9628916—1

CHAPTER 12 – THE B-45 IN COLOR, Part 1, Scale Models

1/72 AEROFORM VACUFORMED B-45 TORNADO KIT NUMBER 72001

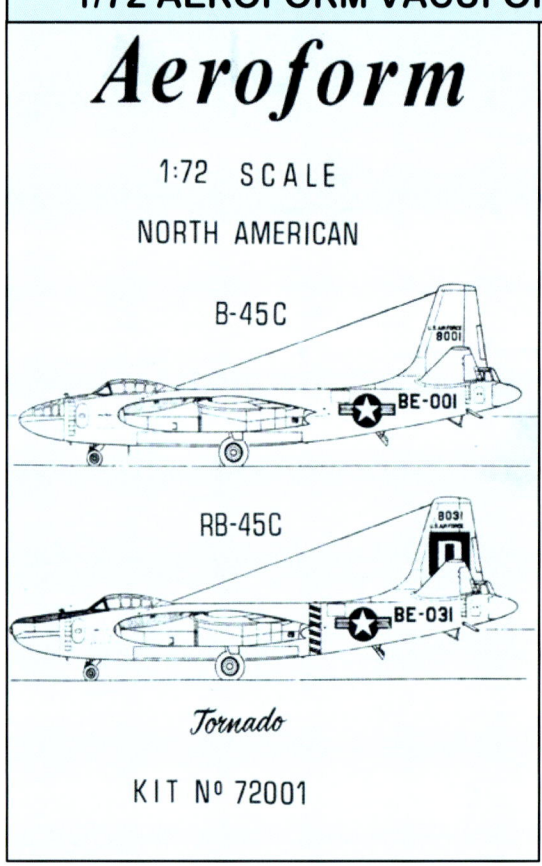

Jones for the two aircraft depicted above.

By Luis Santos: In the early 1970s, a Southern California Company, Aeroform produced a vac-u-form 1/72 scale B-45C/RB-45C, kit number 72001. The kit had 60 vac-u-formed parts plus the clear vac-u-form canopies and a four page instruction sheet. However, the kit did not include a nose for the B-45C as shown above on the instructions. The kit also included decals by Lloyd

I would only recommend this to someone that wants to learn how to build a vac-u-form kit. If you prefer the vac-u-form medium, the Gerald Elliot kit would be a finer model to build. I spent forty plus hours building this model. Most of that time was spent researching and scratch building. The instructions are overly simplified especially in the landing gear bays and the drawings lack adequate details. When comparing it with official North American drawings the kit scales out perfectly. The plastic it is molded in is of fair quality and thinner than expected so the surface details are generally soft and require re-scribing. Due to this and the shortcomings of the vac-u-form process the engine intakes, nose and especially the exhaust areas lack definition and need attention to make an accurate replica. The tires and seats came from my spares box while the cockpit interior, pitot tube, and fuel dump mast were scratch built. The kit's canopy was used as it was clear and thinly molded. The vac-u-form landing gear is only good as a pattern, but luckily they are available in metal from S.A.C. (who kindly provided samples for this review) as aftermarket parts.

1/72 GERALD J. ELLIOTT VACUFORMED B-45 TORNADO KIT

This kit is similar to the Aeroform model. There are twenty white vacuformed parts, two clear vacuformed canopy parts and eleven white metal detail parts (two nose wheels, nose gear, two main gear, two cockpit seats, two instrument panels, two control sticks, and the tail skid. Although a RB-45C is depicted on the box top, no RB-45C nose or decals were included.

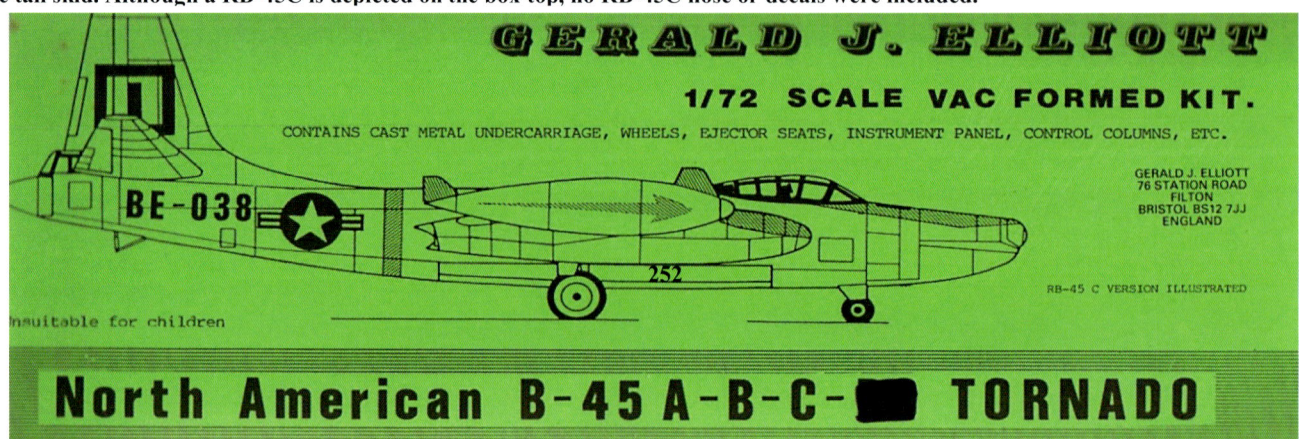

1/159 FROG INJECTION MOLDED B-45C TORNADO KIT NUMBER 388P

The Frog 1958 box scale North American B-45A model kit number 388P is surprisingly accurate for such a dated model. And this was the first injection modeled kit of the Tornado ever producded. It measures in at only 5.5 inches long with a 6.8 inch wing span. The only short fall on parts is that the two aft stabilizing fins on the tip tanks are not included. But is ths scale tthey will not be a challenge. It comes with a clear two-piece display stand and has no landing gear or pilot figures. The kit is molded in 17 silver plastic pieces and three clear canopy parts. Decals are provided for B-45A-5, 47-047, of the 84th Bomb Squadron. The kit was later re-issued by Tri-ang models in France and New Zealand. This series of Frog kits also included the Tornado's replacement, the Douglas B-66.

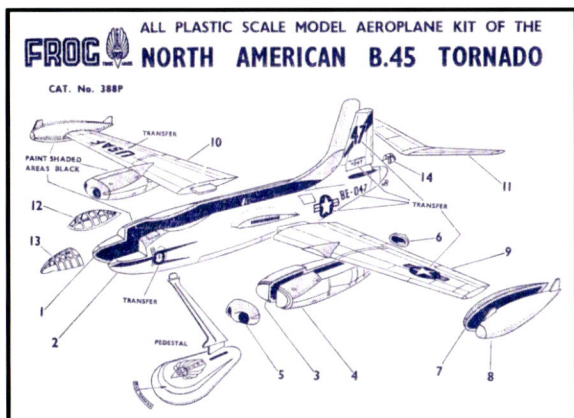

1/72 MACH 2 INJECTION MOLDED B-45A/C/RB TORNADO NUMBER MC-0008

My least favorite B-45 model is the Mach 2 kit number MC-0008. This is mostly because of the type of plastic they use, the poor quality of clear parts, the difficult fit and sink marks. Still I thank this company for producing such a diverse and interesting product line. Their kit was produced in 1993 and was molded in sixty-five grey plastic parts and seven clear pieces. Although parts are supplied for building a B-45A, B-45C, or RB-45C, the model only provides decals for a 84th Bomb Squadron B-45A, 47-080. One big plus for this kit is the inclusion of an Emerson A-6 turret and it's overhead clear parts, a first for the industry.

1/72 VALOM INJECTION MOLDED B/RB-45A/C TORNADO KITS

By Geoff Hays: Valom has issued five different North American B-45 Tornado kits in 1/72 scale that allow you to do all of the unarmed versions straight out of the box. The kits go together quite easily and fit well with very little filling, even without alignment holes. A concern is that the designers did not have good access to tech data and drawings from J35 powered B-45A-1s, J47 powered B-45A-5s, and BACKBREAKER/SECOND CALL-modified aircraft, nor sources for turret equipped RB-45Cs. Changes in the nose, flight deck and bomb bay were extensive on the inside, and significantly different between all four groups. The applicability of the changes to different modifications is not well laid out. All kits contains ninety-three tan plastic pieces on three sprues; eleven clear parts; fifty-three photo etched metal pieces; one transparent instrument panel sheet; and eight resin intake and exhaust parts. The first kit, number 72120, has decals for the J35 powered B-45A-1 preserved at the Castle Air Museum, Atwater, CA in the markings of the 84th Bomb Sq., 47th Bomb Wg.; and a B-45A-1 assigned to the 47th Bomb Gp., 84th Bomb Sq. at Barksdale AFB, LA in 1949. The second kit, number 72121, models two B-45Cs: the first has decals for the C model preserved at the National Museum of the United States Air Force Museum, in the markings of the 85th Bomb Sq., 47th Bomb Wg.; and the second is the initial B-45C assigned to the 4925th Test Wing (Atomic) at Kirtland, AFB, NM in 1951. The third kit, number 72122, models two RB-45Cs. The first decals are for 8013, the first RB-45C to arrive at Yokota AB, Japan during the Korean War. The decals are for early 1952 markings for Det 2, 91st Strategic Reconnaissance Sq. including the shark-mouth on the nose and tip tanks. The second set of decals are for a RB-45C assigned to the 19th Tactical Reconnaissance Sq. at RAF Sculthorpe in mid-1954. This kit also contains a fourth sprue that builds into a Douglas Mk 7 atomic weapon and its handling trailer along with the forward oblique camera station in the nose. The fourth kit, number 72123, builds either an aircraft from RAF Special Duties Flight carrying out OPERATION JU JITSU I on the night of 17-18 April 1952; or JU JITSU II on the night of 28-29 April 1954. Additional pieces for the forward oblique camera station in the nose are included. The fifth and final kit, number 72125, first allows modeling one of the first RB-45Cs to be camouflaged and armed during the Korean War, 8027 at Yokota AB, Japan. Or the modeler may complete one of the very early RB-45Cs from 1951 assigned to the 323rd Strategic Reconnaissance Sq., 91st Strategic Reconnaissance Wg. at Barksdale AFB, LA which flew Peripheral Aerial Reconnaissance Program (PARPRO) missions from RAF Sculthorpe in the UK.

Even with the concerns mentioned, these kits have the potential to become top-flight award winning models, especially if the after-market community comes forward after good research with images, drawings and technical data and provide Emerson and Bell turrets, bomb bay tanks, external Fletcher tanks, bomb bay weapons racks, Mk 5 and Mk 8 weapons, and BACKBREAKER/SECOND CALL mods for the bomb bays, nose and cockpits. Thanks to Valom for producing these kits!

Valom Kit Number 72122 Box Top and Decals

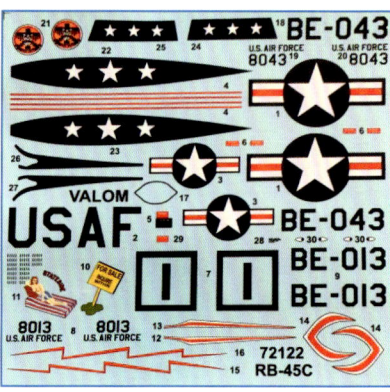

| Valom Kit # 72120 | Valom Kit # 72121 | Valom Kit # 72123 | Vlaom Kit # 72125 |

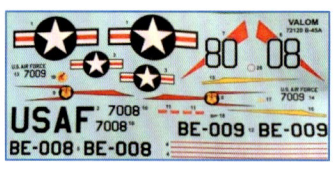  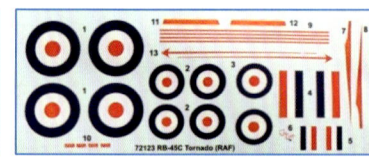 

1/48 COLLECT-AIRE MODELS RESIN B/RB-45A/C TORNADO KIT NUMBER 4873

By Gil Hodges and Fotiou Rouch: Collect-Aire 1/48 scale multimedia kit number 4873 has been the only 1/48 scale Tornado produced to date and remains one of the best kits in any scale available. Available that is, if you can find it on E-Bay and are willing to pay the price. The mostly resin kit provided optional nose sections for the B-45A or the RB-45C. Resin, metal, and photo-etched parts were included to create accurate and detailed cockpits, tail gunners position, wheel wells, gear doors, crew entrance door and engine intakes and exhausts. The Collect-Aire kit comes with decals for seven versions: an RAF Special Duties Flt RB-45C from OPERATION JU JITSU I or II; B-45C-5 8035 assigned to the 84th Bomb Sq. at RAF Sculthorpe; RB-45C 8027, the first RB-45C camouflaged in the Korean War and one of the first two to be armed in the Korean War; RB-45C 8013, the first Tornado assigned to Yokota AB, Japan; RB-45C 8012 assigned to the 19th TRS at RAF Sculthorpe in 1954; B-45A-5 7089 assigned to the 84th Bomb Sq and 7093 of the 85th Bomb Sq. both of the 47th Bomb Wing at RAF Sculthorpe in 1955. Quality control was the most prevelant issue with Collect-Aire kits, and as Gil and Fotios report in this article you might find one that has no problems with warping while another may have issues in multiple areas. The lesson is be prepared for either possibility. The kit is large, heavy and supported by very sturdy metal landing gear. The interiors of the bomb/nav and cockpit stations are well detailed and parts for both bomber and photo versions are provided. The Collect-Aire kit is the only kit that provides a reasonable example of an Emerson A-6 turret, but Fotios points out that aside for the vac-u-formed turret transparency, it is very difficult to finish. A Bell M-7 turret is not provided for the recce versions. Bomb doors and bomb bay details are not provided. Gil reports that "the engineering is straight forward and

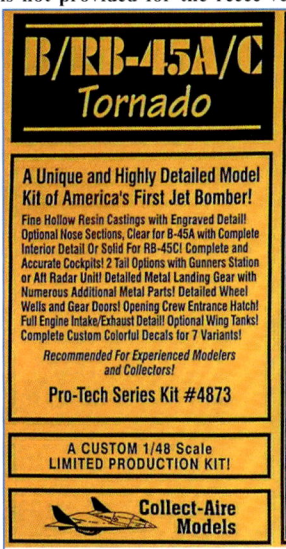 

capable of being handled by experienced resin builders, or builders that have some experience with limited-run injection kits or vac-u-form kits." Fotios' kit was hampered by warpage in large solid pieces of cast resin that caused very difficult uneven shrinkage problems in the nacelle intakes and exhausts. And while the four main fuselage sections were well designed, they experienced uneven curing issues after casting. The solid wings showed no issues of warping, but the fuselage wing-mounts did not match the contours of the wing, as is common on many short run kits. With these comments made, the photo below is Fotios' completed model, showing what an exceptional model is possible from this Collect-Aire kit. Remembering the quality control issues at Collect-Aire, Gil's final note was "If you can find it and afford it, I recommend it." Model and notes by Fotios Rouch; box illustration and notes by Gil Hodges.

CHAPTER 12 – THE B-45 IN COLOR, Part 2, Unit Insignia

The insignia in this section are presented in chronological order of the B-45 unit assignment. The insignia are shown in subordinant order of Major Command Units, then combat Group or Wing, and finally squardons within the Group or Wing. (Air Force Histroical Research Center and the Author's Collection)

BOMBARDMENT UNITS....

47th Bombardment Group, Barksdale AFB, LA, 22 August 1947 to 2 October 1949

Continental Air Command | Tactical Air Command, subordiate to CAC | 47th Bombardment Group | 84th Bombardment Squadron | 85th Bombardment Squadron

363rd Tactical Reconnaissance Group, Langley AFB, VA, 17 October 1949 to 11 March 1951

Tactical Air Command | 363rd Tactical Reconnaissance Group | 84th Bombardment Squadron | 85th Bombardment Squadron

47th Bombardment Wing, Langley AFB, VA, 12 March 1951 to 31 May 1952.

Tactical Air Command | 47th Bombardment Wing | 84th Bombardment Squadron | 85th Bombardment Squadron

47th Bombardment Wing, RAF Sculthorpe, 23 March 1954 to 30 September 1955.

United States Forces In Europe | 47th Bombardment Wing | 84th Bombardment Squadron | 85th Bombardment Squadron | 86th Bombardment Squadron (from 23 March 1954)

47th Tactical Bombardment Wing, RAF Sculthorpe, 1 October 1955 to 31 August 1958 with B-45A-5s. Served with Douglas B-66Bs until 22 June 1962.

United States Forces In Europe   47th Tactical Bombardment Wing   84th Tactical Bombardment Squadron   85th Tactical Bombardment Squadron   86th Tactical Bombardment Squadron

Air Defense Command B-45A-equipped Target Tow Squadrons, October 1949 to late 1958. (No ilustrations for the 2600th nor the 2750th Target Tow Squadrons were found.)

Air Defense Command Biggs AFB, TX   1st Target Tow Squadron, Castle AFB, CA   2nd Target Tow Squadron, George AFB, CA   4th Target Tow Squadron, Yuma and Vincent AFBs, AZ   17th Target Tow

USAF and Government Research and Development Organizations Assigned B-45s, March 1947 to June 1957.

Air Force Flight Test Center, Special Weapons Center, Edwards AFB, CA Kirtland AFB, NM   Wright Air Development Center, Wright-Patterson AFB, OH   Rome Air Development Center Griffiss AFB, NY   Air Proving Ground Command,   AF Eglin AFB, FL

National Advisory Committee On Aeronautics, Langley AFB, VA

RECONNAISSANCE UNITS....

Far East Air Forces, 91st Strategic Reconnaissance Squadron, Yokota AB, Japan, 28 September 1950 to 23 March 1951.

Far East Air Forces, Yokota, AB Japan   91st Strategis Reconnaissance Squadron, Yokota, AB Japan

Strategic Air Command, Det. 2, 91st Strategic Reconnaissance Squadron, Yokota AB, Japan 24 March 1951 to 29 November 1953.

Strategic Air Command, Yokota AB, Japan   Det 2. 2, 91st Strategic Reconnaissance Squadron, Yokota AB, Japan   There was very close coordination between SAC and FEAF's 5th Air Force Bomber Command at Yokota AB, Japan.

Far East Air Forces, 6091st Reconnaissance Flight, 20 April 1953 to 30 November 1953, Yokota AB, Japan; and 6091st Reconnaissance Sqaudron, from 1 Decmber 1953 to 20 March 1956, Yokota AB, Japan. 6091st RS was assigned to the 6007th Composite Sq, (R), 1 February 1954, Yokota AB, Japan. Became the 6007th Reconnaissance Group (Composite) on 1 August 1956, Yokota AB, Japan.

Far East Air Force   6091st Reconnaissance Flight and 6091st Reconnaissance Squadron   6007th Composite Squadron (Reconnaissance) 6007th Reconnaissance Group (Composite)

Operations in the CONUS and PAPRO in Europe by the 91st Strategic Reconnaissance Wing, Barksdale AFB, LA, between 10 February 1951 to 31 August 1951; and from Lockbourne AFB, OH from 1 September 1951 to 5 June 1953.

Strategic Air Command Squadron   91st Strategic Reconnaissance Wing   322nd Strategic Reconnaissance Squadron   323rd Strategic Reconnaissance Squadron   324th Strategic Reconnaissance

91st Air Refueling Squadron   RB-45C 8035 was imaged on the ramp at the 1953 Detroit, MI air show. (B. Kemp via David Menard)

Tactical Air Command, 363rd Tactical Reconnaissance Wing, 19th Tactical Reconnaissance Squadron, 5 June 1953 to 7 May 1954, Shaw AFB, SC.

Tacitical Air Command

363rd Tactical Reconnaissance Wing

19th Tactical Reconnaissance Wing

United States Air Force Europe, 47th Bombardment Wing, 19th Tactical Reconnaissance Squadron, RAF Sculthorpe, 8 May 1954 to 30 April 1955.

United States Air Forces Europe

47th Bombardment Wing

19th Tactical Reconnonaissance Squadron

United States Air Forces Europe, 47th Tactical Bombardment Wing, 19th Tactical Reconnaissance Squadron, RAF Sculthorpe, May 1955 to 7 January 1958.

United States Air Forces Europe

47th Tactical Bombardment Wing

19th Tactical Reconnonaissance Squadron

USAF Research and Development Organizations Assigned RB-45s, March 1950 to June 1957.

Air Force Flight Test Center Edwards AFB, CA

Wright Air Development Center, Wright-Patterson AFB, OH

Air Proving Ground Command, Eglin AFB, FL

CHAPTER 12 – THE B-45 IN COLOR, Part 3, the Aircraft

Photo 1: The second Experimental XB-45, 559480, criuses over the El Paso Mountains north of Muroc Dry Lake during an early test flight out of Muroc AFB. (Craig Kaston via Steve Ginter)

Photo 2: A 17th Target Tow Squadron TB-45A readies for take-off at Vincent AFB (formally Yuma AFB), AZ in late 1957 or early 1958. (B.Hostetter via David Menard) Photo 3: B-45A-5, 7063, was photographed at an open house at Edward AFB in the mid-1950s carrying arctic conspicuity markings. This color was used on this target tow aircraft used for interceptor rocket and fighter gunnery tests at Edwards. (D. Duncan via David Menard)

Photo 4: This image shows the first camouflaged and one of the first armed RB-45Cs to fly anywhere in USAF service. The aircraft flew a long list of highly classified sorties over Communist controlled territory in North Korea, China, Manchuria and the Eastern Soviet Union during the Korean War. (Bob Dorr via Steve Ginter)

Photo 5: As the War began to wind down in mid-1953, northern airfields in South Korea began to see intense use supporting the refueling of departing and arriving reconnaissance missions carried out over the Communist Controlled northern peninsula and the surrounding locales. Here, RB-45C 8038 assigned to the 6091th Strategic Reconnaissance Squadron prepares for engine start at K-13, Suwon AB, ROK, just south of Seoul. (Norm Taylor via Steve Ginter)

Photo 6: This image shows another 6091st RB-45C, 8025, departing to the north from K-14 Kimpo AB, ROK in mid-1953. K-14 soon became the Seoul International Airport and served that role until replaced by Gimpo International Airport at Inchon on 2001. This aircraft also carried shark's teeth on the outboard side of the tip tanks, and on the fairing between the bomb/nav station transparent panels and the radome for the AN/APQ-24. (NMUSAF)

Photo 7: At the arrival of the first copy of a new model of B-45, North American Aviation (and many other companies) would take the aircarft out onto the desert just off the ramp at the North Base at Muroc or Edwards for a color photo session. Usually a good clear or high party cloudy day was chosen and the hour was just after sunsrie. This is where and when this image of the first B-45C, 8001, was taken. This aircraft, was historic in its own right as it was the first US turbojet powered aircarft to carry and drop a live atomic weapon on a target in the Nevada Test Site. (Courtesy The Boeing Aircraft Company)

Photo 8: B-45A-5s of the 84th Bombardment Squadron are marshalled on the ramp at Langley AFB within a few days of the Wing's mass flight across the North Atlantic to the United Kingdom in June of 1954. This flight established the United States' first jet bomber deterrent force in support of NATO. (NARA II via Tommy Thomason)

Photo 9: B-45A-5s, 7076 of the 85th Bomb Sq. and 7082 of the 84th Bomb Sq. begin their second day on Naval Air Station Keflavik Iceland before the third leg of the Wing's relocation to RAF Sculthorpe, Norfolk, East Anglia, UK. (Courtesy the RAF Sculthorpe Heritage Centre via Ian Brown)

Photo 10: 84th Bombardment Squadron B-45A-5, 7030, awaits a long range training mission on the ramp at RAF Sculthorpe. (Steve Ginter)

Photo 11: One of the most unusual aircraft assigned to the 47th Bombardment Wing were the three B-45C-5s. All three started life as standard RB-45Cs, and were converted in all interior respects to match B-45A-5 Atomic bombers. One, 8035, shown here, retained its Bell M-7 turret as originally fitted, and the other two were equipped with Emerson A-6 turrets. (Author's Collection)

Photo 12: B-45A-5, 7093, leads a two-ship formation, known as an "element", upon arrival at RAF Sculthorpe. 7093 will begin a climb and a right hand turn for a 180 degree change of course, known as a "pitch out". As soon as 7093 is clear, the wing-man will do the same maneuver. Once both aircraft have fully changed direction, the landing gear and flaps will be lowered on both aircraft to slow them and they will begin another 180 degree descending turn to line up for landing at Sculthorpe. (Courtesy the RAF Sculthorpe Heritage Centre via Ian Brown)

Photo 13: At least six B-45A-5s assigned to the 86th Bombardment Squadron conduct a formation fly-over or ceremonial fly-past at RAF Sculthorpe or another RAF installation in Norfolk, East Anglia in mid-1955. (Courtesy the RAF Sculthorpe Heritage Centre via Ian Brown)

Photo 14: The 19th Tactical Reconnaissance Squadron arrived at RAF Sculthorpe in April of 1954 as seen in this photo. The Squadron was assigned to the 47th Bombardment Wing on 8 May 1954. It assumed the role of Atomic target reconnaissance for the three USAF Atomic capable units in Europe through 1957. (Steve Ginter) Photo 15: RB-45C, 8040, was photographed during a fly-by of RAF Sculthorpe in the summer of 1955. The 19th TRS was assigned to the 47th Bombardment Wing at RAF Sculthorpe on 8 May 1954. From that date, the 19th repainted its nose trim to Insignia Red from Insignia Blue. (Courtesy the RAF Sculthorpe Heritage Centre via Ian Brown)